GM LS-Series Engines

The Complete Swap Guide, 2nd Edition

Joseph Potak

Quarto.com

First Edition published in 2011 by Motorbooks, an imprint of The Quarto Group,
Second Edition published in 2023 by Motorbooks, an imprint of The Quarto Group,
100 Cummings Center, Suite 265-D, Beverly, MA 01915, USA.
T (978) 282-9590 F (978) 283-2742

27 26 25 24 23 1 2 3 4 5

ISBN: 978-0-7603-7654-6

Digital edition published in 2023
eISBN: 978-0-7603-7655-3

Formerly found under *GM-LS Series Engines, 1st Edition*
Library of Congress Cataloging-in-Publication Data

Publisher: Zack Miller
1st Edition Editor: Chris Endres
Design: Danielle Smith, Cindy Samargia Laun
Cover Image: LS swap into first-generation Chevy Camaro.
Holley Performance Products

Printed in China

On the cover: LS swap into first-generation Chevy Camaro. *Holley Performance Products*

On Contents (opposite): Known for their tremendous LS-swap parts catalog, ICT Billet has pulled a few impressive tricks out of their hat for this LS-swapped Lamborghini. This Electric Green Lambo is equipped with a GM LS3 6.2-liter engine and has twin turbos feeding the Holley Hi-Ram intake and Tick Performance intercooler. The transmission and accessory drive pose challenges for builds like this: ICT Billet definitely had their hands full with such an awesome build.

On Acknowledgments (p4): This super-clean 1971 Pontiac Firebird has an LSA supercharger on top of an SDPC Raceshop 416-cubic-inch LS3 and makes close to 700 horsepower and 700 torque. For transmission it uses a 6L80 with Tapshift. This is a true street car: the owner often takes it out for events such as Holley LS Fest East. Tastefully built, it gives a nod to its Pontiac lineage by repurposing the shaker hood on top of the LSA supercharger.

About the Author: Joseph Potak is a professional engine specialist and specializes in upgrading engine power in the performance aftermarket. He spends generous amount of time on the engine dyno through research and development of new engine performance packages and testing components for maximum power. Joseph's work has been featured in numerous magazines and online media. He's built various setups at the drag strip over the years, including Texas Speed & Performance's 7-, 8-, and 9-second Camaros and many LS/LT swapped vehicles.

Contents

Acknowledgments

With all publications of this caliber there certainly is outside help involved, from the first idea to the final printed product and all steps in between. Many hands and minds are involved, and I would like to thank them all as follows.

First of all I would like to thank my editor, Chris Endres, who worked with me on the first edition of this book. Many thanks go out to Nichole Schiele and Steve Roth for their help in getting the word out on my publications. Thanks to all other behind the scenes Motorbooks staff who have gotten their hands dirty helping compile this edition while dealing with eleventh hour changes.

I must give thanks to Nicky Fowler of Scoggin-Dickey Parts Center, Dr. Jamie Meyer, and the GM Performance Parts contacts who helped with further GM and GM Performance Parts information and images. Thanks to John Spears of Speartech for images, wiring harness, and ECM function information, Ian McDonald of LSXWorks for ECM discussions and information, Wayne Powell of KWiK Performance for accessory bracket pictures and information I also want to recognize Mike Norcia of Ram Clutches and Kevin Winstead for their continued technical support. Equal gratitude goes to the owners, staff, and members of the LS1tech.com for their support.

Huge recognition should also be given to Aeromotive, Ron Davis, Lingenfelter Performance Engineering, Holley Performance Products, Dakota Digital, ICT Billet, Speed Engineering, Mast Motorsports, MSD Ignition, Vintage Air, Hedman Hedders, Moroso Performance Products, Dirty Dingo Motorsports, Car Shop Inc., Canton Racing Products, and Denny's Driveshafts. Final thanks go to Trevor Doelling for letting me use his 1969 Camaro for the 7.0L LS7 T56 swap and images, his 1967 Chevelle for the 6.2L LS3 4L65E swap and images, and his 1968 Camaro for LS-swapped engine bay photography.

Introduction

The GM LS-series Gen III and Gen IV small-block V-8s are some of the most popular late-model engines to modify. The 1997–2004 Corvettes are equipped with the LS1 (and LS6 in 2001–2004 Z06s) 5.7-liter (346-cubic-inch) engines, while the infamous LS1 was housed in the popular F-body platform from 1998 through 2002 models.

These Y-body and F-car platforms first gave the original LS1 its reputation as a menacing street machine powerplant that is easily upgraded, but the Gen III engine is not just a performance-only engine platform. It is the new small-block Chevy from the GM house, and you can find versions of it installed in pickups, Suburbans, Trailblazers, Hummers, Impalas, and all the way to the $100K+ C6 ZR1 Corvette!

As a general rule, if it's a GM product newer than model year 2000 and equipped with a V-8, it is virtually guaranteed to have a Gen III– or Gen IV–derived engine package. GM kept this LS family of engines until the more recent Gen V engine arrived: in 2013 in Corvettes, the next year in light trucks, then in 2015 in Camaros, and in 2017 in heavy-duty trucks. (See the engine chart for identification purposes.) For the sake of simplicity, I classify nonspecific Gen III or Gen IV engines as LS-series for items that apply to the general swap info.

Why does all this matter in an LS-swap manual? The LS-series engines are widely acknowledged as a compact lightweight (in aluminum-block) design that is reliable, powerful, and versatile all at the same time. The dimensions and weight advantages open new doors in smaller cars that want V-8 performance, while the powerful and high-tech features would be welcome in a chassis that may have come with a carbureted 350-cubic-inch (SBC). From daily drivers to high-end street machines, the LS engine has the versatility to meet everyone's V-8 needs and desires. The LS-series engines can cater to everything from weekend cruisers to drag-only applications.

LS-series retrofits and swaps are no new trend. From the onset of the first LS1, people with all different makes of cars have been doing retrofits with this platform. Early in the Gen III engines' life, this was no small feat as there were no swap parts commercially available from any source. Now that these engines have been available for more than two decades, the swaps are much easier due to extensive aftermarket support. Yet many would-be engine swappers are still in unfamiliar territory.

Current aftermarket companies offer mostly bolt-in solutions for installing your LS-series engine into just about any popular chassis. Chances are that as long as you are installing your engine into a conventional chassis, you can find multiple LS fanatical peers who have done the same. GM swaps are, of course, the easiest and most common, while swaps into Ford, Dodge, and Jeep vehicles are also popular. More unique swaps, such as Toyota minivans and Dodge Stealths, will require extra attention and fabrication skills. Many items will be similar, but vehicle-specific mounting, cooling, and fueling are always a chore in unconventional vehicle applications.

Some installation information throughout the manual is going to require generalizations. What may physically fit in one chassis and body may not fit in the next. Research is required of the owner to decide what actually fits or to make it fit, if necessary. We are talking of taking technology and powertrain advancements from our day and sometimes fitting it into cars designed and built before man landed on the moon. The engineer designing your 1965 Impala never dreamed that it would end up with anything other than its original 327-cubic-inch V-8 and Powerglide. Likewise, Ford Mustang designers never imagined that someone would take a turbocharged Chevy Tahoe 5.3-liter truck engine and slap it between the fenders of a Fox body, yet it happens every day.

All LS-series engines have identical exterior dimensions, save for the dry-sump Corvette C6 LS7, LS9, and LS3, ensuring good parts interchangeability and that many wet-sump internal components are similar. The basic design has not varied much throughout the generations, so the majority of exterior components that fit a Gen III 6.0-liter LQ9 truck engine will also fit the Gen IV 5.7-liter LS1 (as an example). This allows a world of interchangeability options between the Gen III and Gen IV truck and car designations. Water pumps, oil pans, flexplates and flywheels, intakes, and valve covers swap back and forth between just about any engine with ease.

From the minuscule 4.8-liter GM truck engine to the 7.0-liter LS7 Corvette Z06 behemoth, there are a variety of engine designs to choose from. They all "appear" the same externally when equipped in a car, so if you are just looking for the LS engine look, any engine will fit the bill. If you are looking for something more performance minded, stick to the cars from the performance lineup. Look for (OEM) LS1, LS2, LS3, LS6, LS7, and LS9 engines. If OEM performance levels are not your thing, you can build up any LS-series engine from just about any donor vehicle to suit your performance needs. In these pages you will see how to harvest these late-model legends for use in just about any vehicle.

Chapter 1
Introduction to LS-Series Engines and Planning Your Project

HISTORY

In 1997 the first Generation III engine hit the streets, powering the C5 Corvette with 345 horsepower and 350 foot-pounds of torque. This basic engine carried over into the next year into the face-lifted 1998 F-body models (Camaro and Firebird) and soon thereafter gained a cultlike following on such message boards as LS1.com, LS1tech.com, CamaroZ28.com, and even swap groups on Facebook, where enthusiasts quickly found the untapped power potential of the Gen III engine platform.

Though the fourth-generation F-body met its demise in 2002, it lived on and flourished under the hood of the Corvette through the 2004 model year. The LS1 also powered the short-lived Pontiac GTO in the first year of production, 2004.

The introduction of the LS6 in the 2001 Z06 came with a 385-horsepower rating, courtesy of a thoroughly updated engine design that featured improved cylinder heads, crankcase breathing, and induction. The following year, the 2002 LS6 was further enhanced with subtle improvements to the camshaft design that was complemented by lightweight intake and exhaust valves.

In 2005, a new version of the LS-series engine was introduced with the new sixth-generation Corvette. Known as the LS2, it shares many previous features such as cylinder heads, crankshaft stroke, and other similarities throughout. LS2 engines are factory rated at 400 horsepower and are found not only in the 2005–2007 C6 Corvettes, but also the 2005–2006 Pontiac GTOs and 2006–2009 Trailblazer SS SUV. The LS2 is considered a Generation IV small-block.

The LS-series engine is produced in many forms and displacements. If you cannot find what you want in the used or crate engine arena, you can build just about anything your heart desires.

A popular swap candidate is the 427-cubic-inch LS7 found in the 2006-2013 Corvette Z06 and 2014-2015 Camaro Z28. With 505 horsepower on tap, this is one engine that will definitely help peel away the pavement. The only drawback for conventional usage is the dry-sump system, which for most swaps must be changed to a conventional wet-sump oiling system.

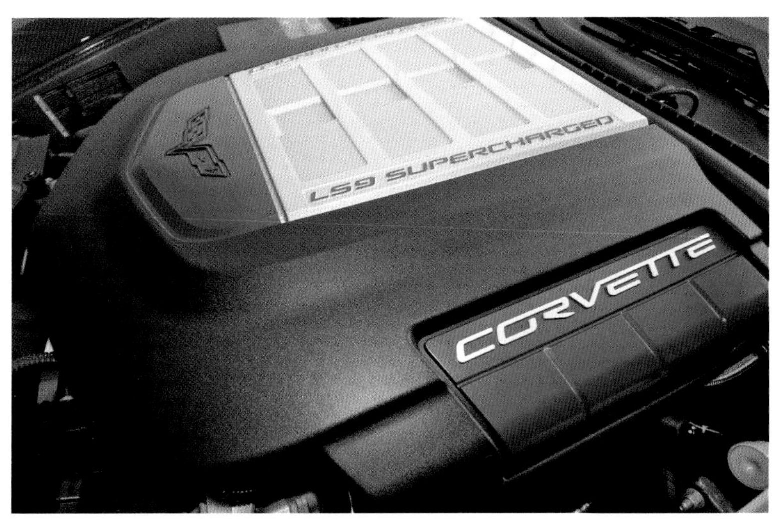

While not a budget friendly choice, the supercharged LS9 engine from the 2009-2013 Corvette ZR1 can set your build apart from all others and add coolness points when your hood is open. With 638 blower-forced horsepower (the highest of all OEM LS engines), it is the factory engine to install when any of the others are just not enough. Much like the factory LS7, the LS9 features dry-sump oiling among other amenities not offered in more conventional LS-series engines.

Speaking of Generation IV, an easy rule of thumb (speaking with performance-oriented cars in mind) is this: 1997 to 2004 = Gen III; 2005 and newer = Gen IV. Starting with 2007, certain truck redesigns began using Gen IV engines.

The Z06 was on brief hiatus in 2005 but came back on steroids in 2006 as the high-tech 7.0-liter (427-cubic-inch) LS7 engine rated at 505 horsepower. Bristling with innovation, the LS7 features a dry-sump oiling system, GM CNC-ported rectangle port heads with titanium intake valves, titanium connecting rods, forged crankshaft, and lightweight hyper-eutectic pistons. Its 505 horsepower makes the LS7 the highest horsepower naturally aspirated pushrod engine GM has ever built.

If that was not enough, Chevrolet reset the bar in 2009 with the re-introduction of the Corvette ZR1. This wasn't the former King of the Hill ZR-1 you remember from the 1990s. This was a Gen IV–based 6.2-liter engine. Much of the LS9 short-block was carry-over technology from the LS7 (such as titanium rods and intake valves), but it was further optimized with the addition of forged pistons. Why forged pistons? Because the LS9 engine for the then new ZR1 was topped off with a 2.3-liter roots-style supercharger making an authoritative 638 horsepower. What could be more perfect in your LS-swap than the LS9?

Additionally in 2009, the Cadillac CTS-V got the supercharged 6.2-liter LSA engine, equipped with a 1.9-liter roots/TVS-style supercharger. While similar to the LS9, it has many differences, with the displacement and supercharger options being the most common points. This LSA engine is a great swap choice. As a bonus, the supercharger makes a perfect donor to other Gen IV engines, and even Vortec truck engines (with cathedral heads) by using intake adapters and brackets from sources such as ICT Billet. Further, the 2013–2015 Camaro ZL1 came with a version of the LSA, with only the top lid and exhaust manifolds as the main differences.

LS-SERIES ENGINE DIFFERENCES

As you may know, the LS-series engines are unlike all other GM engine offerings. The LS blocks are a modern, deep-skirt design produced in either a lightweight aluminum-block casting with integral iron sleeves or a regular full-cast-iron block. The iron blocks were found in the Gen III Vortec and some Gen IV Vortec truck engines, although the later Gen IV offerings are slowly converting to aluminum. Engine blocks such as those used for the L76 and L92 truck engines, which are 6.0- and 6.2-liter engines, respectively, are now aluminum.

Many LS swap aficionados are quite savvy when it comes to sourcing donor engines for their project vehicles, but some recent LS-swap devotees may not know the ins and outs of correctly finding and identifying the engine they desire. Budget may dictate whether you build from a 4.8/5.3-liter Vortec or a 7.0-liter LS7-based transplant. As stated in the chapter text, you can spend from several hundred up to many thousands of dollars for your engine of choice. In this section, I will help you determine which engine you are looking at and how to avoid getting an engine you don't want.

The OEM blocks have the displacement cast or stamped in the front or rear of the block. This is useful for wrecking-yard scavenging, as you can quickly identify displacement, and with a few other identifying marks, the exact application and year of the engine can be determined. The original bad-boy Gen III LS1 is shown.

The Gen IV LS-series enhancements came about in 2005, corresponding to the first 6.0-liter designated aluminum block being released with the new Chevrolet SSR C6 Corvette chassis and the GTO. There are quite a few differences between a Gen III block and Gen IV block, but either can be used with the proper external components in almost any vehicle.

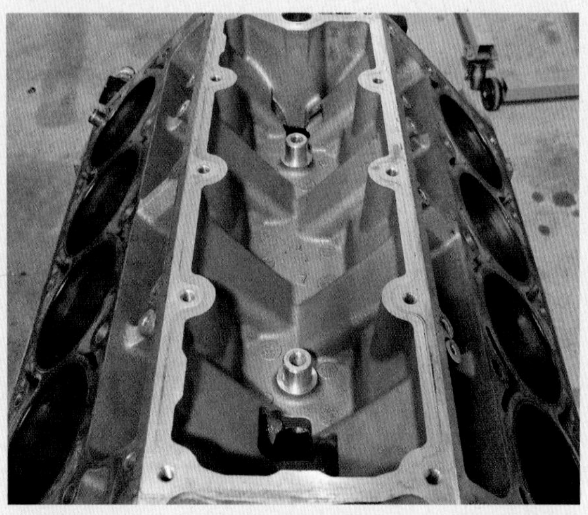

One of the Gen III LS traits is the centrally located knock sensors. All Gen III blocks have the knock sensors in the valley plate area under the intake manifold. The only production aluminum blocks with centrally located knock sensors are the LS1/LS6. Most of the 1999–2006 Vortec LS truck engines are also derived from the Gen III.

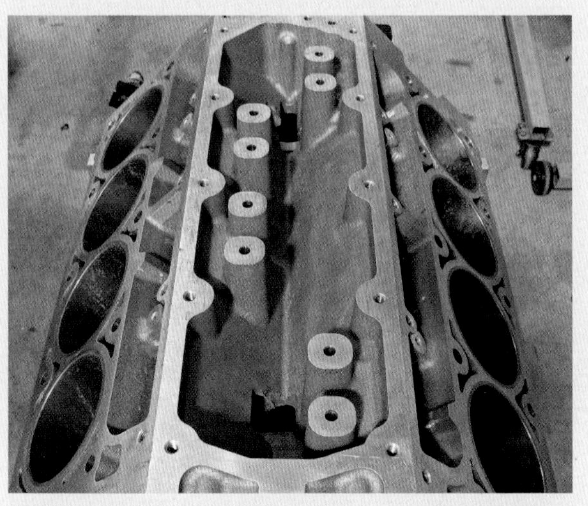

Gen IV blocks do not have centrally located knock sensors; they are located above the oil pan rails on each side of the block. In their place are these lifter oil towers, which are functional in the active fuel management (AFM) engines. The AFM engines have the capability to deactivate cylinders 1, 4, 6, and 7 under cruise conditions for an increase in fuel economy. If not used, these eight lifter-control oil galleries are plugged off using the LS2-style valley plate.

The main (nonmaterial) difference between all engines is the displacement. You have 4.8-, 5.3-, 5.7-, 6.0-, 6.2-, and 7.0-liter factory engines to weed through, and in each of these displacements, there are also some further variations, depending on the year and vehicle in which it was originally installed. The only engine that is largely unchanged internally and externally through its lifespan is the LS7 7.0-liter, as it was only available in the 2006–2013 Corvette Z06 models and the 2014–2015 Camaro Z28.

Iron blocks were available only in trucks such as Tahoes, Silverados, Suburbans, and the like. So, if someone is selling you an "LS1" or "LS2" and you see that it is an iron block, stay clear if it is an actual LS1 or LS2 that you are looking for (remember: alloy blocks are nonmagnetic and can be tested with a $3 pen magnet). Now there are other external identifiers, such as the cylinder head castings, block displacement castings and design, and also in further electronic sensor identifiers. The blocks will not show the engine option code but usually have the displacement cast or stamped into the front on either cylinder bank.

Most LS engines are Gen III–based prior to the 2005 passenger car model year (2006 for trucks). If you are looking for a Gen IV engine, you'll need to find 2005 or newer passenger cars and 2006 or newer trucks.

One simple way to differentiate between a Gen III block and a Gen IV block is that the Gen III used under-the-intake knock sensors mounted through the valley plate. If you spot these knock sensor wells, it is a Gen III block. If the knock sensors are on the side of the block above the oil pan rail, it is a Gen IV design.

In addition, the Gen III valley plate uses 10 mounting bolts and no oil pressure sensor fitting in the cover itself (it is in the block). The Gen IV valley plate will have 11 mounting bolts and an integral M14 metric oil pressure fitting. Gen III has a rear-mounted camshaft position sensor for the machined-into-camshaft reluctor wheel; Gen IV is mounted in the timing cover as the reluctor wheel is part of the upper timing gear.

Keep in mind that if you source a Gen IV engine with the 58×/4× crankshaft/camshaft reluctor wheels, you will be required to use a 58× crank reluctor-compatible ECM or a crankshaft/camshaft signal converter with a Gen III ECM.

If equipped with active fuel management (AFM) or variable valve timing (VVT), they will need to be mechanically disabled by replacing the AFM lifters, lifter trays, and valley plate. If VVT is provided on the engine: a non-VVT camshaft, timing set, and timing cover is required if a mismatching ECM and harness is used. Gen IV engines that do not have AFM or VVT are the LS2, LS3, LSA, LS7, and LS9.

Another change introduced shortly after the Gen IV release is the swap over to the higher resolution 58×/4× crankshaft and camshaft signals. This is easily externally identified by the color of the crankshaft and camshaft sensors. Black is the 24×/1× reluctors, and gray and tan identifies the 58×/4× reluctor wheels. (Note that current Gen IV OEM replacement sensors are black.)

The black sensor is the Gen IV knock sensor, which is only compatible with Gen IV ECMs. The gray sensor is the 58× crankshaft sensor. The Gen III knock sensor would fit in the same location if the block is drilled and tapped to the larger M10 bolt diameter instead of the M8 of the Gen IV. Alternately, an empty side motor-mount hole can be used. Optionally, an adapter from ICT Billet can relocate the knock sensors to the oil pan side rails.

Internally we can see the difference in the two reluctor wheels: Left is the 24× reluctor; right is the 58× reluctor. You can interchange these during a buildup if you have the correct alignment and press tools. If you don't have access to these tools, this procedure is best left to a specialist.

Light trucks were not left out in the cold as these, too, received variants of the Gen III and Gen IV engines. This was something that had never happened with the Gen II LT1 style of engines, showing GM's commitment to the LS-engine platform. The new body style on 1999 and newer GM trucks received 4.8-liter, 5.3-liter, and 6.0-liter engines that are closely related to the LS. All early truck engines are cast-iron blocks, unlike their "LS"-designated car cousins, which are all aluminum. Later-model 2008 and newer 4.8-liter and 5.3-liter truck engines sometimes have aluminum blocks, but the 6.0-liter block in trucks has always remained cast iron.

A noteworthy change to the truck engines starting in 2006 in the Cadillac Escalade and GMC Denali was the addition of the L92 6.2-liter aluminum-block engine. The significance is not due solely to block material or displacement, but mainly to the significant change from "cathedral-shaped" intake ports to "rectangle" port heads, similar to those found on the LS7 engine. This 6.2-liter engine was the prequel to the later production car engines such as the LS3 found in C6 Corvettes beginning in 2008 and the 5th-Generation Camaro LS3 and L99 engines; 6.2-liter blocks are always aluminum alloy.

Gen IV engines have internal engine provisions for integration of GM's active fuel management (AFM) system, where cylinders 1, 4, 6, and 7 shut off during light throttle applications such as cruising at steady speeds. These engines also have the ability to use variable valve timing (VVT) to broaden the engine torque curve across a wider rpm range, and provide an exhaust gas recirculation (EGR) effect while cruising by retarding the camshaft. Some engines use both systems (AFM/VVT), such as the 5th-Generation Camaro L99 engine. These engines can be reverted back to conventional LS configuration by using non-VVT/AFM engine hardware, such as an AFM/DOD/VVT delete kit would provide.

All OEM engines are stamped with the last few digits of the vehicle identification number (VIN) of the vehicle that the engine was installed in at the factory. Sometimes these numbers are useful if other identifying marks prove unreliable in engine identification.

If you want to keep a car's vintage look, some tasteful presentation and parts selection will do the trick. In this Chevy II, the owner painted his 6.2-liter LS3 engine Chevy Orange and added a carb-style intake manifold and "327" badges to the polished valve covers. Top it off with an original-looking chrome dual-snorkel air cleaner assembly and you have classic looks with modern EFI technology hiding right under the covers.

LS-Series Engine Specifications and Ratings

Year Range	RPO Code	Displacement	Generation	Horsepower*	Torque	Compression	Block Material	Cylinder Head Style	Notes
1997 - 2004	LS1	5.7L I 346cid	III 24x	350	365	10.1 - 1	Aluminum	Cathedral	
2001 - 2005	LS6	5.7L I 346cid	III 24x	405	400	10.5 - 1	Aluminum	Cathedral	
2005 - 2009	LS2	6.0L I 364cid	IV 24x I 58x	400	400	10.9 - 1	Aluminum	Cathedral	2006+ C6 is 58x
2007 - 2014	LS3	6.2L I 376cid	IV 58x	430	424	10.7 - 1	Aluminum	Rectangle	
2010 - 2015	L99	6.2L I 376cid	IV 58x	400	410	10.4 - 1	Aluminum	Rectangle	VVT/AFM
2006 - 2015	LS7	7.0L I 427cid	IV 58x	505	470	11.0 - 1	Aluminum	CNC Rectangle	Dry Sump Oiling
2009 - 2013	LSA	6.2L I 376cid	IV 58x	556	551	9.1 - 1	Aluminum	Rectangle	1.9L Supercharged
2009 - 2013	LS9	6.2L I 376cid	IV 58x	638	604	9.1 - 1	Aluminum	Rectangle	2.3L Supercharged
1999 - 2013	LR4	4.8L I 293cid	III 24x	255 - 280	285 - 295	9.1 - 1	Iron	Cathedral	
1999 - 2013	LM7	5.3L I 325cid	III 24x	270 - 295	315 - 335	9.5 - 1	Iron	Cathedral	Gen IV VVT/AFM
2005 - 2007	L33	5.3L I 325cid	III 24x	310	335	9.9 - 1	Aluminum	Cathedral	H.O. Vortec
2007 - 2014	LC9	5.3L I 325cid	IV 58x	302 - 315	330 - 338	9.6 or 9.9	Aluminum	Cathedral	AFM (/VVT)
1999 - 2007	LQ4	6.0L I 364cid	III 24x	300 - 330	360 - 370	9.4 - 1	Iron	Cathedral	
2002 - 2007	LQ9	6.0L I 364cid	III 24x	347	380	10.1 - 1	Iron	Cathedral	
2007 - 2008	L92	6.2L I 376cid	IV 58x	403	417	10.5 - 1	Aluminum	Rectangle	VVT
2010 - 2014	L94	6.2L I 376cid	IV 58x	403	417	10.5 - 1	Aluminum	Rectangle	VVT / AFM
2009 - 2017	L96	6.0L I 364cid	IV 58x	360	380	9.6 - 1	Iron	Rectangle	VVT
2008 - 2009	L76	6.0L I 364cid	IV 58x	355 - 361	384	10.4 - 1	Aluminum	Rectangle	AFM
2011 - 2017	L77	6.0L I 364cid	IV 58x	362	391	10.4 - 1	Aluminum	Rectangle	AFM

* Actual HP and TQ vary with OEM vehicle application

As for vehicles, the patients receiving new LS transplants vary quite a bit. Quite possibly the most popular swap is first-generation 1967–1969 Camaros. While commanding a high price in both restored or pro-touring modes, that doesn't stop fanatics from swapping in just about any LS-based engine into this muscle car.

Another popular swap using a more plentiful vehicle is the S10 truck and Blazer swap. These trucks can be found for $500 to $1,000 not running, so these make a perfect platform for your swap project. The road to an LS-powered S10 is well traveled.

RECIPIENT VEHICLES

The question of which vehicle to use for your LS swap is not for me to answer. Many of the more popular LS swaps require comparatively little effort, as the course of action has been well trodden. Nearly all performance-oriented GM vehicles—such as 1st-, 2nd-, and 3rd-Generation F-bodies; Corvettes; B-bodies; G-bodies; 1955–1957 Chevys; every generation of GM truck; and many less obvious choices, such as vans and S-body Blazers have undergone LS swaps. There is no limit to the type of vehicle that might receive your LS swap. BMWs, Nissan Xterras, Ford Mustangs, Ford F100 trucks, and Saturn Skys have all been LS recipients. Even exotic cars like Dodge Vipers and Lamborghinis aren't immune from the LS swap frenzy.

More obscure vehicles are always going to present a more challenging swap, but that should not deter you. If you keep the basic principles of LS swapping in mind, the swap can be performed by nearly anyone, assuming they follow the general guidelines set forth in this manual. I will next touch on a few major hurdles that all LS swap projects have to clear. This will give you a clear starting point in your quest to cram that LS7 into your grandmother's AMC Pacer.

OEM ENGINE DESCRIPTIONS

4.8- AND 5.3-LITER VORTEC TRUCK ENGINES (1999–CURRENT)

As these age, they will become more popular for LS retrofits and swaps, but at the present the 1999–2005 Gen III Vortec engines are more widely available, much more so than even the original Gen III LS1. These are the most economically obtainable "LS" engines to find.

The 4.8- and 5.3-liter engines share the same block and bore size and are externally identical. The only difference is internal, as displacement is decreased by installation of a shorter-stroke crankshaft in the 4.8-liter (3.267 inches versus 3.622 inches) and a longer connecting rod. Differentiating between these two is a slightly harder task but not impossible. The Gen III 5.3-liter uses slightly dished pistons. The 4.8-liter engines have a true flattop design. Gen IV 5.3-liter engines from 2007 and on use the same flattop piston design as the earlier 4.8-liters for an increase in compression and the better LS2/LS6 243/799 casting design cylinder heads for a small bump in rated power.

All trucks use the tall "truck" intake manifold. Earlier 1999–2002 trucks retained EGR. The 2003 and newer did not require the EGR system. It is plugged off at the manifold. The 2003 also marked the first year for returnless fuel systems for truck usage. The Gen IV 4.8- and 5.3-liter truck engines for 2007 and on have other additional features such as active fuel management and variable valve timing for a broader torque curve.

As for installing these engines into passenger cars, the major components required would be the LS1/LS6/LS2, MSD or FAST intake manifold for cathedral heads and the car-style oil pan hardware required for your specific swap project. Once car external equipment is installed on a 4.8- or 5.3-liter engine, it would appear externally as the other LS-based engines instead of looking like a "truck" engine setup. To the untrained eye, it would be difficult to determine if it were a 4.8- or a 6.0-liter, which is one of the benefits that comes with swapping these Vortec engines. The smaller displacement Vortec engines are much more plentiful and always an economical way to obtain an LS-appearing powertrain in your swap vehicle than the true LS-series engines.

5.7-LITER LS1 AND LS6 CAR ENGINES

First introduced in 1997 and carried through to 2004 with the fifth generation of the Corvette platform (C5), the 5.7-liter LS1 was the newest high-tech engine of its time and is still regarded as such to the inexperienced. With a black composite intake manifold, aluminum architecture throughout, and individual coil-per-cylinder, they look intimidating to those unfamiliar with the LS-series engines. The 1998–2002 F-body (Gen IV Camaro or Firebird) received the LS1 as well and is a popular late-model car to modify as well as the C5.

The 1997–1998 LS1 uses perimeter bolt valve covers and coil packs that bolt on to the valve covers themselves. The 1999 to current engines come equipped with center-bolt valve covers and a coil pack mounting rail for convenience. The 2004 Pontiac GTOs also came with the 5.7-liter LS1.

LS1s are an all-aluminum external structure. OEM blocks will be marked 5.7-liter on the front and rear of the block. C5s were drive-by-wire, equipped with an electronic throttle body, whereas Gen IV F-bodies and 2004 GTOs are cable throttle actuated. The 1997–2000 C5s use the EGR-less LS1 intake manifold; whereas, the 1998–2000 F-bodies were equipped with the less desirable LS1 intake manifold with EGR, the EGR tube being introduced directly behind the throttle body. These are some extra identifiers with the engine removed: The F-body has a conventional rear-sump oil pan, the C5 uses the unconventional appearing batwing oil pan, and the GTO uses a front-sump design with a forward pan-mounted oil filter.

Other identifiers are that the 1997–1998 C5 used a feed and return line fuel injector rail, meaning the fuel pressure regulator is located in the injector rail. All others are returnless simulated systems, where the fuel pressure regulator is near or in the fuel tank and a single pressurized feed line feeds the single inlet fuel injector rail.

The 2001–2004 marked a high point for the 5.7-liter LS engine with the re-release of the Z06 nameplate. The C5 Z06 received a revised engine named the LS6. The LS6 features improved crankcase breathing windows, quieter coated piston skirts, reduced-tension piston rings, higher-rpm-designed camshaft, lightweight valves—all for a much-improved top end. The LS6 received the desirable "LS6" heads, casting number 243. The 243 heads are a highly sought-after casting and variants were in production on the later Gen IV 4.8- and 5.3-liter truck engines.

In 2001 all LS1 and LS6 engines received the "LS6" intake manifold, which with its semi-deceiving nickname is not "LS6-only" as it implies. The 2001–2004 Gen III LS engines all received this intake. Some identifying characteristics are the lack of EGR provisions and the flat bottom instead of the LS1-rounded style. This manifold requires an LS6 cylinder head coolant bleed pipe and rear air bleed plugs instead of the spider coolant tube design that connects to both front and rear of the cylinder heads.

6.0-LITER VORTEC: IRON BLOCK

In addition to the 4.8- and 5.3-liter regular truck engines, the "new body" heavier duty trucks would get the largest Gen III at the time. The 6.0-liter LQ4 was first delivered in 2000 with an iron-block, cast-aluminum dished pistons, and unique cast-iron LS heads. The cast-iron heads would last only for the 2000 model years before being shelved in favor of aluminum 317 casting heads that came in 2001.

One nuance with the early 2000–2001 6.0-liter iron blocks (and very early 1999 manual transmission, 4.8-liter trucks) is that the rear crankshaft flange is lengthened for use with the conventional pre-existing Gen I–style 4L80E transmission and torque converter. This crankshaft will allow easier Gen I–style transmission interchangeability without additional hardware, but this is not enough of a reason to look for one of these as it deems it incompatible with all other LS-style transmissions, unless you get a custom torque converter built. It will not mesh with an LS1 transmission due to the extra length of the LS torque converter dimensions. While it's best to avoid these engines, if you end up with one of these LQ4 engines with a long crankshaft flange, you could always swap the crankshaft with the later model to use it with all other LS transmissions.

Late 2001–on 6.0-liters came with the standard Gen III–length crankshaft. Other than the obvious changes as stated, the 6.0-liter iron blocks remained fairly consistent in external design from 2001 to 2007 as the LQ4.

The 2002 brought a new face to the 6.0-liter lineup with the LQ9 engine, which is a higher output version of the LQ4. LQ9s came with true hypereutectic flattop pistons and full-floating piston pins. The remaining components are identical to the similar LQ4. Both Gen III iron-block variants were produced through 2007.

6.0L LS: ALUMINUM BLOCK

When you say "6.0-liter aluminum block" you think of one motor: the 400-horsepower LS2. The LS2 was introduced in the 2005 Corvette C6 and carried on into other performance oriented vehicles. The LS2 is the first Gen IV engine released, although many now well-known Gen IV features such as VVT and AFM are not included. The 2005 C6 and 2005–2006 GTO/SSR/TBSS LS2 engines shared one major thing with the Gen III engine electronics and that is the 24× crankshaft reluctor wheel and sensor.

The camshaft sensor, while relocated into the timing cover, still has the same function and sends the same 1× cam signal to the ECM as the LS1 Gen III making the first LS2 engines more compatible with a wide variety of early ECMs. These will require only some wiring extensions for knock sensors and cam sensors. The LS2 ended C6 life in 2007, but carried on in several other vehicles such as the 2006–2009 Trailblazer SS and 2006–2007 Caddy CTS-V.

The aluminum 6.0-liter also goes by a few other RPO engine codes in other platforms, such as L76 and L98. These variants use square port cylinder heads instead of cathedral intake runners. Vehicles that use these engines are the Pontiac G8 and the 2007 and newer trucks/SUV's. The L76 and cast-iron L96 6.0-liter supersedes the LQ9 and LQ4 6.0-liter truck engines as a top-end facelift while keeping the same displacement. The L98 and L76 short blocks are virtually identical to the LS2 engine in function and design. Many variants have AFM and/or VVT, which are easier to segregate and identify in chart form.

6.2-LITER VORTEC AND LS FAMILY

The 2006 brought a new engine to the GM lineup with the 6.2-liter L92, available first in the Escalade and Denali SUV trucks. It is a Gen IV–designated engine; the difference is basically a 0.065-inch increase in bore size as compared to the prior 6.0-liter engine. The 6.2-liter is the first LS-based engine to use the VVT functions to advance or retard the timing at different engine rpm. The 2006 L92 is a 58×/4× crank/cam signal engine, so to use it without the VVT provision, you will need to replace the timing cover, camshaft, and upper timing gear with non-VVT hardware.

The 6.2-liter engines are found in many current vehicles such as the Corvette C6, late-model trucks, and the 2010 and newer Camaro as the LS3 and L99 engines. The LS3 is the more performance-oriented engine, while the L99 features VVT and AFM systems for a broad acceleration powerband. During steady-state cruising, the L99 has the ability to shut down cylinders to gain extra fuel mileage in four-cylinder mode.

Positive displacement supercharger versions of the 6.2-liter engine are found in the 2009 and newer Cadillac CTS-V and Corvette ZR1. Both were available as GM crate engines if you fancy factory-boosted setups. The LS9 is the only factory-produced LS engine to use forged pistons, while its other exotic internal engine materials are similar in design to the notoriously stout LS7.

To identify a 6.2-liter engine, you can read the engine displacement stamped into the front of the engine block. The 6.2 engines all feature the rectangle intake ports and can either have the tall truck intake manifold or the low-profile passenger car intake. The truck engines usually have VVT chain components and camshaft. Note that all 6.2-liter blocks are made of aluminum alloy.

7.0-LITER LS7

One of the most recognizable names in the Gen III and IV stable is the LS7, as found in the 2006 and newer Corvette Z06. LS7s are the behemoth of the LS world with 427 cubic inches of displacement. The LS7 features a race-inspired dry-sump oiling system, which stores the engine oil in an external canister instead of the engine oil pan. The dry sump system ensures no lack of engine oil pressure during extreme racing conditions or cornering and greatly resists oil slosh that would uncover the oil pickup tube normally.

LS7 engines have a longer crankshaft snout to facilitate the extra room for the LS7 two-stage oiling system (engine oil pressure and engine oil return). Due to the extra snout length, it limits some of the front accessory interchangeability without converting it to wet sump, although it shares the same accessory spacing with the LS2/LS3 C6 engines. The dry sump system may cause a setback in that the Z06 oil canister may not physically fit your engine compartment. An aftermarket or custom tank may be required to fit the engine bay where your tank will reside. A wet-sump conversion is usually performed on these for swap projects. The LS7 does not use the VVT or AFM features, as these are left unmachined and disabled.

Many other vehicles receive LS swaps, so few chassis are left untouched. This 1980s model Malibu received a transplant that has way more power from the LS engine in stock form than the early 1980s engine could have dreamed about.

Here's a 1978 Malibu built by Mast Motorsports Shop Manager Damon Sampson and serves as a test bed for some Mast Motorsports projects. This Malibu was saved from an early death at the crusher with the talking power of $300 cash. Damon's Malibu has gone 10.40s at 131 miles per hour at one of the LSX races using a 454-cubic-inch LSX-based engine buildup.

Chevy IIs are lightweight and fast when equipped with a conventional SBC engine. Add an LS-engine package and then lookout! Nowadays, the lighter and more powerful LS alloy engines drop right into early muscle cars with ease, as if the cars were made for the engine from the get-go. *Tom Munsch*

CHOOSING YOUR LS SWAP ENGINE

There are many decisions to be made once you decide to swap an LSX into your car. The route will be determined, in part, by the engine assembly you procure for your swap. Factors to consider are desired engine displacement, automatic or standard transmission, carbureted or EFI, and the inclusion (or not) of engine accessories. Some external choices can be made after the fact, but the big decisions need to be made before even turning the first bolt.

For example, the EFI engine harness and engine computer will need to match the generation of the engine. You cannot use a 1998 F-body engine harness with an OEM LS7 engine without some matching engine modifications. This is because all Gen III engines used 24×/1× crankshaft and camshaft reluctor tooth counts, while the 2006–2007 and newer Gen IV engines are equipped with 58×/4× crankshaft and camshaft reluctor tooth counts. In all cases, the engine needs to use equipment matching its own generation or will require the use of a Gen IV–Gen III conversion box. If you are starting from scratch, it is usually best to just have the right engine and ECM combination without adding extra harness adapters.

Reluctor identification is easy if the engine is OEM built. GM made the 24×/1× crank/cam sensors black, while the 58×/4× crank/cam sensors are tan and gray. While it is easy to differentiate engine family between Gen III or Gen IV, some early Gen IV model years received 24×/1× setups—mainly 2005 Corvettes and 2005–06 GTO LS2-equipped vehicles. You can convert either reluctor configuration to the other, but this will require disassembly of the engine. Gen III engines can also be built with the newer equipment and used with Gen IV 58× electronics. If you have a specific, one-off combination in mind, there are a handful of aftermarket basic and custom engine harness conversion builders, as well as several GM Performance Parts options for crate engines.

As you can see, many of the decisions you will make are based on the engine choice, whether you buy a used pullout engine, crate engine, or build one from the ground up. Many used engines may come with the wiring harness, and sometimes even the transmission is included to lessen the decision load on you. Many LS-swappers will use everything from the donor pullout vehicles, trying to keep everything intact. This is usually a decent way to do an LS swap, as sometimes starting with nothing at all will not show the true costs involved. Things like engine sensors, coil packs, accessories, and bolts can nickel-and-dime you to death. When you purchase a take-out powertrain assembly, these items and more are usually just left on the engine. This is especially helpful if you are starting from nothing. If you have a source for these remaining parts or already own the parts needed, you have a little more leeway as to which engine assembly you want to locate, buy, or build.

Factory Engine Materials and Stock Displacements

Engine	Family	Material	Bore	Stroke	Size
LS1/LS6	Gen III	Aluminum	3.898"	3.62"	346ci/5.7L
LM7	Gen III	Cast Iron	3.780"	3.62"	325ci/5.3L
LQ4/LQ9	Gen III	Cast Iron	4.000"	3.62"	364ci/6.0L
LS2/L76	Gen IV	Aluminum	4.000"	3.62"	364ci/6.0L
LS3/L92/LS9	Gen IV	Aluminum	4.065"	3.62"	376ci/6.2L
LS7	Gen IV	Aluminum	4.125"	4.00"	427ci/7.0L

Engine Build Displacement Combinations

Bore	3.62" Stroke	4.00" Stroke	4.10" Stroke	4.125" Stroke
3.905"	347ci	383ci	393ci	395ci
4.000"	364ci	402ci	412ci	415ci
4.030"	370ci	408ci	418ci	421ci
4.065"	376ci	415ci	426ci	428ci
4.100"	383ci	422ci	433ci	436ci
4.125"	387ci	427ci	438ci	441ci

Engine displacement formula = bore × bore × stroke × pi/4 × 8

The Mopar purists will hate this one. Yes, you can add a Hellcat crate engine from the Dodge house to your ride, but this Plymouth Superbird owner chose to use GM Silverado donor parts to build his own version of this classic car. And who can blame him? "LS_the_world" should be the hashtag here.

Lime Time is a first-generation Camaro built by John Wargo at the Custom Shop in Flanagan, Illinois. Under the hood is an LQ9 6.0-liter powerplant from an Escalade. The car was completely restored and converted to LS power in a timeframe of six weeks for the 2009 SEMA show. This, amazingly, was a budget buildup for the powerplant setup. Using KWiK Performance brackets and a used engine saved the builders a ton of money. *KWiK Performance*

The engine you choose dictates which throttle body, water pump, PCM, and engine harness is required. The throttle body and water pump may seem to be little things, but those minor external engine parts are as diverse as the vehicles they go to, so they must be selected to match.

WHERE TO FIND AND BUY AN LS-SERIES ENGINE

The Gen III/IV GM "small-block" engine can be found in anything from sports cars to soccer mom SUVs and cargo vans in the used car or salvage yard marketplace. There are plenty of other sources, and depending on if you want a new crate engine with a GM dealership warranty or just something completely custom like a forged internals 427-cubic-inch LSX buildup, the sources and costs vary.

USED PULLOUT ENGINE ASSEMBLY

The salvage yard engine "pullout" can be hit or miss, if you know what you're looking for you can score big. If not, you may only get an average deal. Sometimes you can get the engine/transmission complete with wiring and all sensors, which if you are starting from nothing at all, could be advantageous.

The unknown variables here are the maintenance records of the engine and transmission and how the vehicle was used. Did it pull a trailer all its life, or did it live in a dusty environment with little care for oil changes and proper air-filter replacement? I'm not trying to scare anyone away from these engines, as the vast majority are perfect swap donors and known for longevity, but if seeking one of these as your powerplant, be alert and make sure you are getting what you pay for.

The 4.8- and 5.3-liter used engines alone can be anywhere from $350 to $750, depending on mileage and completeness. The aluminum flavor of performance engines in the same condition can cost anywhere from $1,500 to $2,500 in the LS1 version, with each newer and larger engine costing a tad more. LS2 pullouts are in the $3,000–$4,000 range. Used LS3 engines seem to be in the $4000–$4500 range depending on condition. Often as a pull-out powertrain, the engine will come as an assembly with the transmission; expect this to add about 50–60 percent to the pricing.

WHAT DO YOU GET WITH A USED ENGINE?

Typically, a used engine will be complete with exterior engine covers, such as valve covers, oil pan, and the front and rear covers, depending on whom and where you get it from. Some salvage yards sell the engine in take-out form, meaning that they tried their best to keep as much attached as possible. This usually excludes engine accessories and small sell-able electronics like the starter. Often the coil pack assemblies, intake manifold, fuel rail, and injectors will be in place, and if you are lucky it will come with an engine wiring harness. Take note that the harness could have been cut for removal or unbolted, depending on the procedures used for removal.

The inclusion of the engine harness can be a blessing, but it really depends on which intake manifold you are going to be using. For instance if you buy a truck engine with the tall truck intake manifold and it comes with the harness, you are likely not going to want it with a car intake manifold as the truck harness drapes over the intake manifold in a weird, truck-specific routing pattern. You don't notice this in a truck when installed with the beauty cover intact, but put a truck harness over a car intake and it will look like Frankenstein built it—not to mention the difference in truck-car injector connectors or the extra wiring required to use the truck harness. If you have a truck swap planned and have the late-model truck intake or harness and remainder hardware, by all means use what you have access to. The Vortec truck intake manifold looks at home in any truck, but would look awkward and out of place on most car swaps. Performance of the truck intake is geared toward the lower rpm ranges that trucks are intended for, although they hold their own when compared to LS1 and LS6 intakes.

Sometimes you may find an engine for a once-in-a-lifetime deal of a price. For instance, a family member needed some coil packs for his 1973 Camaro LS1 swap. We located a complete police Tahoe 2007 5.3-liter engine, from the harness to the oil pan, for $200. Turns out coil packs are almost that price used, so it was a no-brainer decision to pick up this engine assembly. If you see a complete truck engine for $300, even if you do not use the engine short-block itself, parts that transfer to other engines such as ignition coils, sensors, timing cover, rear cover, and cylinder heads and rocker arms quickly add up. Buying the assembly for the parts included can save you some money, and then sell what you do not need if you have extras.

The cylinder heads were quickly sold for $400 since they were the same as LS6, so instantly we had free coil packs, $200 extra cash to spend on the swap, and a 5.3-liter short block and a ton-o-parts leftover. The 5.3-liter ended up only needing a connecting rod ($10), new heads, head gaskets, and head bolts. Older 5.3-liter non-LS6 heads are all over, and I already had a stock pair ready to install. It ended up being another completed engine for minimal cost. There is less than $100 invested in the entire engine, not counting the surplus of funds from the cylinder head sale. That is what being resourceful can do for your LS swap.

ECM/PCM CONTROLS AND HARNESS

The OEM PCM system and harness is amazingly flexible for the majority of LS swappers. Many retrofits use a modified (into stand-alone) F-body harness and reprogrammed PCM. The use of the OEM-based system and sensors adds reliability, simplicity, and ease of use, not to mention the accessibility of factory replacement components. For both stock and highly modified LS-based engines, the factory controls can handle most anything thrown at it. You can have a harness custom built to your exact dimensions or specs by a company specializing in such swaps such as Speartech.

If you purchase a take-out powertrain setup such as an LS1/4L60-E combination with the OEM engine wiring harness, Speartech can convert it to a stand-alone setup as well. This service can save approximately 40 percent over the stand-alone LS harness outright. We will look at all harness options and help you decide which setup best suits your needs and requirements.

Stock engine wiring harness modifications are required if you want to reuse an existing harness. Several companies offer this service, which involves removing unnecessary wiring and adding the required circuits to convert a stock engine harness to a stand-alone harness. This usually means using two power wires to simplify the installation. *Speartech*

LS ENGINE MOUNTING

Have the engine figured out? That is the easy part. In fact, even engine mounting is fairly straightforward in most commonly swapped vehicles. However, this can be one of the most frustrating aspects of mounting an LS-based engine into an oddball chassis configuration. There are a variety of adapter-plate designs to mount to most SBC-style mounts. The most popular for many 1960s to 1990s rear-wheel-drive (RWD) GM vehicles have a simple adapter plate design that adapts to the OEM SBC mount. Both engine-mounted clamshell-style and the frame-mounted clamshell mounts can be used with this adapter plate system. Dirty Dingo, ICT Billet, Holley Performance, Trans-Dapt, Speed Engineering, and a handful of other manufacturers all offer these mounts.

For more unique swaps, there are a few manufacturers that offer vehicle-specific mounting solutions. If you are swapping into a Jeep Wrangler, 280Z, 350Z, RX7, or Pontiac Solstice, there are swap kits available. For those vehicles where no bolt-in options are available, the retrofitter would have to fabricate a mounting system for that specific vehicle. With the popularity of the LS-based engines, you may not have to wait long for a bolt-in answer for every common swap possible as companies are adding swap components all the time. If you really want to take the road less traveled, some fabrication skills can come in handy for the more obscure LS swaps. Weld-in kits from Dirty Dingo, many off-road shops, and Speed Engineering are available, all of which include most of the items needed for fabrication.

Engine motor mount adapters are required to mate the four-bolt square block mounting to the incompatible three-bolt flanges of traditional GM mounts. There are several ways to do this. The most common is the plate setup, which adapts a SBC mount to the LS engine-mounting flange. *ICT Billet*

FUELING REQUIREMENTS

The fuel system is one of the major costs of building an LS retrofit project car. You can go the cheap route for a few hundred dollars, or you can spend thousands of dollars building a custom-fuel system. Your horsepower needs dictate how much fuel is required. The OEM fuel system will dictate how sophisticated the new fuel system will need to be. Later-model EFI cars can be much easier to adapt than early low-pressure carbureted vehicles because the newer cars have EFI-compatible fuel tanks and lines from the factory.

Most stock and lightly modified LS swaps can use an OEM fuel setup such as the Aeromotive 340-liter-per-hour in-tank fuel pump or the Walbro 255-liters-per-hour pump, either internal to the fuel tank or externally mounted along a frame rail or subframe. Companies such as Aeromotive and Holley Performance also have drop-in fuel hangers for a slew of popular swaps. The OEM PCM and fuel system is calibrated for 58 psi for the bulk of LS engines, which requires the use of a fuel pressure regulator of some sort. This can be had as an aftermarket EFI regulator and corresponding high-pressure fuel line/fittings, or by using the economical 1999–2003 Corvette C5 fuel filter/regulator assembly with proper EFI swap adapter fittings and attachments. This filter includes a bypass regulator internally that regulates fuel pressure to the proper 58 psi setting and bypasses excess fuel pressure by returning fuel to the fuel tank via a 5⁄16-inch outlet. Additionally, it usually comes equipped with a mounting bracket attached. The use of 3⁄8-inch steel for EFI or -6AN braided fuel line is adequate for this setup, which mimics the OEM C5 system in function. This is an ideal setup for a mostly stock engine or driver build; it offers easy maintenance and, because it is OEM, it is serviceable in just about any town with a parts store.

For the higher horsepower and a big power-adder when the 255-liter-per-hour won't cut it (above 500 rear wheel horsepower usually), you can either double up with two Walbro fuel pumps and a larger fuel feed line (-8AN or larger feed) or use of a specific high-capacity internal or external fuel pump, such as an Aeromotive Stealth, A1000, or Eliminator pump. The only drawbacks to either is that the cost jumps significantly due not only to increased cost of the fuel pump itself, but also the hardware required for custom length AN fuel lines, custom or sumped fuel tank, AN return lines, and a specific EFI fuel pressure regulator. There is much more involved than shown in this small preview. Stay tuned for complete fuel system building and modification.

Fueling is a big deal, especially if coming from a mechanical fuel pump system and a carburetor. The fuel system can be one of the expensive aspects of the LS-engine swap, but just like eating out at a restaurant, there are some economical options and there are plenty of ways to spend more money on premium selections.

COOLING SYSTEM REQUIREMENTS

Your car's existing cooling system is something that will need to be addressed. If you are using a 30-plus-year-old radiator that barely provided adequate cooling for your 305-cubic-inch SBC, now is the time to upgrade.

It is no surprise that the LS-series of engines makes way more engine power and torque than its predecessors. This increase in engine power and resultant heat can quickly overtax the OEM system. As a minimum upgrade, the radiator should be a three-core copper radiator or, better still, one of the aftermarket fabricated-aluminum radiators. An OEM radiator from an LS-equipped vehicle can often be adapted into your project, depending on current radiator dimensions and the fabrication required.

In addition, the LS passenger car engines do not have provisions for a mechanical clutch cooling fan. Many truck engines do have the clutch fan or at least provisions for adding one, but for all LS swaps, a high-CFM dual-fan setup and proper wiring installation would be highly recommended. Either a donor dual-fan F-body, C5, or late-model truck fan setup would be completely adequate, and often a better choice than some of the aftermarket universal options. These are often overlooked areas until the last minute, but are important considerations. Many swappers have retrofitted the complete F-body radiator and fan setup into their project vehicle with good results, taking advantage of OEM reliability to upgrade their aged cooling systems.

The cooling system is another area that needs to be addressed. In many platforms the OEM radiator will not provide sufficient cooling duty and an upgraded aftermarket radiator will be required. Other than trucks, all LS engines came with electric cooling fans. There is no easy way to have a mechanical fan without using the truck accessory bracket assembly, intake manifold, and water pump, so if you had a mechanical fan prior to your swap, count on replacing it with either a large single or high-capacity dual fans if you are using the lower profile car-type intake manifolds.

TRANSMISSION COMPATIBILITY

The LS engines have come paired with a multitude of late-model transmissions since their introduction. The 4L60-E, Borg Warner T56 (now Tremec), 4L80-E, and the six-speed automatics such as the 6L80/6L90 have proven themselves to be stellar transmissions for swap usage. These OEM LS-mated transmissions are usually top choice of LS swappers due to their desire to follow the trend of adding late-model performance, economy, and drivability to their older muscle car.

It need not be complicated, though. The fact is that all GM SBC and BBC V-8 RWD transmissions have identical block-mounting locations. For the builder wanting to add LS performance to his 1979 Camaro, this means he can adapt his TH350 or Saginaw four-speed for use with his new LS engine, thus saving transmission upgrade costs incurred and keeping some tried-and-true old-school reliability. Powerglides, TH350s, TH400s, and even late 1980s and 1990s versions of the 700R4, including the LT1/SBC 4L60-E, are all capable performers behind an LS engine and can be adapted (refer to the transmission discussion later in this book).

The most challenging part of a transmission retrofit will be the crossmember. Due to the wide variety of engine-mounting locations and options, there may not be a ready-made crossmember that works with your engine-mount setup, transmission choice, and original crossmember design. As with engine-mounting solutions, bolt-in options are quickly filling the needs of LS swappers. If there is not one available, this is where you might have to bust out the welder and make something work. You may be able to modify your OEM crossmember with a new mounting location. Companies such as Competition Engineering offer weld-in crossmember kits, if all else fails.

SWAP COSTS

I will not lie, the expenses involved in swapping an LS-series engine into your ride may not be cost-effective as compared to just installing a new, similar original engine. If you blew up your engine in the only means of transportation you own, then this swap may not be for you—at least not without proper planning. In some instances, an LS can be an economical choice, though, especially if you have a rare factory engine.

I have witnessed LS swaps completed for less than $1,500, but many stock engine retrofits end up in the $5,000 to $7,000 range. The big portion of the cost is the engine itself, but the many ancillary swap parts do add up. I would add up all costs involved before jumping in head first to make sure you can afford the swap and that it can be done in a timely fashion—on your schedule.

If you are an LS fanatic and can't stand to not have one on the cheap, then just buy a running LS1 F-body or one of the many truck/SUV offerings to curb your appetite. When you tire of that, pull the engine out and prepare your retrofit. Likely, the F-body would be more economical to purchase than performing the swap itself from nothing. Although, as vehicles age, nicer cars are becoming more rare or less affordable.

Transmissions are one area that allows a lot of flexibility. The LS engines use the same SBC bell housing bolt pattern that has been used forever. Shown here is a 6L80-E from a 2010 Camaro; SBC gurus will spot the same bell housing bolt pattern quickly.

As far as timelines: If you plan to spend five years on it, costs are a little less important as compared to the builder who wants his completed in a month or two. You can go ridiculously expensive on many parts, or you can go budget style and find the best deals on everything. Both recipes will be outlined in this manual. While the destination is similar, the route you take is up to you.

LS-SERIES SWAP PLANNING

The actual grunt work of the swap does not take a whole lot of time, and the mechanical portions are actually surprisingly quick to install when you have the right parts lined up. If you worked on the project a few hours a day, you could have it done in three to four weeks or less. If you spend one day a weekend on it, expect to invest three to four months or more.

CRATE ENGINE ASSEMBLY

A GM Performance Parts dealer such as Scoggin-Dickey Parts Center (www.sdparts.com) would be the go-to place for a factory crate engine package. Who else can manufacture an engine intended for older swap projects—and back it up with a GM warranty?

The GMPP crate offerings vary from a 5.3-liter engine all the way to the 454-cubic-inch LSX bulletproof iron block, all with varying power levels depending on displacement and GMPP's camshaft specification. Furthermore, SDPC can even perform further performance work to your crate engine such as ported cylinder heads or an aftermarket camshaft before the engine leaves their facility.

Many of the GMPP crate engines are derived from the factory production line such as the 5.3- and 6.2-liter LS3 before being diverted into "crate-engine" designation. Because of this there are times you will score big and get an engine oil pan and flexplate or clutch, depending on which automobile the engine was originally destined for. This is not the case with the GMPP LSX iron-block–based buildups, as these are not production applications.

In addition to these crate engines, don't count out the factory replacement Goodwrench engines. These are brand-new versions of engines that would come in production cars, the difference being you don't get all the extras such as intake manifold, coil packs, or other engine externals that you would possibly get with a used engine. You have to specify exact year and application when ordering. Refer to the engine description topics if this is your desired route. For exact crate engine details or newer crate engine packages not covered in this manual refer to the GM Performance Parts website: www.gmperformanceparts.com.

Here are a few of the more popular GMPP crate engines offered:

LS327/327–P/N 19165628: This crate engine package is based on the production 5.3-liter iron-block truck engine, specifically the LM7. It uses a 24×/1× crankshaft/camshaft reluctor wheel design, suited for Gen III–style ECMs. The crate engine reflects this design in using the truck oil pan, cathedral 5.3-liter casting heads, and stock 5.3-liter camshaft: This setup is a reliable proven package that will idle smoothly and run perfectly fine with low-octane fueling with the 9.5:1 compression ratio. This engine has a good V-8–style power curve, using either an EFI setup or carbureted setup and makes 332 horsepower and 352TQ at the flywheel.

LS376/480–P/N 19171224: This 6.2-liter crate engine is based on the factory EFI LS3 but with a notable upgrade: GM installs the LS1 Hot Cam, which is 219/228 duration at .050 inch, .525/.525-inch intake and exhaust valve lift, respectively, and on a 112-degree lobe separation for that recognizable camshaft idle lope while helping to increase power another 50 horsepower above the stock LS3. This isn't a huge cam for the 6.2-liter displacement, so it still retains almost all of the LS3's drivability but has a good power increase at the same time. You may notice some light cam surge below 2,500 rpm due to the camshaft. The LS3s, as mentioned elsewhere, use rectangular intake ports, similar but different than the LS7 intake ports. Due to the 10.7-1 compression ratio, the minimum fuel quality recommended is Premium 92 octane. This is a Gen IV–crate engine and thus has the 58×/4× crankshaft/camshaft reluctor wheels requiring a Lingenfelter Gen IV to Gen III signal conversion box, or a Gen IV 58×/4× designed ECM and harness.

LS376/515–P/N 19171225: The 515-horsepower version of the LS3 also features a camshaft swap, this time to the larger ASA camshaft: 226/236 duration at .050 inch, .525/.525-inch intake and exhaust valve lift, respectively, and on a 110-degree lobe separation. While this engine can be installed with a Gen IV 58×/4× ECM and harness, the LS376/515 comes with a carbureted single plane intake manifold and a Holley 4160–style 770-cfm carburetor with dual feel float bowls, vacuum secondaries, and electric choke. The use of a carburetor requires a low-pressure electric fuel pump that delivers a minimum 50-gallon per hour and 6-psi of fuel pressure such as the Holley Blue inline fuel pump. This engine also requires an ignition control unit as available from GM as P/N 19171130, or through an aftermarket company such as MSD Ignition.

LS9 6.2L–P/N 19201990: This is the powerplant of the exclusive supercharged ZR1 Corvette. It is now available as a complete crate motor, so you don't have to shell out $110K+ to have a ZR1-powered supercar: You can build your own. Now don't get me wrong: the ZR1 engine, like its C4 ZR1 LT5 counterpart, is still an expensive item in itself, but in an engine retrofit application, how cool would it be to have a factory supercharged ZR1 LS9 engine in a 1963 Corvette?

The LS9 consists of a boost-friendly short block with forged 9.1:1 compression pistons, a camshaft with 211/230 duration at .050 inch, .562/.558-inch intake and exhaust valve lift, respectively, piston oilers, titanium connecting rods, forged crankshaft, and topped off with a Roots-type Supercharger for a grand total of 638 horsepower. The compressed intake air is cooled via an integral intercooler, which uses a separate cooling system, reservoir, pump, and front mounted heat exchanger. Inlet intercooler coolant temps are recommended to be kept below 95 degrees F for optimal engine performance and below 175 degrees F for engine safety. This requires an intercooler pump that pumps a minimum of 5.5 gallons per minute.

(continued on page 22)

CRATE ENGINE ASSEMBLY *(continued from page 21)*

The ZR1 requires an external dry sump tank for engine oiling much like the LS7 dry sump system. In retrofit applications a custom oil tank and oil separator will likely need to be additionally fabricated if the ZR1 oil tank does not physically fit.

ZR1 engines require the GMPP ZR1 engine accessory kit due to the wider accessory belt used to drive the blower itself. The part number for this system is 19243524 and includes the 2009 Corvette ZR1 front engine accessories.

An additional difference is the fuel system. The ZR1 Corvette uses a fuel pressure control module that varies fuel pressure to three different pressure specifications used depending on conditions. Below 1,500 rpm, the fuel pressure is commanded to 36.3 psi, between 1,500 rpm and 5,200 rpm, the fuel pressure is increased to 72.5 psi, and above 5,200 rpm, the fuel pressure maxes out at 87 psi. The fuel system must be capable of flowing minimum of 75 gallons per hour at 87 psi or engine damage is possible.

LSX 376 P/N 19171049: This build is not based on any production engine and is a true crate-engine setup using GMPP items. It is delivered fully assembled with LS3 heads and a 204/211 duration at .050 inch stock LS3 camshaft with .551 inch intake and .522 inch exhaust valve lift netting 450 horsepower and 444TQ N/A.

The core of this engine is the stout LSX iron block using a 6.2-liter crankshaft and connecting rods. Dissimilar from other naturally aspirated crate engine designs, this particular one uses forged 4032 material 14-cc dished pistons resulting in 9.0:1 compression, which is extremely gas-pump friendly, requires 87 octane naturally aspirated and 92 octane with forced induction. Due to the lower compression ratio and forged pistons, this engine would be more durable than most others for use with a mild blower or turbo setup. Use of the LSX 376 and a CTS-V blower or the Magnacharger blower kit would be pretty impressive in an older ride and rival ZR1 power at costs far lower.

Perfect "drop-in" new car swap candidates would be anything factory equipped with a 58×/4× crank/cam reluctors and requiring boost-friendly setups on applications such as 2010 and newer Camaros, 2006 and newer Corvettes, and 2007 and newer trucks. With a signal conversion box, this extends the application to all other GM Gen III–controlled vehicles. Keep in mind that newer cars are required to meet EPA guidelines, so this engine may not be a legal swap by the letter of the law, depending on your location.

LSX 454 P/N 19244611: As the biggest bad-boy engine in the GMPP crate engine lineup, the LSX 454 leaves nothing on the table for power or cubic inches. At 454-cubic inch displacement using the GMPP LSX iron block bored to 4.185 inches, it puts a new twist on the term "small-block."

Coming with a monster (by GM standards) camshaft with 236/246 duration at .050 inch, .635/.635-inch intake and exhaust valve lift, respectively, six-bolt LS7-derived GMPP cylinder heads, and 11-1 compression via forged pistons, this Goliath puts down 620 horsepower and 590TQ in carbureted mode without breaking a sweat. It drops about 40 horsepower when fitted with EFI using the stock LS7 intake, mainly due to the intake not able to feed the larger cubic inches of the engine. Other internals included with the engine are a forged 4.125-inch stroke crankshaft with an 8-bolt flywheel flange and forged steel connecting rods.

GM has quite the selection of factory-style engine assemblies from mild replacement engines to wild 454-cubic-inch monsters. Choosing the LS engine for your powerplant has never been easier than today. Imagine if you were to do this swap before 2005, when you only had the truck engines and the LS1 and LS6 5.7-liter engines. Now you can have anything. *Courtesy of General Motors*

E-Rod LS3 Crate Engines: The E-Rod crate engines are the first 50-state legal retrofit engine available off the shelf. Sure you can take a new engine and put it in an older car with every emissions control device, but the E-Rod kit does it with ease without having to research what to do with this or that. It comes with everything required in the crate.

What makes the E-Rod emissions compliant are the charcoal canister, purge solenoid, nonventing PCV system, functional catalytic converters, and rear catalyst-monitoring O_2 sensors. The combination of such is much like what you would find in something like a 2010 Camaro with an LS3. By using the retrofit-intended engine harness like the other GMPP engine harnesses, it makes the conversion to the E-Rod a hassle-free installation.

The 6.2-liter LS3 is the current crate engine package available through the E-Rod program, with other crate engines likely to follow. The original E-Rod LS3 is set up for an automatic transmission but is now also available for a manual transmission with a slightly different ECM calibration. Be aware that an external transmission controller is required for the automatic version as the E67 ECM does not control transmissions. The necessary engine sensor outputs are in the bulkhead connector for an easy install of the transmission controller.

Many of the remaining components are the same items you would find in the Gen IV GMPP retrofit harness kits such as the DBW throttle pedal, MAF element, pre-calibrated E67 ECM, and the wiring harness itself. The difference is that this kit, as compared to buying the crate engine and harness separate, comes with everything that is required to make it emissions legal, including the engine itself.

The LS3 engine comes with a 24-month, 24,000 mile warranty, while the external supplemental components carry a 12-month, 12,000-mile warranty. For a price-tag of about $10,000 through an authorized GM Performance Parts retailer for the entire E-Rod engine package, you can have the same 430-horsepower as the Corvette and Camaro and further cut down on exhaust emissions, a huge feat that only new-cars pricing usually come with.

The E-Rod emissions-legal crate engine program addresses a market need that the LS engines accomplish easily: The need to still meet emissions requirements while having a hopped-up hot-rod. While you can swap everything from a new donor car and meet the letter of the law, no one really did that. Now you can buy the kit over the counter from your GM Performance Parts dealer that provides a fully compliant emissions-legal LS3. Amazing. *Courtesy of General Motors*

This can be a winter project to do in the down season or something that you can just do as time allows over a year's time. If you have a garage or shop, you are in a more advantageous position, as you can just stop working when you are done and come back to the project when ready or as more parts arrive. The best course of action is to have a plan (write it down!) and follow the order of installation with the components. If you find something missing or that you have to order it, make a note of it and move on to the next nonrelated part.

If you have to borrow or lease garage space for your LS swap, you need to have everything lined out or a trailer handy to move your project out of the way if you get in a time crunch. If doing this at home, the best suggestion is to have plenty of organized room in your garage. A two-car garage is certainly preferable. You can put the car on one side and the engine and remaining parts squeezed into the other size, waiting on your free time.

As far as tools, it really depends on the car used, but the engine itself is entirely metric, having a plethora of metric sockets, wrenches, and even spare M6, M8, and M10 nuts and bolts is recommended. Depending on the car model year and brand, it may use standard-sized fasteners for things like engine mounts, fuel tanks, and transmission crossmembers, or it may use metric throughout if it is more modern.

CUSTOM-BUILT TURN-KEY ENGINE ASSEMBLY

Well you have looked through the stock engines and the crate engines, and you just don't see anything that you can't live without. You may want more power. Or a built engine for a specific application or power-adder. This is where a custom engine build would better suit your needs.

Sure, they cost more sweet moolah, and for that price you get to customize your project with whatever engine size or style your heart desires. You are customizing your entire car with the LS swap, so why not make the engine a little stronger?

To start with, you will need to establish a goal, usually a horsepower number or displacement size, or sometimes just something better than what you currently have is enough. Some of the main decisions are block design and material, which then leads into engine size, and from there you can tailor cylinder heads and a custom-specified camshaft. After the block game plan is drawn out and considered, then the engine shop will machine, balance, measure, and assemble the short block, and then further into a long block.

The drawbacks to building an engine from scratch are that you don't get all the little amenities that a used engine may come with, such as timing cover, oil pan, coil packs, sensors, valve covers, and so on. You will either have to procure these from the used marketplace or a GM dealership. These items often nickel-and-dime you to death. I've even seen firsthand where someone bought a complete engine that was damaged beyond repair and salvaged these external components off it for a huge savings. This can amount to up to several hundred dollars that could be saved, depending on what you can use.

Typically, you can build anything from a forged 347-cubic-inch LS6 that makes 550 horsepower to a 454-cubic-inch resleeved LS7 that makes 700 horsepower. Costs for an assembled long block can start at about $10,000, and the sky is the limit depending on options. Usually large displacement engines cost anywhere from $12,000 to $15,000, depending on how they are optioned and which cylinder heads are used and so on. Adding a supercharger and/or a billet front accessory drive will easily put that engine into the mid- to high-$20K range. Custom engine builds vary a lot depending on exactly which parts are used. I recommend using an LS-specific shop that builds a lot of LS engines. These specialty shops have gained the familiarity with what is necessary for machining the blocks and know the assembly requirements necessary for an engine to survive. Your local machine shop can likely build a killer SBC 350, but if they have not built several successful LS performance-based engine buildups, I strongly suggest looking elsewhere.

You can find LS engines in all kinds of vehicles. Here, a Mazda RX-7 originally equipped with a rotary engine now sports a nice (and reliable) LS swap. The gold valve covers in the other image may appear to be a Ford 390 from a 1960s Thunderbird, but you'd be wrong: there's an LS swap hiding from the Ford purists enclosed under all that gold paint. Both swaps are examples of non-GM LS builds that are basically wolves in sheep's clothing.

SUMMARY

The Gen III and Gen IV engines are a great engine package deserving of retrofit attention. Lightweight, efficient, versatile, powerful, and plentiful are some of the best traits to come out of this lineup. If your old engine kicked the bucket or is on its last leg, there is little reason to not look into swapping your vehicle over to one of these engine designations. Join the LS-swapping frenzy and see for yourself what the automotive community has been shouting about for the past two decades. The power and performance of the LS engines are now available in the vehicle of your choice.

Perfect sleepers to LS swap: 1980s Caprices, especially with a Magnuson Supercharger providing the boost. The front accessory drive always poses a challenge with LS swaps and supercharger kits. As shown here, an accessory drive courtesy of Wagner provides the supercharger belt drive and the main accessory drive kit.

DIY: BUILDING YOUR OWN ENGINE

If you have built a few engines and have a quality machine shop to use, building your own engine can be a satisfying experience, especially so if you have tackled the rest of the car or have done everything with your own two hands, such as a restoration. I wouldn't recommend your first engine build to be LS-based, but it is possible. It's just a little better to have a working knowledge of how things go together before jumping headfirst into these engines.

To build the engine, you need to first obtain all the parts for at least the short block so that proper machining can be accomplished. This would include the block and pistons mainly, but shortly after the block is machined, the rotating assembly (crank, rods, bearings, and pistons) would need to be balanced to finish the machining steps before mock-up. From that point, it's a matter of checking clearances and, in short, carefully assembling all the components.

If building your own engine sounds like something you would enjoy, I will shamelessly plug one of my other manuals, *How to Build and Modify GM LS-Series Engines,* which covers every step of a typical street engine buildup from start to finish, including certain steps that are specific to LS-series engines. If you have an LS engine that is in need of a buildup, the same guidelines apply as do to a brand-new engine buildup.

In this book you will find examples of accessory drives, air intake kits, and performance modifications. This 1969 Camaro is equipped with a clean LS swap and a slew of tasteful upgrades from Detroit Speed. I am positive it performs on the track as well as its workmanship shows here.

Chapter 2
LS Engine-Mounting Solutions

Properly locating and mounting your LS-series engine into the desired chassis can be either a headache or as painless as spending 10 minutes bolting on conversion motor mounts. The deciding factor is how unusual your LS swap candidate is. As you probably know, the more popular LS swaps such as GM rear-wheel-drive muscle and sports cars are a road well traveled that allow many options for engine mounting. Many companies offer these solutions, and it's rare that there's only one choice unless it is an unconventional build. Obscure swaps such as Pontiac Fieros and Mini Coopers will present a more challenging bolt-in conversion, although that certainly does not mean you cannot fabricate what you need yourself. It is not unusual for mounts to be custom-fabricated, even when bolt-in conversion parts are readily available.

THE ISSUES

As stated earlier, the LS-series engines are 99% identical externally. This allows a multitude of engine swap possibilities between vehicles originally equipped with LS engines, much like the original SBC's well-known versatility.

While the LS engines are classified as being in the Small-Block Chevrolet (SBC) family as Gen III and Gen IV variants, they do not share engine-mounting similarities with any other prior SBC generation of engines. The original Gen I SBC used three motor-mount bolts in an upside-down triangle shape throughout its many decades of production. This design was shared only with its larger brother, the Big-Block Chevy (BBC). Gen III and Gen IV engines use several differing mounting techniques based on the various chassis in which they were installed. Depending on application, LS engines may use three or four of the square pattern motor-mount bolt locations on each side of the block.

Of the popular LS conversions the easiest to swap seems to always be GM vehicles. This makes a lot of sense being the LS engines are GM-designed. Other makes, such as Ford Mustangs and a variety of popular and ideal LS-swap candidates of foreign vehicles, also have motor-mounting options available now. It is amazing how many differing applications are endowed with the LS engine package.

There are a large number of engine-mount adapter plates available to fasten conventional SBC engine mounts to the LS engine in a few differing locations (depending on the adapter or conversion manufacturers goals). The non-vehicle-specific universal adapters usually have a fixed location such as standard and 1-inch forward or back. Dirty Dingo has unique slider engine mounts that have adjustable slots on the engine side and retain the factory frame-mounted clamshell for

Third-generation F-bodies such as this 1991 Camaro 1LE's make perfect swap candidates. Many of the required components—including engine mount adapters and brackets—are readily available for a true bolt-in engine swap process. These Dirty Dingo slider mounts (P/N DD-3550C) allow 2½-inch movement forward or ½-inch rearward adjustment from the original SBC location, offering a precise, strategic mounting position for oil pan, exhaust, and transmission clearance.

OEM Corvette 1997 and newer C5 and C6 engine mounts use fluid-filled rubber mounts to dampen engine shakes and rattles. These mounts may be useful to use in street rods or pre-1960s vehicles where some fabrication is already required.

The 1998–2002 F-body engine mounts use clamshell engine side mounts and cast-iron frame-mount pedestals. You can likely use the engine side clamshell mounts in any application where the frame-side mount is fabricated.

100% front-rear adjustability in engine placement. Another option is to purchase vehicle-specific engine mounts made for your exact swap candidate. These vehicle- specific mounts will usually employ a frame-side mount and a matching engine-side mount to locate the engine centerline left-right and forward-back to match a certain header or specific oil-pan clearance.

For the more hard-core swappers with the required know-how and equipment, fabricating custom engine mounts is not out of the question. Fabricated mounts can range from the basic everyday swaps to the more obscure unique vehicles. If you have the ability to build your own mounting equipment, that may give you the advantage if something else does not quite fit as precisely as you require.

Depending on which transmission is selected, you may need to relocate, modify, purchase, or build a new transmission crossmember. Of course, if you had a TH-350 or another SBC GM transmission and choose to retain it, you may not need to modify the transmission mount at all. This would delay the onset of new gray hair or possibly lessen habitual tool throwing.

Other issues relating to the engine-mounting location are header and accessory clearance. Often unexpected interference between items such as steering shafts, steering gearboxes, oil pans, and even the body or subframe can pose problems. These are all issues that you can either deal with in the trial-and-error installation by testing close-appearing parts or by procuring full kit-form LS-swap items.

ENGINE-MOUNTING SOLUTIONS

In many rear-wheel-drive GM platforms, simply using the adapter plate technique suffices for 9 out of 10 swaps. These adapter plates can be purchased from retrofit companies such as Dirty Dingo Motorsports, Hooker, and Speed Engineering. They are all the same basic design being laser cut from aluminum or steel plating and machined into a completed product from that point.

Some header kits come with their own required engine-mounting adapter plates, which allow proper clearance when using these plates with the same manufacturer's header kit. One such example would be the Hooker LS1 headers for 1967–1969 Camaros. The Hooker mount adapters set the engine in a specific location to allow optimal clearance for use with the Hooker LS1 conversion header kit and Holley oil pan.

Another option for motor mounting is to use the 1998–2002 LS1 F-body engine-side "clamshell" mounts in conjunction with custom-made or purchased frame-side bolt-in engine mounts. This style works best with GM vehicles that have either factory frame-side solid or clamshell mounts. This is because the OEM non-LS engine mounts are ditched in favor of the new frame-side mounts that mate the GM LS1 F-body mount to the vehicle. This technique alleviates the adapter plate from the equation, as you are now using only an engine side mount and a frame side mount, offering a cleaner installation. An example of this is Hooker BlackHeart engine-mount bracket solutions.

Avoiding the use of adapter plates is also a trend with a few retrofit companies. These specialists offer both custom frame side and engine side mounts, skipping the OEM components altogether. The mounting kits supplied by

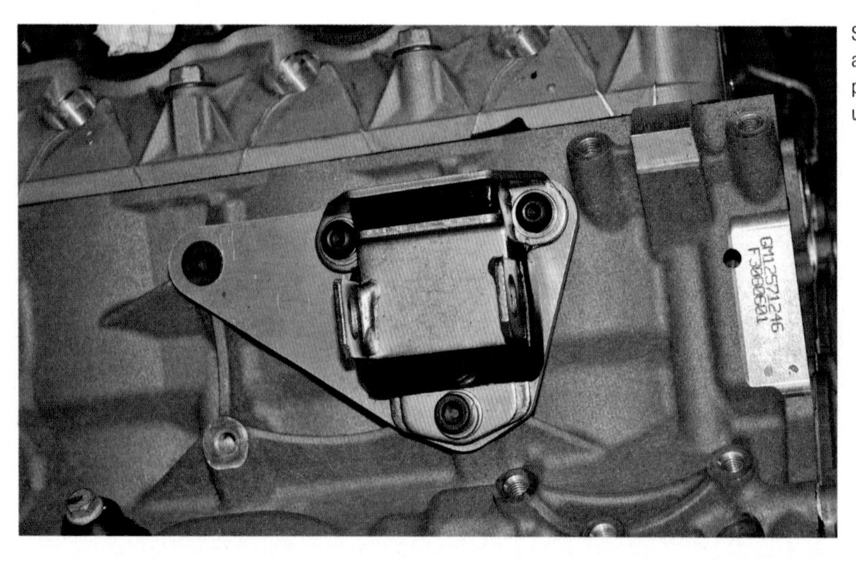

Some subframe kits meant for LS1 conversions also adhere to the motor-mount plate setups. This aluminum plate setup is for the total cost involved subframe when using an LS based engine.

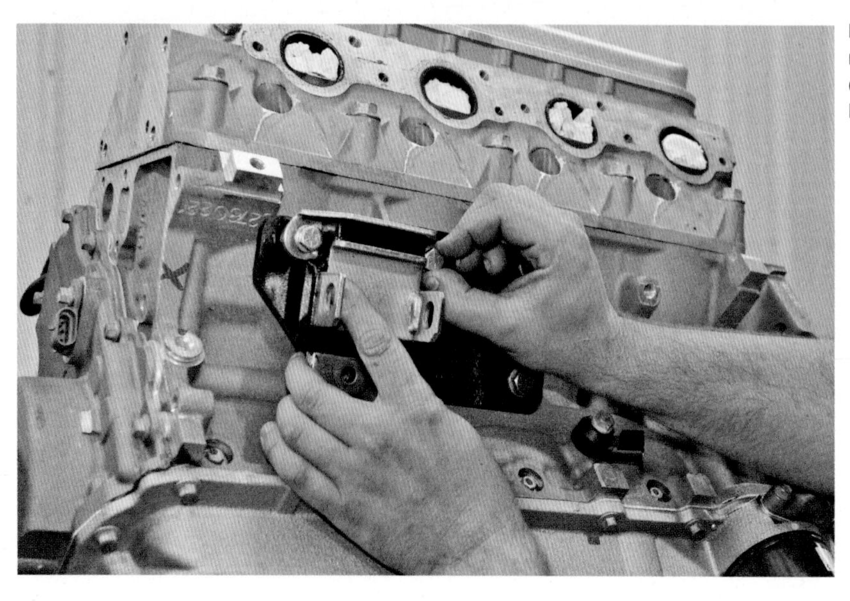

In some instances, 1-inch setback motor mount plates are required. Whether for increased oil pan sump clearance or exhaust header clearance, the mounts shift the engine's location in the chassis back exactly 1 inch.

Without the frame-mount pedestals, you can see the clamshell-style engine side mount. Some people have had good results earlier on in the LS swap timeline by using these mounts and some cutting and welding skills for the mating frame-side mount. These days you can buy most applications as bolt-in motor mounts, but fabricating is always an option.

such companies are specifically paired to that exact LS swap candidate. Custom-dimensioned mounts also allow swap-specific headers to be used from the same source for those who do not like to do the trial-and-error installation routine. Typically, the new engine mounts are polyurethane bushing equipped so as to dampen engine shakes and rattles (better than solid mounts, but not as good as OEM engine-mount dampening). These might be ideal for someone unfamiliar to the swapping routine and not against spending the extra coin for convenience.

There are instances and engines that may require the engine to be shifted farther forward or back in the chassis. There are also minor adjustments, higher and lower, for items such as the header, oil pan, or accessories. Some kits locate the engine based on the OEM engine's mounting location, which may be ideal for front-mounted engine accessories but is sometimes not ideal for some oil pan clearances. It also weeds out certain header choices at times. The engine mounts that shift the engine farther back (similar to something like the OEM SBC location) allow increased oil pan sump clearances and improved header clearance but at a cost of limiting things like OEM air conditioning compressor choices. This is mainly due to where the air conditioning (A/C) compressor is located, as the LS location is a low-mount that fits very few swaps without fabrication work. If OEM A/C is a requirement, then keep these engine-mounting factors in mind. Notching the subframe or using aftermarket accessories and mounting solutions are a requirement if you have chassis-to-accessory interference and do not wish to cut, weld, or fabricate to shoehorn your parts into place. Keep in mind that an alternative top-mount bolt-in A/C compressor kit is often the easiest solution.

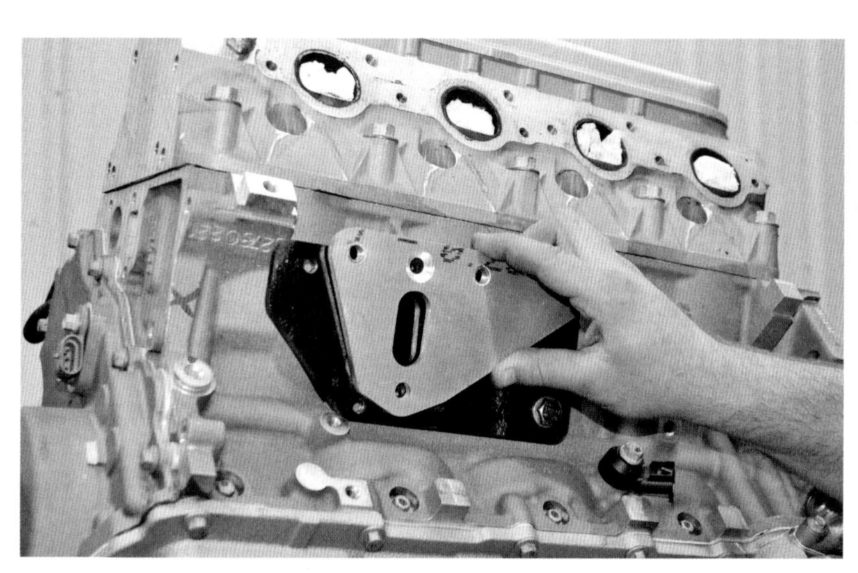

Comparing standard location motor mount plates to the 1-inch setback mounts; you can easily see the difference. One isn't really better than the other, but if you need the extra room, the 1-inch will help out. It seems that most applications require standard location mounts; at least attempting these first will give you a good point of reference.

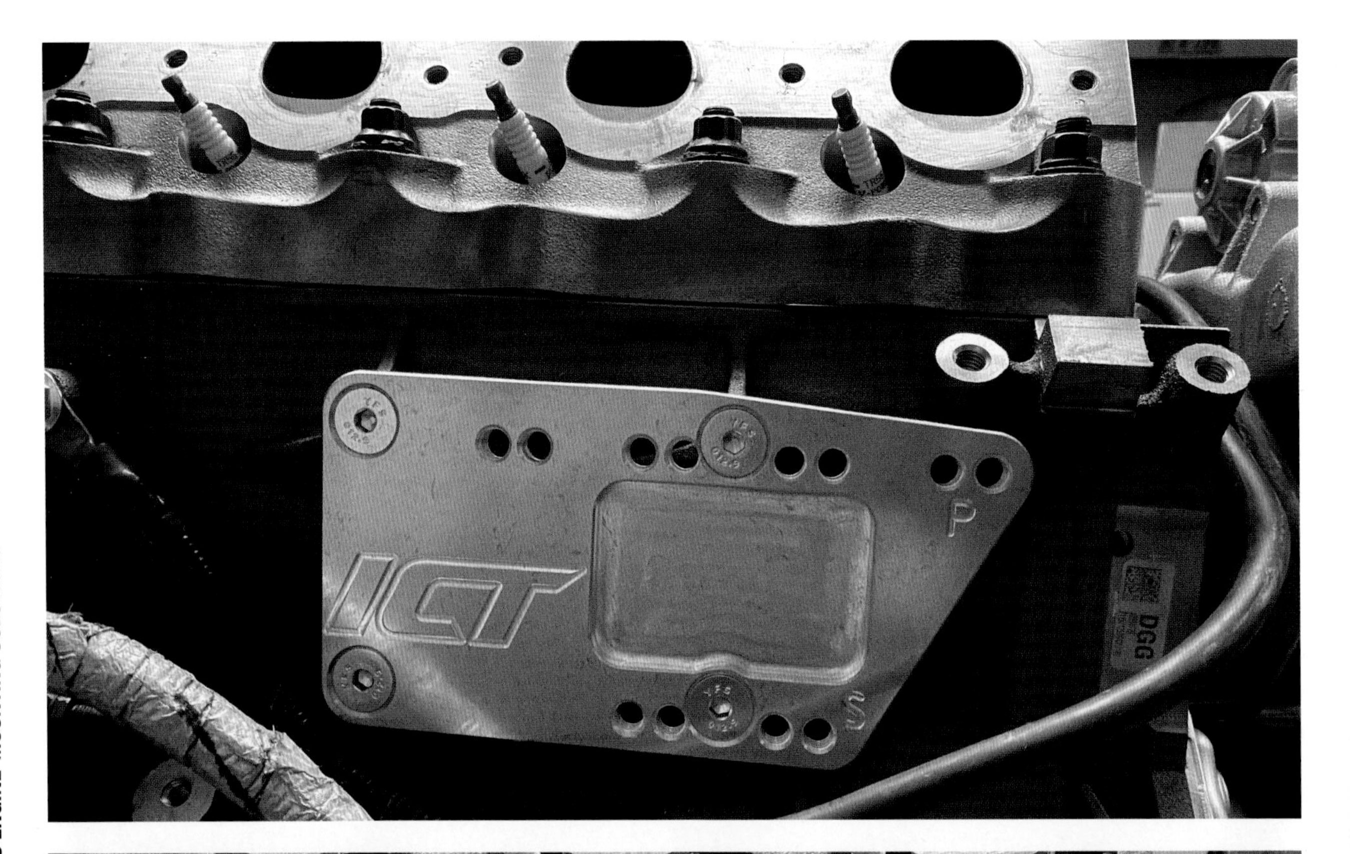

The ICT Billet SBC-LS adapter mounts (P/N 551628) allow for either original frame-side rubber mount adaptation or engine-side motor mount adaptation. They have several mounting points to move the engine forward or rearward depending on application needs. Shown here in my 1967 Camaro along with the tall replacement Prothane (P/N 7-504-BL) engine mounts. The taller version of the mounts help with extra oil pan to crossmember clearance if needed.

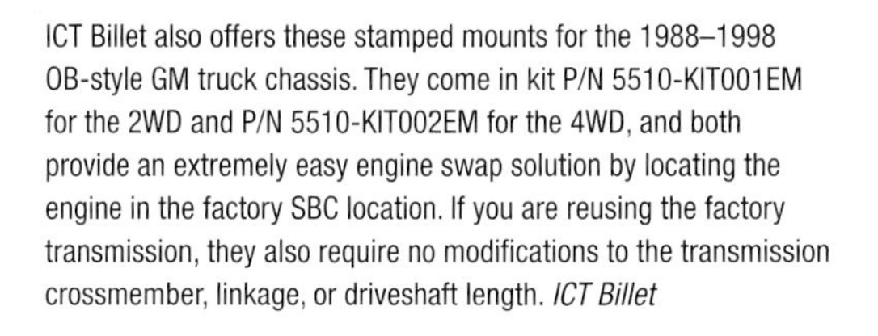

ICT Billet also offers these stamped mounts for the 1988–1998 OB-style GM truck chassis. They come in kit P/N 5510-KIT001EM for the 2WD and P/N 5510-KIT002EM for the 4WD, and both provide an extremely easy engine swap solution by locating the engine in the factory SBC location. If you are reusing the factory transmission, they also require no modifications to the transmission crossmember, linkage, or driveshaft length. *ICT Billet*

Sometimes you may find that the powertrain will fit better if it just had a little more vertical clearance at the transmission tunnel area. The easy answer is to just flip and swap the motor-mount plates from side to side. This will give approximately ½ inch extra clearance at the top of the engine and transmission.

Flipping the motor mounts upside-down does make it a little harder to align, but any extra clearance that helps avoid interference is a good thing. When coupled with a 1-inch setback mounting plate, the engine is now 1 inch further rearward and ½ inch lower than when using a standard location mounting plate.

When swapping LS-series engines into early pre-1968 GM A-body vehicles, you would be better off using '68 and later style "tall and narrow" style frame and engine side mounts. Pre-assemble the engine mounts on the engine and then set the engine in place, then level and adjust as necessary for oil pan, exhaust, transmission, and steering clearance. At that point you can mark and drill the frame for new frame-side motor mount bolt holes.

The BRP mounting kit locates the engine in a proven location that BRP has found to work during actual vehicle testing. This LSX-powered Bandit Trans-Am is an employee buildup being performed at Mast Motorsports. Second-generation F-bodies are extremely popular LS-swap candidates, along with the other non-LS F-body generations. *Mast Motorsports*

SPECIFIC ENGINE MOUNTS

TRANS-DAPT: WWW.TDPERFORMANCE.COM

TD offers motor-mount adapter plates in the simplest form. They allow you to adapt your SBC engine mounts to the LS engine in the factory location and optionally allow a setback of 1 inch, if desired. These are cut from steel plates and available with or without early-style replacement GM 3-bolt rubber motor mounts. These mount plates can adapt both early and later-model GM V-8 engine side mounts to most SBC-equipped vehicles and are very inexpensive.

CAR SHOP INC.: WWW.CARSHOPINC.COM

Car Shop has a few different engine-mount solutions for several different vehicles. They have a selection of motor-mount plates to adapt to both early V-8 GM cars and to set the engine back 1 inch in such vehicles. Specific engine mounts available for S-10 trucks, C3 Corvette LS swap applications, and weld-in style for street rod buildups. The 1-inch setback mounting plates are helpful if you have a ton of firewall-to-engine clearance.

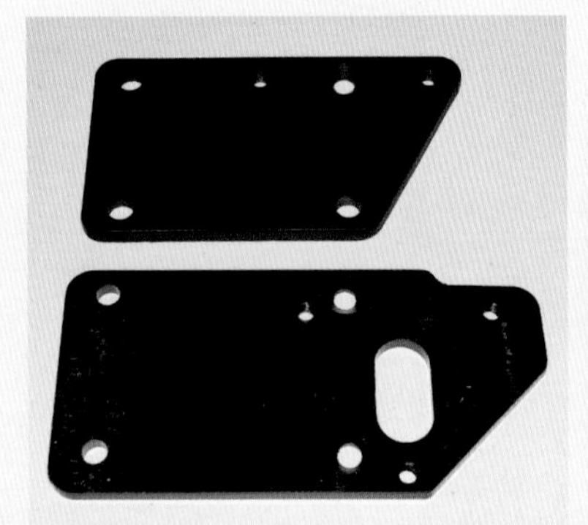

One thing that has made LS swap life easier was the introduction of the now-commonplace adapter plates. There are many versions and manufacturers. The top plate is the Trans-Dapt mount adapter in the standard mounting location. The bottom is the Car Shop 1-inch setback plate (with motor-mount relief hole), which sets the engine 1 inch farther rearward than the standard location mounts.

Carshopinc.com has a few popular motor-mount plate designs. The standard location and 1-inch setback is the most commonly used with GM V-8 chassis. The flat, flush-mount plates are used with stock 1970 and newer stamped-steel motor mounts (with rubber-frame side mounts). The little stamped dimple on the rear of the steel motor mounts will need to be ground and blended smooth. *Car Shop Inc*

Carshopinc.com also manufactures adjustable engine mounts for LS1 S-10 conversions. The adjustable nature allows fine-tuning and tweaking to get the engine exactly where it needs to reside, if minute adjustments are required. *Car Shop Inc*

Using adapter plates works well for both engine-mounted clamshell mounts and frame-mounted clamshell mounts. Many builders will purchase new OEM-style rubber mounts or the energy suspension polyurethane motor mounts at the same time. Poly mounts significantly firm up the engine mounting compared to OEM mounts. This allows less deflection under acceleration.

Some engine mount solutions are specific to the platform, as in this Saturn Sky LS swap. These mounts are from Kappahaus and use OEM C6 engine mounts for the isolator. Many of the chassis-specific engine mounts place the engine in a designated location with minimal adjustments and require complementary headers, oil pan, and accessories, making a matched swap setup of components.

ICT Billet universal adapters (P/N 551628) can be used with engine side mounts, or they can be used with frame-side OEM GM mounts when adapting the factory-stamped steel SBC mount. Such mounts are typically found in 1970s–1990s GM V-8 RWD vehicles. *ICT Billet*

DIRTY DINGO: WWW.DIRTYDINGO.COM

Dirty Dingo also has quite a selection of specific engine mounts for a wide variety of vehicle applications. The DD mounts water jet cut from 0.250-inch thick plate steel and offer about 1 inch of forward and up to 2 inches of rearward adjustability via slotted mounting holes. Using these mounts, you can dictate where exactly the engine sits in the engine bay. Some of the DD mounts are designed to be solid and in some applications are intended to be used with Moroso SBC solid-frame mounts, and many have also been designed to use SBC OEM rubber frame mounts. The adjustability is helpful if you need extra oil pan-to-steering clearance or header-to-frame or body clearance. Once the proper location is found, you simply lock the bolts down, securing the mount solidly in place.

STREET AND PERFORMANCE: WWW.HOTRODLANE.CC

One of the pioneer companies that were the go-to people for LS-series engine swaps from the beginning was Street and Performance, which has a full selection of motor-mounting solutions to mount your LS-based engine into most GM V-8 applications using SBC frame mounts.

BRP: BRPHOTRODS.COM

Brewers Performance manufactures complete LS-swap kits, including motor mounts, oil pans, and headers for specific chassis as an all-in-one package. The BRP mounts normally come with both frame and engine side mounts and are poly bushed to help dampen engine harmonics and vibrations. Many of Brewers current mounting kits are intended to be used with the Hummer H3 engine oil pan. Older kit designs were equipped for the 1998–2002 LS1 F-body oil pan when used with front-steer rack-and-pinion systems, and the mounts pushed the engine farther forward and upward to clear and allow OEM A/C compressors to fit better, before a few aftermarket accessory drive systems became available.

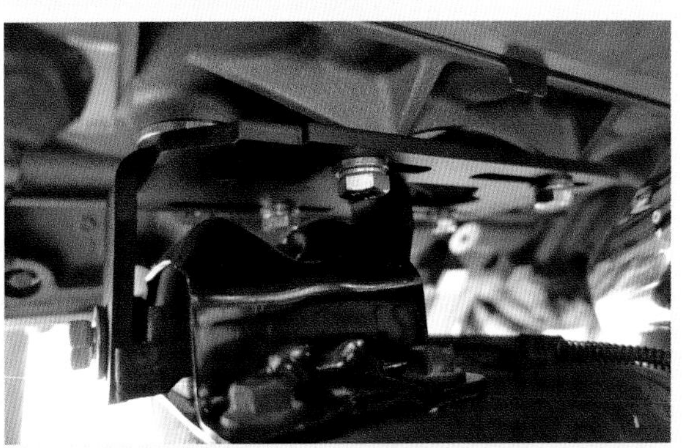

The Dirty Dingo adjustable slider mounts helped set the oil pan for optimum clearance in my 1991 Camaro. Since I used the factory subframe, minimal adjustment was needed once the engine was in place. I did move the engine ¼ inch rearward before locking down the mounting bolts, which assisted the 4th-Gen oil pan clearance with its snug fit.

HOOKER HEADER LS MOUNTS: WWW.HOLLEY.COM

Hooker LS retrofit headers require hooker-specific engine-mount adapters for the early Camaros and Chevelles (and variants). These set the engine back roughly an inch to allow proper header clearance relative to the steering box. These are really only required with these specific headers but will probably work for a few other swaps that need the engine set back just a tad further. Keep in mind that when the engine is relocated further rearward, the low-mount F-body alternator may interfere with the subframe or motor mounts themselves.

MOTOR-MOUNT FABRICATION

Some of the more resourceful LS engine swap gurus have manufactured their own engine-mounting solutions. This mounting bracket mimics the three-bolt SBC stamped-steel mount but uses the LS1 four-bolt–style mounting pattern. *Andrew Mattei*

The fabricated four-bolt LS1 mount setup fits snugly on the factory SBC frame-mounted engine mount. If you have the skills and equipment to do so, a custom-mounting setup such as this can help position your engine exactly where you want it. *Andrew Mattei*

Andrew used a core mock-up LS1 block to test his motor-mount designs on a second-generation Camaro. Looking closer at this image, the oil pan has ample clearance, but the A/C area of the subframe will need to be notched for the A/C compressor, like most. Also, if you take a look at the GTO alternator, you can see that the pitman arm is close. A swap to the F-body alternator should improve clearance. *Andrew Mattei*

Custom applications such as off-road rigs or oddball classic cars and trucks may not have bolt-on solutions. This is where a custom mount may need to be fabricated. Luckily, several companies provide a starting point to build from, such as these weld-in Speed Engineering engine mounts. Tip: Build the frame templates from cardboard and tape first, then transfer this pattern to the metal tabs before cutting and welding.

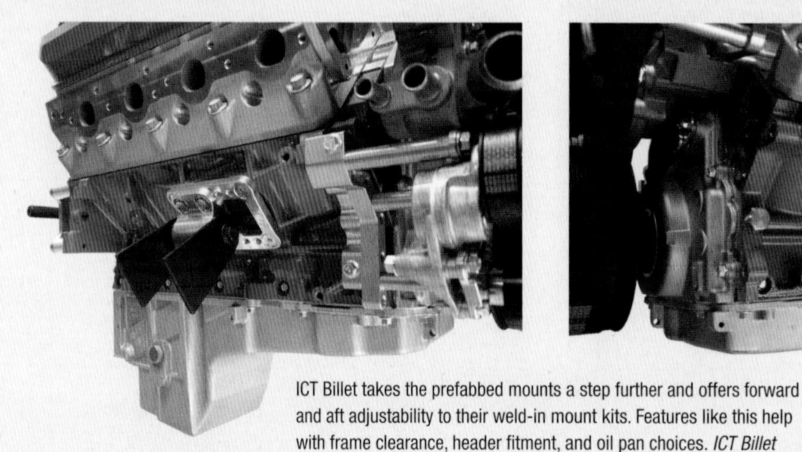

ICT Billet takes the prefabbed mounts a step further and offers forward and aft adjustability to their weld-in mount kits. Features like this help with frame clearance, header fitment, and oil pan choices. *ICT Billet*

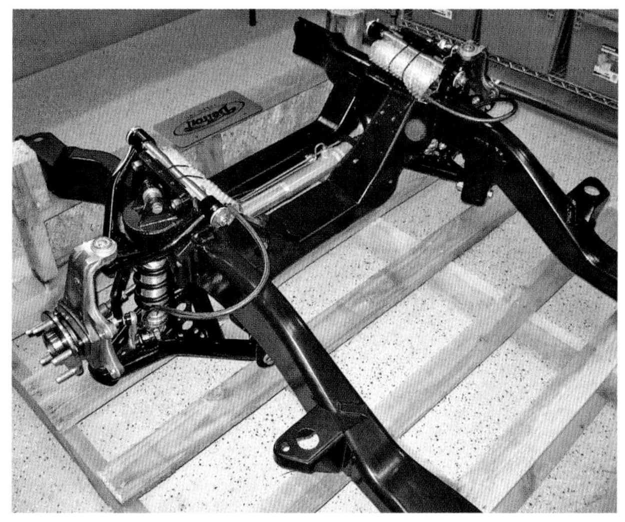

If you have an early muscle car such as a 1967–1969 Camaro or Firebird, there are subframe options available. If you do not want to use the stock subframe and want a dramatically improved high-tech setup, you can get something like the Detroit Speed and Engineering Hydroformed front subframe. It is equipped with C5/C6 braking provisions, front-mount rack-and-pinion steering, and an improved suspension design. *Mast Motorsports*

The transmission mount location is one thing many people forget when performing a swap. Thankfully, with crossmembers available from many manufacturers, this is no longer a problem. This Hooker LS-Swap T56 Crossmember (P/N 71222995HKR) for the 3rd-Gen F-body is CAD designed for fitment and also has an integral torque arm mount for ease of true bolt-on installation.

SUBFRAME OR FRAME REPLACEMENTS

Tubular subframes save weight and open up the underside of the car. Problem areas for components such as oil pans and header clearances quickly become a nonissue with a tubular subframe. UMI Performance, for example, has a direct bolt-in design modular K-Member for third-generation Camaros that can be optioned for LS engine frame mounts, OEM or coil-over springs, and UMI front lower control arms. These K-members shave off 20 pounds of weight and are a great upgrade for street, drag, or autocross events.

Cars such as the Ford Mustang have available replacement subframe choices that allow bolt-in LS-engine packages. Since it is a non-GM swap, other areas need attention that sometimes would not need much thought. Transmission mounting is one area, as these can differ greatly in the non-GM vehicle. Of course, in a non-GM car you can expect to replace the transmission and possibly buy or build a new transmission crossmember. This is due to non-SBC–style transmissions not physically bolting to the LS bell housing flange.

Subframe or frame replacement can be an option for even older vehicles such as F100 Ford pickups and first-, second-, and third-generation Camaros and Firebirds. Even kit cars can be tailored around the LS engine's dimensions. There are a handful of companies that upgrade suspension, brakes, and steering for older vehicles, so while upgrading the powertrain with the LS-swap, you can also upgrade other systems at the same time by replacing the frame or subframe with newer drivetrain technology.

Roadster Shop has added full-frame swaps to its well-engineered SPEC chassis lineups. If you are restoring an

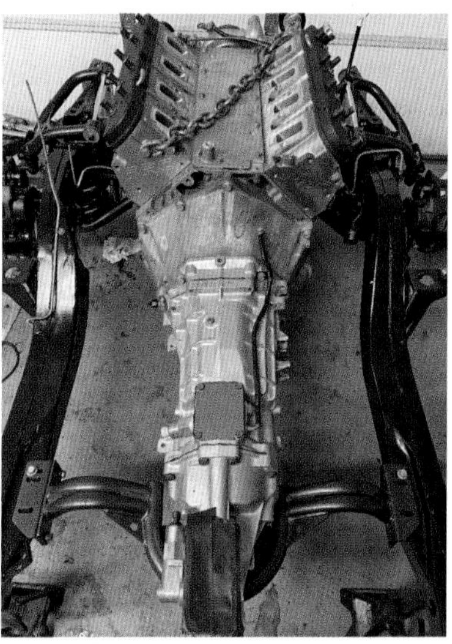

Many suspension companies also have chassis-specific bolt-in crossmember kits. Shown here is the BMR Fabrication T56 crossmember for the 1st-Gen Camaro. Note that I had to drill two holes in the subframe for the crossmember to allow for oil pan clearance as I moved the engine further rearward on the ICT Billet adjustable mounts.

older vehicle and need to address multiple chassis upgrades, Roadster Shop's upgrade solution covers everything that directly bolts to the frame itself. Suspension, brakes, brake lines, and mounts are already installed and ready to bolt under your classic car. It's not a cheap solution if you only need motor mount adapters, but it is definitely a value for builders who need everything at the next level. These SPEC chassis replace your floppy factory leaf-spring setup with modern sports car coil-over springs and underpinnings while allowing you to retain your classic's timeless body styling.

One of the easiest "bolt-in" swaps as far as the mechanical side of things is the 1993–1997 F-body. Though these years were not equipped with the LS engine, the same chassis was still in use during the 1998–2002 model years, which were LS-powered. To retrofit an LS-based engine to this application, all that is needed for engine mounting is the newer LS-compatible front subframe, which handles bolting the engine in as if the car originally came equipped with the LS engine. Wiring and fueling is a different story, but the physical labor portion of the installation is quite easy.

TRANSMISSION CROSSMEMBERS

Transmission mounting may be accomplished after the engine is properly located and addressed as a second priority after engine mounting. In some tight and complex installations, the transmission can be installed at the same time to take into account interference that may be easily remedied by rethinking engine location.

In most GM rear-wheel-drive swaps, transmission tunnel interference and the transmission-mounting crossmember are the most important things to be addressed. GM vehicles are, of course, the easiest, as these often use similar transmission mounting lengths and locations, or at least are close enough to be modified to accept the LS or other GM-compatible transmission choice. For the other popular swap choices, or those for which full swap kits are available, much of the homework has already been accomplished and popular transmission options can be as easy to mount as making a phone call and busting out the credit card.

Swaps that deviate from the more common Tremec T-56 or GM 4L60-E transmissions are the ones that will take some extra thought and planning. These more common transmissions have undergone years of swap development. The six-speed and eight-speed automatics such as the 6L80/6L90 and 8L90 require location-mounting solutions. G-Force Performance Products offers many conversion crossmembers that make installing these more modern transmissions a bolt-in affair. In other obscure-vehicle setups you will need to either fabricate a transmission crossmember from scratch or build off the crossmember that was supporting the OEM transmission. Many people have been able to use a universal crossmember from places such as a DIY G-Force Crossmember or Competition Engineering to fit their powertrain's dimensional needs.

Some vehicles, such as GM trucks that have a full frame, simply need the old crossmember relocated to its new location. These lucky builders only have to bolt in the transmission mount, then mark the frame where the crossmember holes line up, drill new framehole locations, add bolts, and—*voilà*—they're done.

As discussed earlier, if you are mating an OEM transmission to a GM chassis, you do have it much easier with transmission mounting. You will also avoid the requirement to have the corresponding ancillary components such as redesigning and fabricating the shifter linkage or adding provisions for electronically controlling the transmission, as long as the engine-to-transmission mounting point is retained.

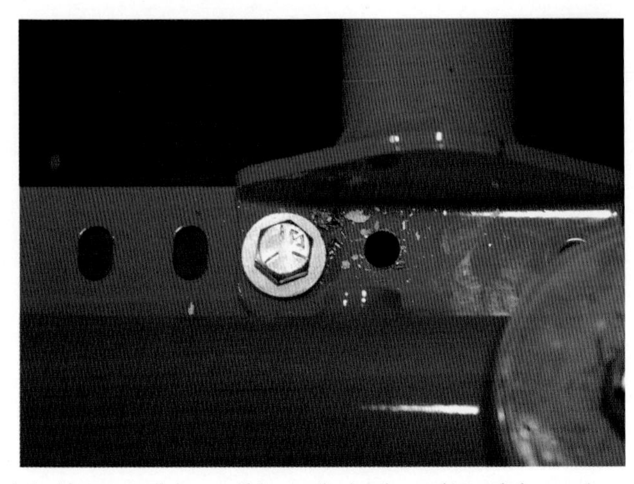

Many transmission crossmember kits are available, both bolt-in and weld-in. The aftermarket subframes usually have multiple mounting locations and transmission mount provisions due to the many different powerplant designs that are used. Even with multiple crossmember mounting locations, sometimes the transmission length is longer than what the subframe provides for. In this case, it was a matter of drilling 2 extra holes for the crossmember.

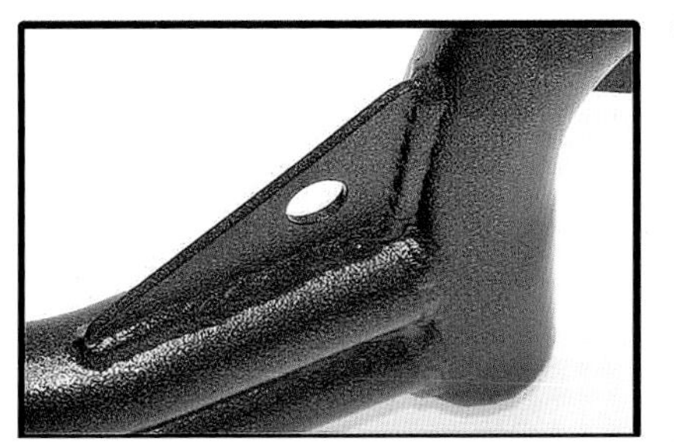

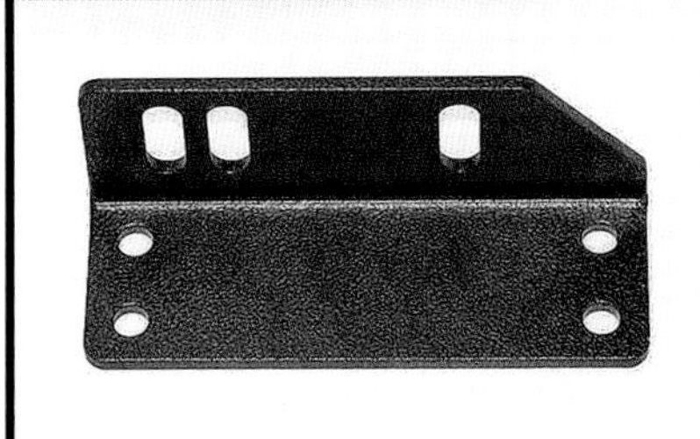

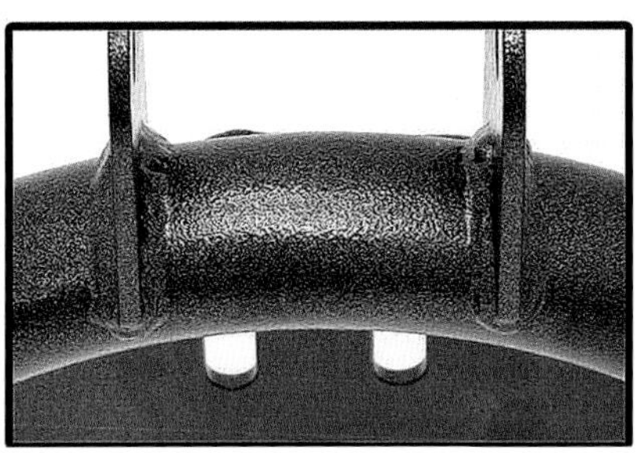

BMR Fabrication also offers a T56 mounting kit for first-generation F-bodies. The BMR kit comes with the beefy transmission crossmember, transmission mount, and hardware. A hydraulic master cylinder mounting plate is also available. *BMR Fabrication*

As with the engine mounting solutions from many specialty LS-swapping companies, if you have the common "LS-style engine in a GM vehicle" swap, all of these items can be purchased and bolted into place for a non-hassle installation. Often, using these solutions are easiest as these builders have already been down this path.

Manual T56 six-speed–style transmissions in conventional LS swaps usually have crossmembers available in kit form. These typically include the required components to retrofit the OEM F-body clutch hydraulic system. You will quickly find out that the most popular swaps are the easiest due to manufacturer support and others that have paved the way beforehand. On the swaps where fabrication is required in every step along the way, consider that if you can build your own motor-mount fixtures, you can likely fabricate or build a new transmission crossmember fairly easily.

Mechanical linkage transmissions are best left in their original location to simplify and prevent binding the original linkage design, but this is difficult to do and does not play well with items such as new exhaust or motor mounts. If you retrofit the clutch mechanism to a hydraulic setup, this is not so important, as the flexible hydraulic hoses allow much more flexibility for transmission location.

Some six-speed manual options do require special effort to function well. The TR6060 from 2010–2015 Camaro is a very capable transmission choice, but it makes for a more difficult swap compared to the common T56 choice. The bell housing is a special depth to allow for the factory twin-disc clutch usage, and it also has a remote-mounted shifter, which typically needs to be discarded for an aftermarket, swap-friendly shifter assembly.

CLUTCH

Attaining proper clutch operation can sometimes be tricky. Many older muscle cars use a mechanical linkage and clutch fork to disengage the pressure plate, which in turn release the clutch disc. Since the mid-1980s, mechanical linkage was superseded by hydraulic clutch technology in most GM vehicles. The hydraulic clutch system works much like a hydraulic brake system. Input is provided via the clutch pedal, which transfers the fluid pressure from the master cylinder into the clutch actuator or slave cylinder via a flexible hydraulic line. It is a fluid leverage system, with the inside piston diameter of each cylinder affecting how much travel is required and how much pedal pressure is required to disengage the clutch. Refer to Chapter 7 for more in-depth transmission discussions.

Both aftermarket and stock windage trays fit the same way and install almost mindlessly. Afterward, you can install the pickup tube. When installing a new pickup tube, make sure to use a new oil pump O-ring and some slick engine lubricant on the O-ring to alleviate possible low oil pressure problems from an ill-fitting pickup tube O-ring.

Once the pickup tube is in place and secured, you can install the oil pan. It's easiest to do this on an engine stand, but you can also install it suspended from an engine hoist, although it is a little more tedious. You can use some RTV silicone on the four corners where the rear cover and timing cover meet the oil pan flange to alleviate oil seepage. This is a GMPP muscle car oil pan kit being installed.

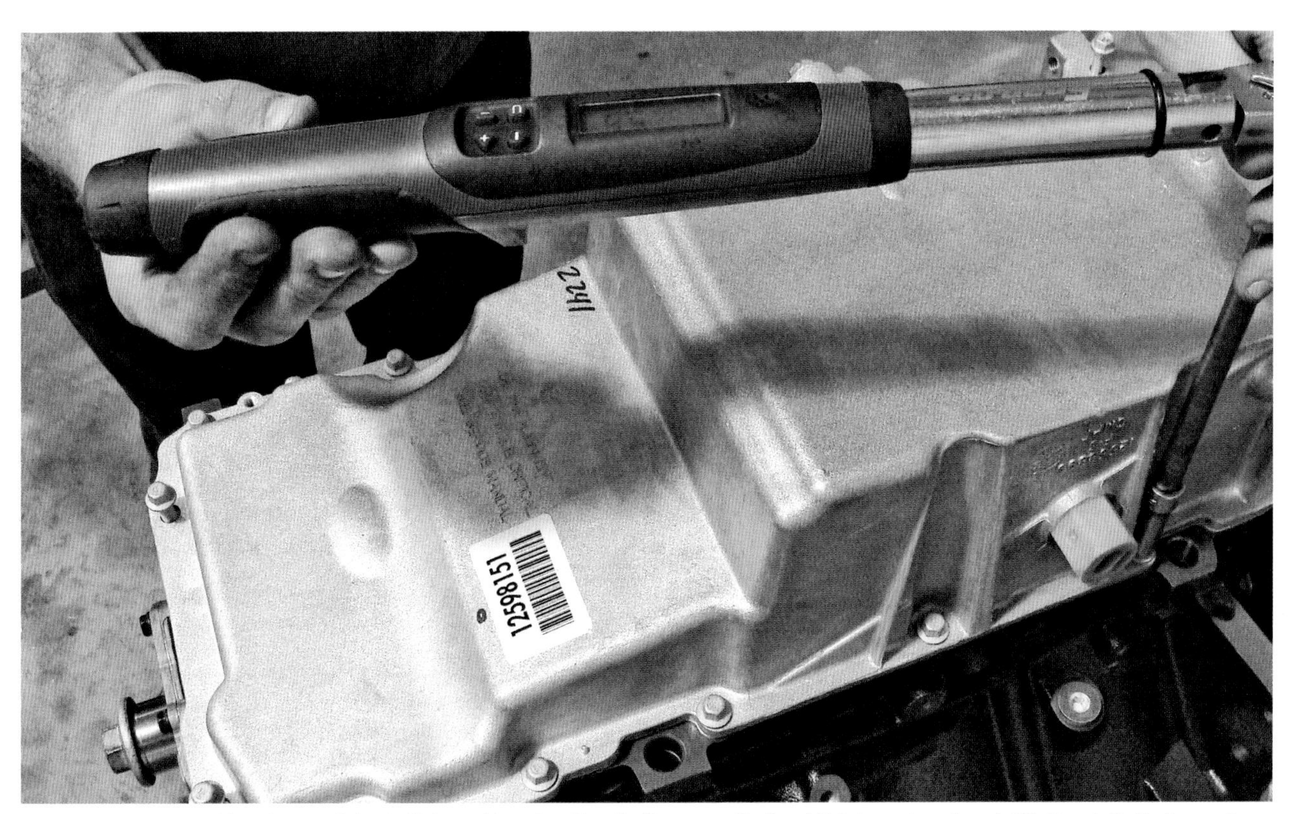

The oil pan must be torqued from the centrally located bolts, working outward to a final torque specification of 18 foot-pounds on the main M8 oil pan bolts. The two smaller diameter but longer M6 rear cover bolts get torqued to 106 inch-pounds. An F-body oil pan is being installed in this image.

Once the oil pan is in place, nothing more needs to be accomplished. The LS6/LS2 CTS-V oil pan is shown installed here.

The OEM F-body clutch master cylinder may be retained if suitable mounting or adapting is provided. Aftermarket universal master cylinders can also be adapted to function properly. Depending on the distance between the master cylinder and the clutch actuator, you may use the OEM braided hydraulic line, an aftermarket line, or a custom-built -3AN or -4AN pressure line for unconventional setups. Some swaps can be accomplished reusing the OEM donor lines.

PILOT BEARING SITUATION

GM made a change to the pilot bearing design in some 2005 model year vehicles, specifically C6s, but other models may be affected. The input shaft is shorter on 2005 and newer transmissions and C6 torque tubes. Correspondingly, the pilot-bearing design has to match; hence, there are two different design pilots. The early smaller bearing has a larger outside diameter and matches the size of a torque converter centering hub, as that is where it resides when pressed into the rear of the crankshaft. The LS1 early smaller bearing (P/N: 14061685) is the same dimensions as the Gen I SBC pilot, but is a roller bearing. The newer Gen IV–style pilot bearing (PN: 12557583) has a larger outside diameter and matches the size of a torque converter centering hub, as that is where it resides when pressed into the rear of the crankshaft. The newer pilot bearing is utilized in such cars as the Corvette C6 and 2010+ Camaro, but also allows older four- and five-speed SBC-style manual transmission usage.

Engine design does not dictate which pilot bearing is required; it depends on which transmission is used. The more popular bearing is, of course, the older design, but the later-model pilot bearing has an additional perk: the ability to allow the use of all conventional three-, four-, and five-speed manual transmissions with the proper flywheel and clutch setup. The pilot bearing location takes up the distance that the shorter LS crankshafts originally took away from the rear crankshaft flange as compared to SBC engines. The difference between the rear crank flanges is about .400 inches.

Warning: If you use the later-model pilot bearing with the LS F-body T56 transmission, you will cause extreme engine and transmission wear due to excessive pressure exerted on the crankshaft thrust surface and input shaft of the transmission. Make sure to match the correct pilot bearing design to the right input shaft length. It is a simple thing that can otherwise have dire consequences to the life of your engine and transmission.

(continued on page 47)

SWAPPING GEN III AND GEN IV ENGINES IN GEN III/IV OEM VEHICLES

The newer Gen IV engines are so desirable and such a hit that even people with older Gen III and early Gen IV engines may want a taste of the new stuff. Not to worry, these engines are similar externally with only minor changes through the different years. There are some pitfalls though, and we are going to cover these here.

First, all external components will bolt up as the OEM engine was equipped. This includes motor mounts, the crankshaft damper (except LS7, LSA, and LS9), all engine accessories, starter, flexplate and flywheel (except LSA, LSX 454, and LS9 with different flywheel bolt counts), and exhaust headers and manifolds. The few small relocated items that you will have to install are the Gen III knock sensors to the outside of the engine block, extend the Gen III camshaft sensor harness to the timing cover–mounted camshaft sensor, and if using a Gen IV intake manifold, you will also have to extend the MAP sensor harness. If using a Gen IV engine with a bolted-in MAP sensor, you will have to adapt back to the older clip-in design for proper MAP scaling also.

If using an iron block in place of an aluminum block, there is one small annoyance with the negative battery cable ground location. The stock bolt-hole location for the negative battery cable above the starter is deleted on iron blocks—simply finding another ground point on the block is required. Use either a motor-mount bolt hole or an A/C bracket bolt hole that is not used. On C5 Corvettes, the extra A/C bolt hole will interfere with the C5 aluminum engine-mount bracket. You can either grind the non-used block bolt hole location or the motor mount itself. I prefer the motor-mount grinding, and since it is on the passenger side, it is not under as much load as the driver's side.

MODIFYING 1998–2002 F-BODY OIL PAN FOR SUSPENSION AND STEERING CLEARANCE

Oil pan modification is an unpopular project now that there are numerous aftermarket cast and fabricated oil pans along with bolt-in rack-and-pinion steering choices. The 1998–2002 LS1 F-body oil pan can be fairly flexible for use in passenger cars such as 2nd-, 3rd-, and 4th-Gen F-bodies—and several GM truck applications without modifications—but in mid- to late-1960s cars with stock steering in rear-steer applications, the oil pan sump interferes with factory steering movement and must be replaced or modified.

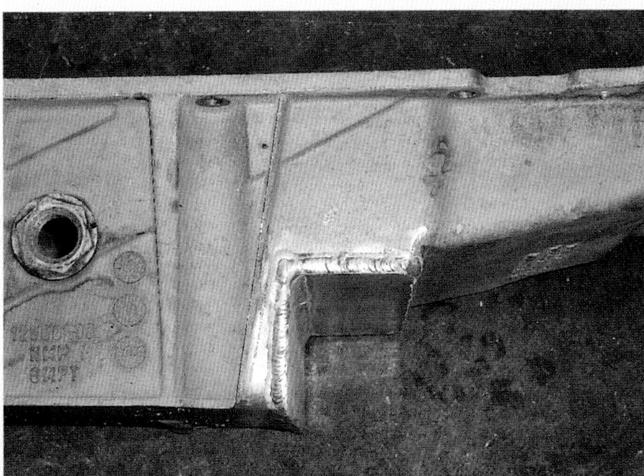

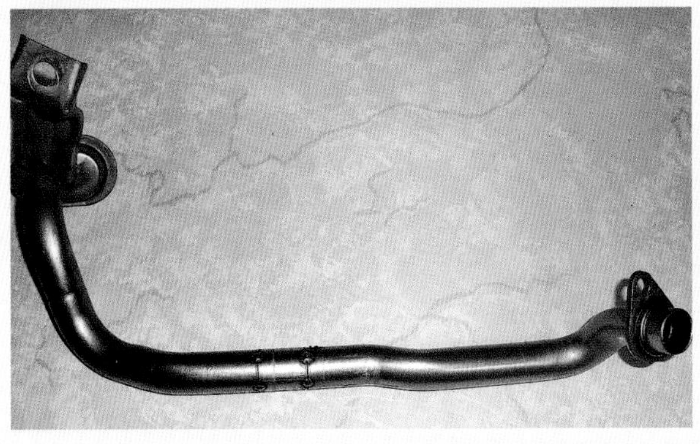

F-body oil pans can be shortened to clear rear-steer applications by notching the oil pan about 2½ inches to allow for steering movement. In addition to notching the oil pan, the pickup tube must be extended a minimum of 1 inch rearward to compensate. I cut a 1-inch section out of an old pickup tube and grafted it into the F-body pickup tube to give it the necessary length. Also, the pickup tube brace must be manipulated and tweaked to be mounted to an alternate main-cap stud. If you have the fabrication skills to do this, take a crack at this, but with the advent of LS swaps and parts come easier bolt-in solutions. Companies such as Holley now have cast aluminum oil pans that have mostly obsoleted modifying stock oil pans for clearance requirements.

LS-SERIES AND VORTEC OEM OIL PANS

1997–2004 5.7-LITER LS1 AND LS6: CORVETTE C5

The C5 oil pan is quite different than any other conventional oil pan. It uses large wings to the left and right sides of the main sump to add oil capacity and help keep the oil pickup tube covered with oil during hard cornering. Some swaps can be tailored around this pan, but many cannot use the C5 "batwing" oil pan due to frame, header, and body clearances. There are a few kit cars that use the C5 oil pan, such as the GTM Supercar. If you purchase an engine with this pan, it is sometimes much better to go another route in the oil pan choice, depending on your chassis.

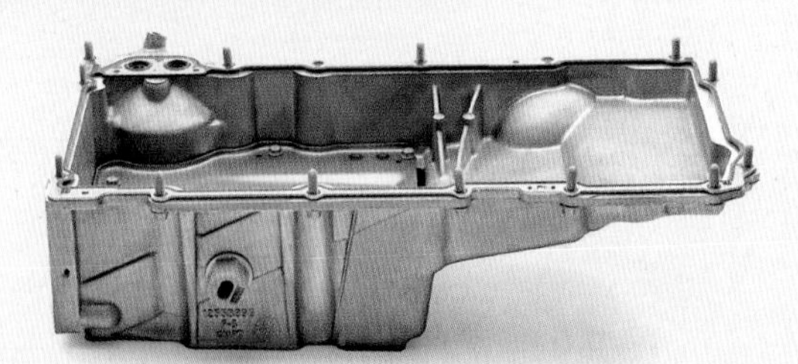

Oil pan selection goes hand in hand with engine-mounting strategy. You have to choose the oil pan based on a number of things such as frame and steering clearances and ground clearance. The most common oil pan to use is the 1998–2002 F-body oil pan (shown). *Courtesy of General Motors*

The C5 "batwing" oil pan is shown with the mating motor mounts. Rarely will such an oil pan be suitable for any retrofits unless there is an unusual amount of lateral space. Something like a rear-engine kit car may work, or possibly sand-buggies, but most passenger cars outside of an actual C5 would not be able to use this pan.

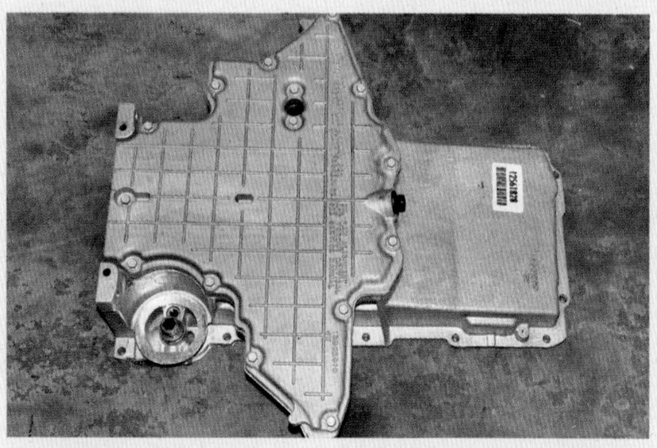

The C5 "batwing" oil pan features a few traps and baffles that allow the C5 Corvette to corner effectively without losing oil pressure. A nice feature, but the dimensions of the oil pan make it rather useless for swapping purposes.

1998–2002 5.7-LITER LS1: CAMARO AND FIREBIRD

This is the LS swapper's first choice in oil pans. The LS1 is GM's firstborn Gen III engine and is found in the popular F-body chassis. This pan is the first OEM pan, so it naturally is the pan many obtain for their LS swap. This pan can fit in some chassis without modification, or can be cut, notched, and welded to fit. Interference often occurs at the front of the sump, where contact occurs with steering and frame components.

2004–2006 5.7- AND 6.0-LITER LS1 AND LS2: PONTIAC GTO

GTOs use a front-sump system that is specific to the GTO chassis. The front-sump pan may be helpful in some import vehicle swaps or others that requires a front sump for clearance. The oil filter is relocated to the front, and the dipstick tube is located in the oil pan, versus the other OEM pans that go through the block dipstick hole.

2005 AND NEWER 6.0- AND 6.2-LITER LS2 AND LS3: CORVETTE C6

In 2005, the Corvette received a facelift, becoming the sixth generation of the Y-body chassis. In addition to getting new externals and drivetrain upgrades, the C6 received a revised oil pan. This oil pan ditches the "batwing" design of the C5 chassis and makes use of a more conventional rear-sump system. The C6 uses a longer sump instead of wider, hence the oil capacity is not much different than before. Since the C6 pan has a longer sump, it does cause interference in vehicles where the F-body oil pan was already a tight fit. This pan can also be modified by cutting, notching, and welding. A good use for the C6 oil pan is on vehicles with aftermarket subframes where pan clearance is generous. This pan is the same width and depth as its F-body counterpart, but with a longer longitudinal sump length, has a larger oil capacity than the F-body pan.

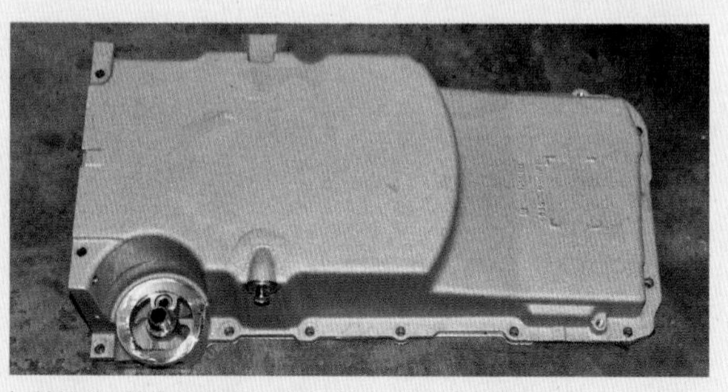

In 2005, the C6 Corvette brought revisions to the C5 Corvette chassis. They are similar chassis but do have many differences. One of the new features of the C6 is the conventional-style oil pan for the LS2—with no wings. This pan has an extended sump, so many vehicles would need a notched oil pan. This pan is unchanged for the newer C6 LS3 engines.

1999+ 4.8-, 5.3-, 6.0-, AND 6.2-LITER: SILVERADO AND GMC TRUCKS

The truck oil pans are a deep rear-sump design, intended to clear the four-wheel-drive front axle, when so equipped. All truck pans are similar dimensions and fit perfectly in older truck applications getting a new LS engine. The sump on these pans can be shortened as required if ground clearance is insufficient. Typically, you want the pan to hang no lower than the lowest part of the frame. Some lowered vehicles may need extensive oil pan modification depending on engine-mounting designs.

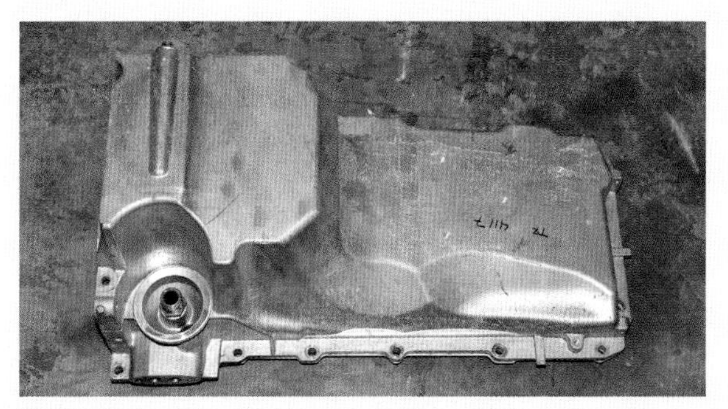

The 1999 and on Silverados, Sierras, and SUVs receive the truck version of the oil pan, with an extended-depth sump. These are suitable for full-size truck applications, such as the earlier non-Vortec and non-LSx trucks.

2004–2007 5.7- AND 6.0-LITER: CADILLAC CTS-V | 2009+ 6.2L CTS-V

The Caddy oil pan is a little deeper than the F-body and C6 pans, but shares similar designs such as the rear-sump design and width dimensions. The CTS-V pan can be fitted to vehicles that have ample room for the sump-to-ground clearance measurements. The deep sump limits this pan's application potential a bit for lowered vehicles, but offers much more clearance than truck oil pans.

The 2004–2006 CTS-Vs with the LS6 or LS2 engine and the 2009+ CTS-V LSA engines are also equipped with a conventional-looking rear-sump oil pan. This oil pan is a new favorite for LS swaps in cars, as the sump length is slightly shorter than the LS1 F-body pan and only a smidge deeper. BRP Muscle Rods offers motor-mount kits that allow the use of the CTS-V oil pan.

2010+ CAMARO 6.2-LITER L99 AND LS3

The 2010 Camaro oil pan has a higher oil capacity and is equipped with a water-to-oil cooler on the left side. The L99 and LS3 oil pan features much larger mid-sump dimension and oil capacity (holds 8 quarts of oil) than comparable car pans. These pans use the newer PF-48 oil filter design with an internal bypass (a high pressure oil bleed-off). These oil pans have minimal swap opportunities aside from the OEM donor, but do fit random off-road type applications such as Jeep Wrangler LS swaps.

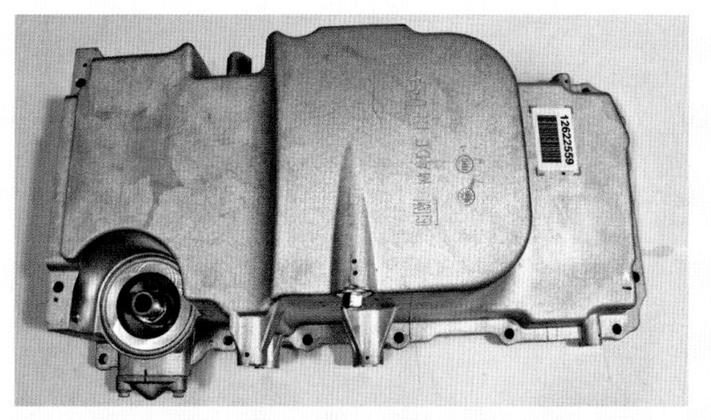

The 2010–11 Camaros and 2008–09 G8s came equipped with this style of engine oil pan. The pan is more of a mid-sump pan but offers a large capacity, around 8 quarts. This pan can utilize the factory fifth Gen Camaro oil cooler, which bolts to the side of the oil pan and utilizes engine coolant to stabilize the oil temperature.

2008+ 5.3-LITER: H3 HUMMER "MUSCLE CAR OIL PAN"

The little brother H3 Hummer uses a unique oil pan that has an interesting trait. It allows some flexibility to use an unmodified oil pan in some of the earlier GM chassis. The drawback to this pan is the slightly deeper sump and lessened oil capacity of 5.5 quarts. While these pans fit quite a few chassis applications, they often do not work well on lowered GM cars, as they sit a little lower than a SBC steel oil pan.

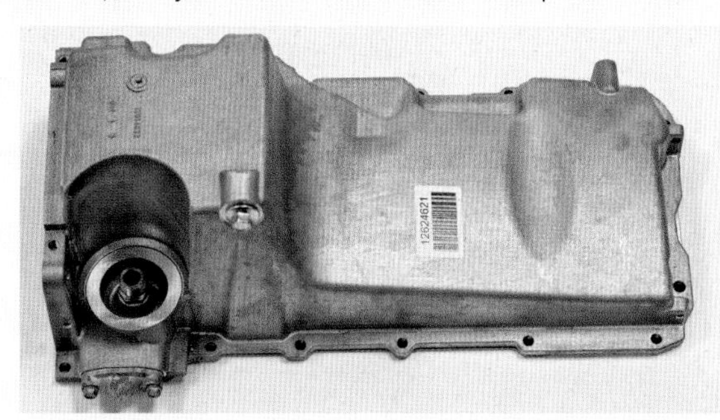

The 2008 Hummer Alpha H3 engines use a short rear-sump design that has the potential to fit a wide range of applications. The pan's drawback is that it is still a truck-style pan; it still has a deep sump as compared to the other passenger car oil pans.

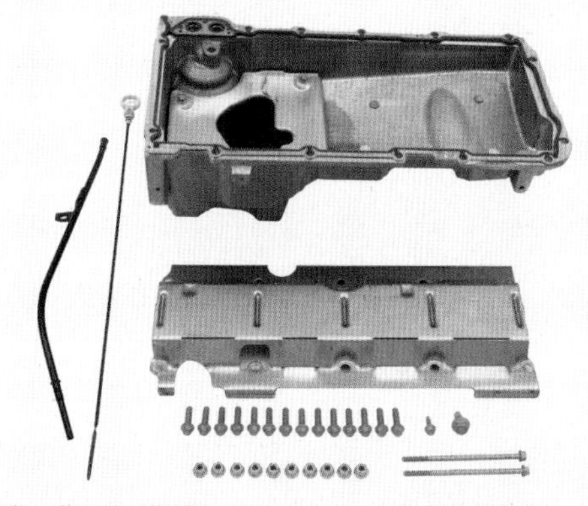

GMPP also offers the H3 LH8 oil pan as a economical kit, including everything necessary to add the oil pan to a crate engine. Be wary of ground clearance with cars, but this pan should be quite useful in S-10 LS swaps. *Courtesy of General Motors*

(continued on page 46)

 (continued from page 45)

6.2-LITER: C6 ZR1

The dry sump LS7 and LS9 oil system keeps a constant supply of oil to the engine under all conditions. This system isn't really meant for power but rather to address oil starvation under extreme cornering, accelerating, or stopping conditions. Oil is not accumulated in the oil pan. It is all scavenged and pumped into the reserve tank using a two-stage internal oil pump—one for return and one for engine oil pressure. The large reserve tank holds all engine oil other than what the engine uses for actual lubrication.

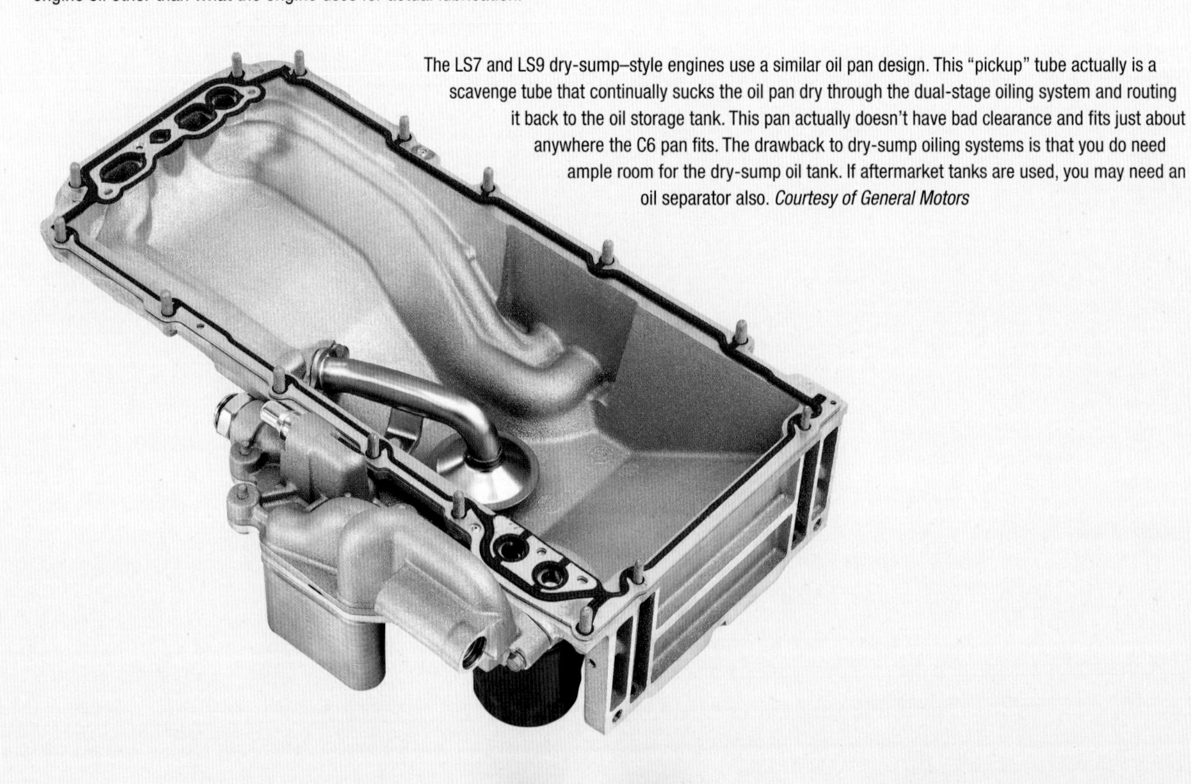

The LS7 and LS9 dry-sump–style engines use a similar oil pan design. This "pickup" tube actually is a scavenge tube that continually sucks the oil pan dry through the dual-stage oiling system and routing it back to the oil storage tank. This pan actually doesn't have bad clearance and fits just about anywhere the C6 pan fits. The drawback to dry-sump oiling systems is that you do need ample room for the dry-sump oil tank. If aftermarket tanks are used, you may need an oil separator also. *Courtesy of General Motors*

GM Performance Parts offers a kit to connect -12AN oiling lines to the stock LS7/LS9 oil pans easily. Anytime the dry-sump engines are in a non-C6 chassis design, the oil tank will be located farther away and likely in a completely different area of the engine bay, so the need for AN lines are an obvious requirement for any non-stock application. *Courtesy of General Motors*

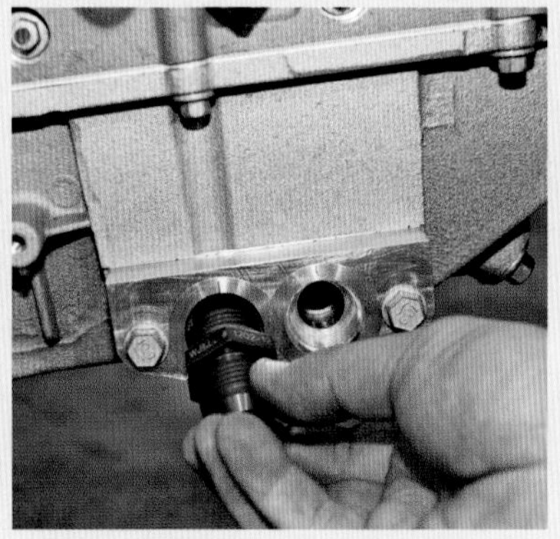

To install the GMPP dry-sump AN adapter kit, you simply bolt the mounting block to the oil pan, then add the fittings. Once the engine is mounted, you will need to build custom length AN lines to your oil tank.

(continued from page 42)

ENGINE OIL PAN SELECTION

The oil pan you select depends on the chassis getting the engine transplant. Choosing the correct oil pan depends on a few factors that influence the decision. You need to consider engine location, subframe interference, ground clearance, and if you race your LS-swapped vehicle, oil capacity may also come into play.

The LS1 F-body oil pan is popular because of its low-profile clearance. In non-lowered truck swaps, you can usually use the LS truck engine's oil pan. This is advantageous due to the increased ground clearance and similar sump design. In a wide range of GM passenger car applications, the F-body pan is used and may need to be cut, welded, and modified for use. Some other cars

(continued on page 51)

One thing to note with active-fuel-management engines is that the lifters can only have so much oil pressure before experiencing unintended disengagement problems. GM included a secondary pressure valve in the VVT and AFM engines to allow a specified maximum 75 psi oil pressure to the special AFM lifters. If you put on non-AFM oil pan onto an AFM engine, you are likely to have unintended lifter disabling issues.

After installing the oil pan and motor mounts, the engine is ready to set into place to check clearance dimensions. Grab a pal to operate the engine hoist as the engine is lowered into place. Line up the motor mounts one by one (using a pry bar or alignment punch as necessary).

Aligning the motor mount through bolts can sometimes be difficult when using poly mounts. The engine has to be at the correct angle and location for the bolt to slide in without much difficulty. Once one bolt is through, go to the other side and use some elbow grease and patience to work on aligning the second mounting bolt.

HOLLEY AFTERMARKET LS CAST OIL PANS

Holley Performance has several LS swap–specific cast-aluminum oil pan solutions. These are all rear-sump designs with plenty of ground clearance, while also allowing clearance for things like 1st-Gen F-body rear-steering tie-rod movements. When you buy a Holley oil pan, it comes with the sump plate and a new oil-pump pickup tube in addition to the oil-cooler bypass plate and oil-filter fitting, all of which need to be installed. These base oil pan kits do not come with bolts or an oil pan gasket. Obviously, a new gasket is preferred unless you are building a new GM crate engine for your vehicle—these "new" oil pan gaskets can generally be reused.

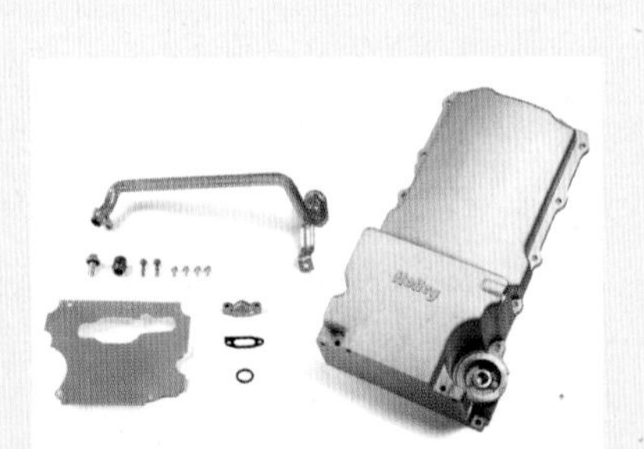

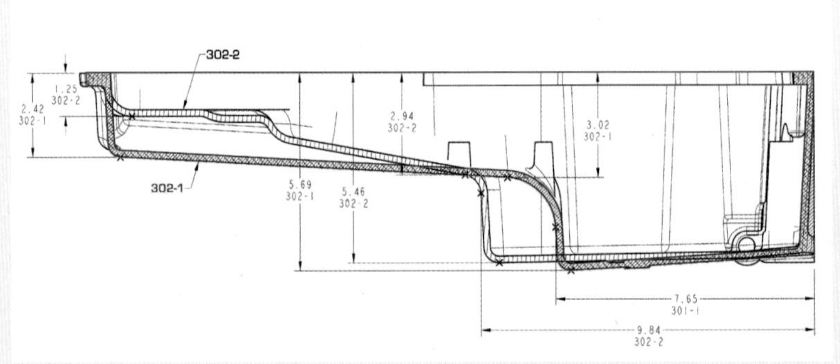

The Holley oil pan kits have filled a need for LS swaps in a big way. They help with subframe clearance, ground clearance, and, in some cars, improved steering clearance, while still providing sufficient oil capacity and strength. Holley 302-1 oil pan and measurements are shown here. The 302-1 is for up to 4-inch stroke crankshafts and can use the full-length GM 12611129 windage tray. *Holley*

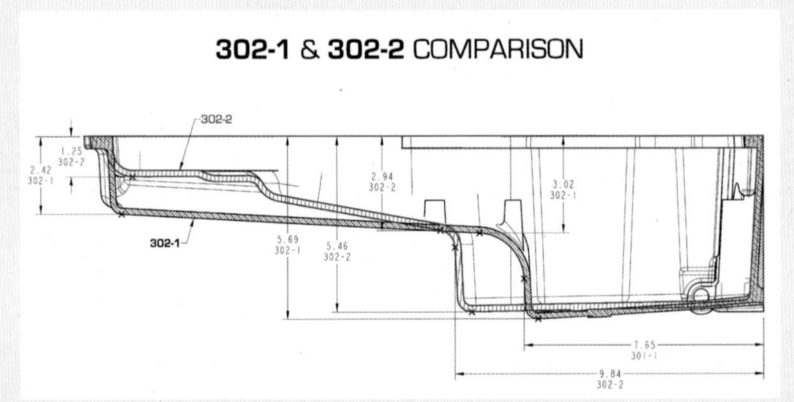

The Holley 302-2 (P/N 302-2BK shown) oil pan kit is ideal for OEM takeout engines and crate engine builds. This oil pan fits stock 3.622-inch stroke engines and provides the most generous front clearance from oil pan to subframe. This oil pan requires a 4th-Gen F-body windage tray (GM 12558253) with a very minor modification for pickup tube clearance. *Holley*

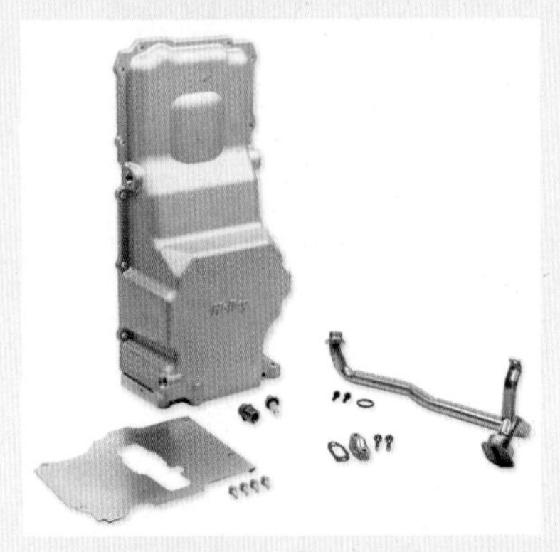

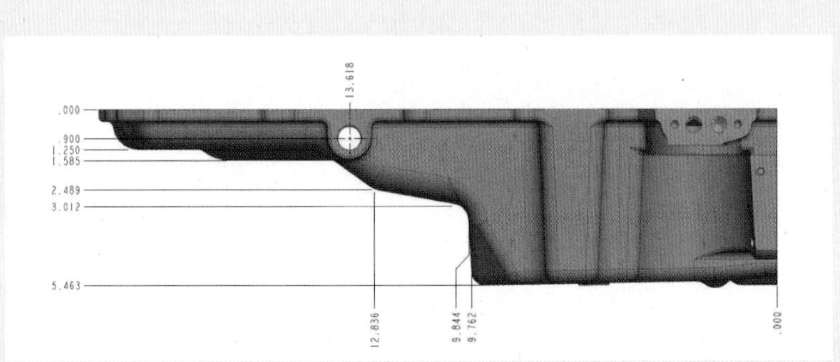

The most versatile oil pan addition is the Holley 302-3, which is kind of a one-size-fits-all design. It allows up to 4.25-inch stroke and has dual ½-inch NPT turbo oil drain back fittings while retaining sufficient front subframe low-profile design clearance. This pan requires a more modified windage tray using either the 4th- or 5th-Gen windage tray cut down for clearance. The template illustration is included with the oil pan instructions. *Holley*

Holley P/N 302-2 is meant for stock 3.622-inch engines such as the 4.8-, 5.3-, 5.7-, 6.0-, and 6.2-liter. It will not clear the long 4.00-inch stroke crankshafts due to minimal front internal rotating assembly clearance (but this provides supplemental crossmember clearance). This oil pan utilizes the 4th-Gen F-body windage tray and requires minimal modifications for the oil-pickup tube clearance, which can be done with a hacksaw or a small reciprocating saw.

Holley P/N 302-1 is the longer-stroke version of the LS swap pan. This uses a full-length windage tray with minimal modifications, but compared to the 302-2 oil pan it has less front crossmember clearance and is not ideally suited for standard SBC-friendly stock subframes like the 1st-Gen Camaro. While it can be made to work with tall motor mounts, this oil pan has minimal room for error if you can squeeze it into such applications.

Holley P/N 302-3 is the latest oil pan design from Holley. It accommodates both short- and long-stroke builds and offers optimal front-subframe clearance. This uses a much modified and cut-down 4th- or 5th-Gen Camaro windage tray due to the internal interference of the oil pan structure. This oil pan is also the only Holley LS swap pan with ½-inch National Pipe Tapered (NPT) turbo oil-drain holes integrated into the casting. This pan is likely the most versatile for GM LS swap applications as it fits long- and short-stroke crankshaft engines, while providing ample crossmember-to-pan clearance.

MOROSO AFTERMARKET LS OIL PANS

Moroso makes race-bred oil pans for many engines. It is no surprise that they have a pan intended for LS swaps (part number 20141). This is a fabricated-steel oil pan with external oil filter lines. The oil pan has a deeper front-sump design that allows a full-length windage tray to be used and also allows clearance for up to 4.125-inch stroke crankshafts. Internally, the pan has trapdoor baffling to control oil slosh and uses a specific length pickup tube to match the sump location. This pan fits many GM vehicles such as the 1967–92 Camaro, 1968–78 Nova, 1965–72 Chevelle, 1953–96 Corvette, 1978–88 G-Body, and GM S-10/S-15 two-wheel-drive Blazers and trucks.

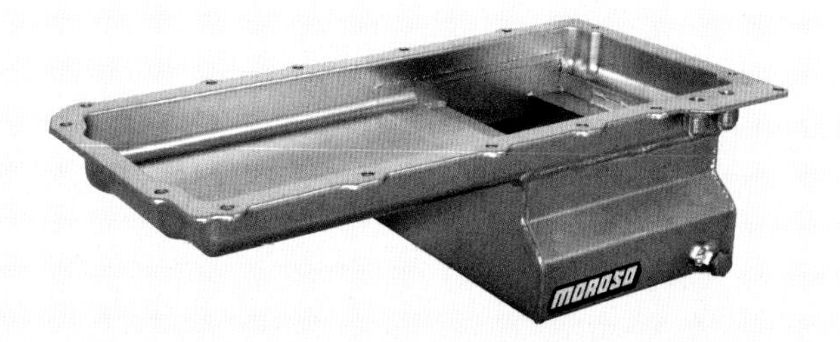

Many aftermarket oil pan companies are fabricating pans for the more popular LS transplant applications. Moroso is one such company. Moroso oil pan part 20140 holds seven quarts and is known to fit the 1968–1972 Novas, 1965–1972 Chevelles, and 1967–1969 Camaros. Use the Moroso part 24050 pickup tube. *Moroso*

Another Moroso design that helps the retrofit crowd is the part 20141 oil pan, another seven-quart oil pan. This pan is known to fit 1967–1992 Camaros, 1968–1978 Nova, 1965–1972 Chevelles, 1953–1996 Corvettes, and 1978–1988 G-bodies (Monte Carlo, El Camino, Regal, Cutlass, and so on). Use with Moroso part 24050 pickup tube and part 22941 windage tray. *Moroso*

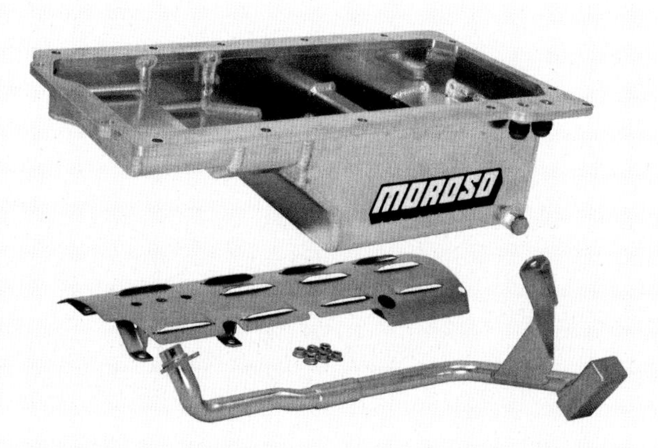

Moroso also offers a direct replacement 1998–2002 F-body oil pan kit that can possibly fit other swap applications such as those that the OEM F-body oil pan would fit. The Moroso part 21150 oil pan is a little shallower, but has a longer sump like the OEM F-body oil pan. This oil pan kit comes with the required windage tray and pickup tube. *Moroso*

The slight installation drawback to most fabricated and aftermarket oil pans is the requirement of an external oil filter. This adds some extra oiling lines and the requisite mounting location. Once installed, the drawbacks are gone, as there are a few benefits to the remote oil filter such as easier servicing and a small cooling effect from having the oil filter away from engine heat. *Moroso*

MILODON AFTERMARKET LS OIL PANS

Milodon's LS retrofit oil pan allows installation of LS engines into early pre-1975 Camaro, Chevelle, 1968 and newer Novas, and 1955–57 Chevys. This pan uses heavy-duty steel and double-thick pan rails but is somehow still five and a half pounds lighter than the factory aluminum oil pan. The increased oil capacity and increased oil control of the Milodon pan allows for consistent oil pressure under heavy usage. The flat front surface of the sump keeps oil from sloshing into the crankcase on deceleration, while the internal baffles in the rear of the sump help to keep the oil pickup tube supplied with oil during heavy acceleration.

A helpful tip for helping to determine ideal engine location is to install a bare donor block and take a few measurements at different points for comparison to oil pan dimensions. In this A-body application, the F-body LS1 oil pan sump dimensions were too tight and against the oil pan front corners, the OEM oil pan would require modification to use. The owner chose to install an aftermarket oil pan kit from Milodon to alleviate clearance issues.

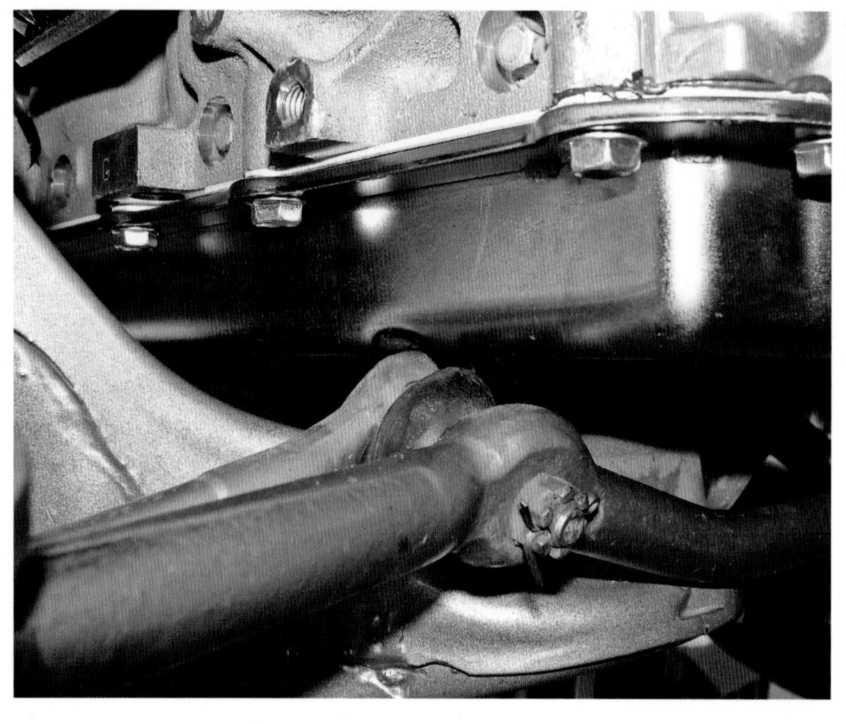

A-body LS swaps also suffer from insufficient steering clearance on the inner tie rods. You can help this clearance by keeping the engine as high as possible in the engine bay by spacing the motor mount plates higher by ¼ inch, or if more clearance is necessary you can dimple the corner of the pan as shown, carefully making use of the back-end of ball-peen hammer.

CANTON AFTERMARKET LS OIL PANS

Canton offers both steel and aluminum drag race–bred oil pans designed for installing LS blocks into earlier chassis. Either pan directly fits A, F, Y, and 1970s X body cars using the common motor-mount adapter plates. The fabricated aluminum version is significantly lighter than the OEM cast-aluminum pan designs. Either oil pan design works with OEM GM 12558253 windage tray. This pan requires an LS remote oil filter adapter or as a plus for the builders who want stock oil filter locations. A factory oil filter mounting kit is available. Custom-built sump configurations to your dimensions are also available as a special order.

Canton LS1 swap oil pan in steel (P/N 13-274) and aluminum (P/N 13-274A) are basically the same, except of course, one is made from a stamped-steel core with a 16-gauge fabricated steel sump, and the other is made from .100 aluminum. Pans fit A, F, Y, and 1970s X-body cars. They have a 6½-inch deep, 9-inch long, 11¾-inch-wide sump, and a 6½-quart capacity. They have trap door baffles for oil control and a magnetic drain plug for collection of metal debris. *Canton*

Canton oil pans have an option to use either the remote oil filter mounting location, or to use an on-pan oil filter adapter (P/N 22-631). Most of the "bolt-in" builders would prefer the oil filter "on pan" for simplicity, but either design is quite functional and they both filter oil. In the end, these are two different ways to do the same thing. Kudos to Canton for thinking of the street builders with this on-pan filter adapter. *Canton*

(continued from page 47)

with generous clearance can use the C6 Corvette oil pan. Another possible OEM-pan solution is the LH8 oil pan from the Hummer H3. It does have less ground clearance than the F-body oil pan and smaller oil capacity but clears many of the "rear-steer" platforms, such as early Nova and first-generation F-bodies.

The OEM pan selections must use matching oil dipstick tubes and indicators, oil pump pickup tubes, and the correct OEM windage tray for proper internal clearance and pickup tube mounting. The pan must also be installed and aligned correctly, being squared with the bell housing flange surface.

Note that the majority of existing oil pans will use the AC Delco PF-46 oil filter up until about 2007 year vehicles. Certain 2007 and newer vehicles, and almost all new replacement oil pans, such as the Holley 302-1, -2, and -3, will use the AC Delco PF-48 filter. It has a larger diameter thread and internal oil filter bypass. These do not interchange and need to be used with the matching oil pan fitting. Remember, though, that all LS engine pans are interchangeable on any LS block with matching dipstick/windage tray hardware, with the exception of the LS7 and LS9 factory oil pans, as these are specific dry-sump engine oil pans and best fit on dry-sump engines.

Problems sometimes rear their ugly heads after spending an hour or so aligning motor mounts. In my 1967 Camaro 5.3-liter installation, I used the LH8 oil pan kit. Not too surprisingly, the sump extended quite a bit below the subframe. This will likely cause a ground-clearance problem, so a swap to the Holley 302-2 cast-aluminum LS swap pan is my likely solution.

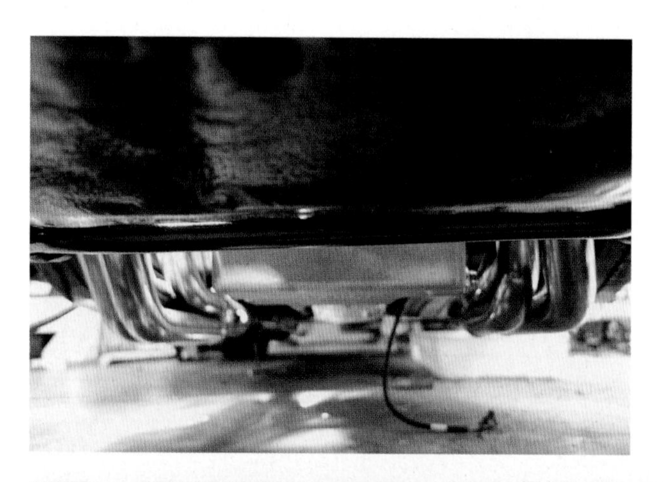

Originally this Camaro had the GM muscle car oil pan with a 6.0-liter LS2. When I built the new 427ci LS engine, I installed the Holley 302-1 oil pan for extra crankshaft stroke clearance and swap compatibility. The Holley oil pan choices afford steering clearance, and they also provide much-improved ground clearance on lowered swap vehicles—exactly where the GM muscle car pan falls short. *Holley*

Chapter 3
Engine EFI and Ignition Controls

So you have your LS engine, your transmission, and you may even have it bolted into your car. You may now be wondering how to get it running. Some folks likely have the donor harness from the pull-out engine, while some builders need to source a complete engine harness if they are doing a new build.

In this chapter we will discuss the electrical portion of the swap, including engine control modules and the complete required engine wiring harness. These are the critical elements that allow the Gen III and Gen IV engines to operate as EFI engines. The basic engine controller operates the fuel injectors, ignition system, fuel pump, and cooling fans by using feedback sensors that measure crankshaft location, engine load, and airflow. It has the capability of trimming air/fuel ratios to ideal stoichiometric during steady speed driving.

While most of the information in this chapter is related to the OEM EFI controls, there are many paths to achieve the same end goal of controlling the engine. The aftermarket has provided many supplemental options to stand-alone harnesses and controllers. The ECU and harness we will cover is one of the most popular aftermarket choices: the Holley Terminator X and Terminator X Max.

Installation of the engine harness might seem like a daunting task, but it really is not all that complicated when using the installation process I call "process of elimination": Install all known connections and then deal with the remaining ones last.

OEM CONTROLLERS AND HARNESSES

If you purchased a complete take-out LS engine from a salvage yard or from a private party sale, you may have received the engine wiring harness and other external components with the engine. This is good, as many of the OEM harnesses can be modified to work in any chassis, both old and new, which involves adding a fuse box to power and protect the harness, an OBD II DLC connector, and removing any unnecessary engine or chassis-specific wiring.

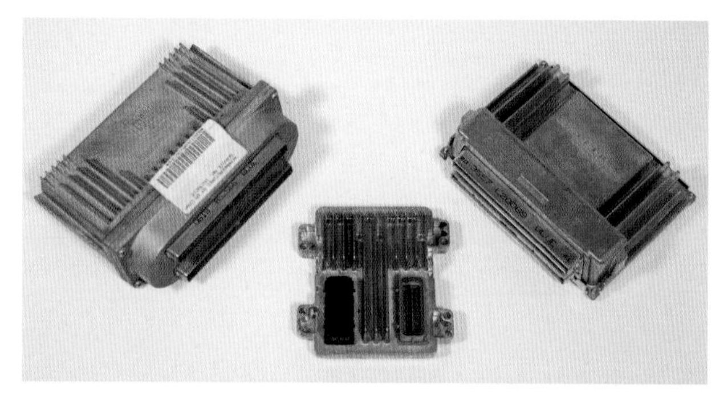

Engine controllers vary in function and external design, depending on which generation and year of engine you are using. To the left is the 1998 F-body ECM, at center is the 2006 and newer E38 Gen IV ECM, and to the right is the 1999–2002 F-body ECM.

Almost all LS vehicles have discrete engine-control harnesses with connectors to mate the engine harness to the chassis harness for power and sensor feedback. If you did not receive a harness with your engine, don't fret, as there are many options for obtaining a harness, from economical to premium.

If you are looking for a harness to use in your swap project, then you need to ensure the harness you purchase matches your genre of engine. There are two basic styles of engines: the earlier pre-2007-era Gen III with the 24× crankshaft reluctor and the later 2007–on Gen IV with the 58× crankshaft reluctor.

With the proper internal engine parts and modifications, you can use either a harness and PCM variant, although it is usually easiest to get the harness that matches your engine. The alternative is swapping crankshaft reluctor wheels and timing sets.

There are a few hybrid applications as well, but with some careful identification, these are easy to identify and weed out. An example of this would be needing Gen III–compatible controls for an early Gen IV engine, such as the 2005 Corvette LS2 and 2005–2006 Pontiac GTO with the

The Vortec trucks have a nice engine cover from the factory. There is a reason for this. The engine harness is located directly over the intake manifold, causing an ugly scene.

The reason the truck harness is less desirable is its routing, shown here. If you are performing a truck LS swap, then there is little reason to not use the truck intake or harness; but if you are swapping an LS engine into a car and are looking for an aesthetically pleasing setup, the truck harness is not for you. Usually truck harnesses are good for truck swaps, while car harnesses are good for both trucks and car swaps.

24× reluctor wheel. These use the E40 ECM and Drive-by-Wire system along with the 24× controls. Additionally, the factory automatic transmissions would have their own separate transmission control unit, whereas earlier 24× ECMs are capable of controlling both automatic and manual choices with just a calibration change.

58× CRANKSHAFT SIGNAL CONVERTER

Even though many 2006–2007 and newer factory and GM crate engines are the 58× crank signal, there are ways around this that do not require internal engine teardown. Lingenfelter has a 58× to 24× signal-conversion box (P/N TRG-002) that plugs in between the Gen III 24× wiring harness and your 58×-equipped engine. This signal box converts the 58× crankshaft and camshaft signals to match the 24× PCM. This is great for builders who peruse wrecking yards and may already have sourced an older OEM harness with which they would like to use a newer engine package. This module is even more helpful with 1997–2006 LS-based vehicles with stock harnesses and PCMs who want newer engine options. A nice feature of Holley EFI systems is the ability to mix and match 58×/24× hardware. You can calibrate the Holley electronic control unit (ECU) to utilize a 58× crank with a 1× camshaft signal, or 24× crank and 4× camshaft. This allows some flexibility in parts choices, which means a mismatch need not impede progress. Additionally, you can choose between DBW or DBC with either engine family, if you have the Terminator X Max ECU.

HARNESS OPTIONS

Many aftermarket wiring harness companies are able to modify the OEM engine harness for an easy, few-wire hookup for older cars. This is a good option to consider instead of trying to isolate the needed wiring among multiple GM bulk-wiring connectors.

One such company that is popular for the LS-series engine wiring harness retrofit and building is Speartech. Speartech can modify your existing harness or build a new harness to suit whatever your project requires. For example, Speartech can rid your harness of unnecessary emission or original chassis-specific wiring, add in features such as 4L60-E, 4L80-E , or 6L80 wiring, and then further recalibrate your PCM and TCM for the LS project vehicle. Speartech-converted harnesses and new harnesses and connectors also come with labels. This makes the end user's life much easier by directly informing them where each connector is to be routed or the item to which it is to be connected.

Regardless of where you choose to source your harness, the end result should be and usually is the same once the engine is in a running state. The decision should be made considering many factors. Budget is one of the main ones, custom harness dimensions is another, and sometimes just following in others' footsteps is a requirement. As with anything, price usually dictates how much service after the sale is available. If you get in a bind, you can usually shoot off an email and many wiring harness companies will help as much as possible, given the limitations of long-distance troubleshooting.

GM Performance Parts (GMPP) is another popular alternative that offers many Gen IV engine harness options for the late-model crate engines available through their many retailers. Most of these crate engines are Gen IV blocks using 58x crankshaft reluctors, hence the reason for the mainly Gen IV intended usage. The GMPP engine controller kit comes with the complete harness, pre-programmed powertrain control module (PCM), oxygen (O_2) sensors, mass airflow (MAF sensor), drive-by-wire (DBW) throttle pedal, assembly-line diagnostic link (ALDL) connector, and an adequate instruction manual.

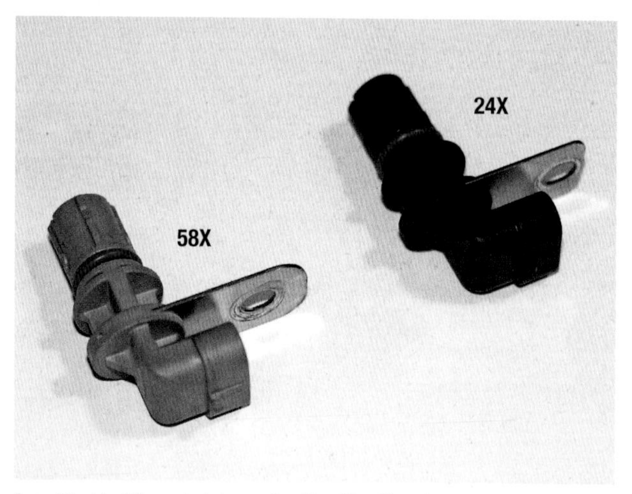

One of the big differences between Gen III and Gen IV engines is the crankshaft reluctor wheel. I hate to sound like a broken record about the reluctor wheel business, but it's better to be annoying and right than to find out you have the wrong reluctor wheel two minutes after your swap project doesn't start up. At left is the gray 58× Gen IV sensor; at right is the black 24× Gen III sensor.

The GMPP harness comes equipped with a fusebox and fairly generic engine wiring. Some harness modifications may be required for certain applications, such as if you need to relocate the fusebox to the other side of the engine bay or lengthen or shorten specific sensor wiring. The wiring pigtails are generously proportioned, so a good amount of flexibility is allowed between vehicles.

Another option is to convert your own OEM harness to your application. This requires a wiring harness pin out, some connectors, fuses, and some time. You could go hard-core and remove all unneeded wiring to clean up the harness, or you can just tap into the required power and ignition wiring and leave the unused remaining wiring in place. The OEM harnesses are not complicated, especially the F-body harness, which is one of the simpler harnesses to adapt into older vehicles.

THROTTLE CONTROL AND COMPATIBILITY

The OEM PCM has the capability to control the electric fuel pump, operate electric fans at programmed temperatures, and operate with aftermarket speedometers and tachometers. Donor vehicles such as 1997 to 2004 C5s and Vortec-based engines such as 2003 and newer pickups, 2000 and newer Tahoes, and 2001 and newer Suburbans with engines that come with drive-by-wire electronics can be fitted into swap vehicles also. Many of the 2000–2002 trucks and SUVs with electronic throttle bodies also are equipped with traction control; whereas cable-drive throttle bodies are not. The earlier "Warren" PCM requires a drive-by-wire TAC module in addition to the compatible throttle pedal assembly and throttle body; while the later, smaller Gen IV PCMs have the throttle control integral to the

(continued on page 60)

EMISSIONS CONTROL SYSTEMS AND WIRING

Many LS swaps do not keep full functionality of the emissions-legal systems. Whether it is from an unwanted feature standpoint or due to the fact that pre-1975 cars (some places say 25 years and older) are not held to the same stringent emissions requirements that the LS-equipped newer vehicles adhere to. Systems that are often disabled are the catalytic converters, catalyst monitoring rear oxygen sensors, air-injection system (if equipped), and exhaust gas recirculation (EGR) systems (if equipped). The LS engines can run perfectly fine with or without any of these systems. Their use really just depends on if you are required to have them or if you just want them functional.

In many LS swaps, LS-swap-specific exhaust headers are used. They require a custom-fabricated exhaust to be built after the header collector to the rear of the vehicle. These aftermarket header systems rarely offer catalytic converter options, and there aren't readily available exhaust systems with cats. You can add catalytic converters to the vehicle, as there are aftermarket cats available in various configurations if you are not retaining the OEM LS configuration. Due to the non-emissions header designs, things like AIR and EGR are not easily bolted on if so desired. Drilling the headers and fabricating EGR and AIR fittings are an option if these systems are to be retained.

As for wiring, all LS-equipped vehicles with full emissions controls use four oxygen sensors, the front two O_2 sensors monitor the unburned air/fuel ratio and allow the ECM to correct the rich or lean condition in real time. The post-cat or rear O_2 sensors monitor catalytic converter health and efficiency in a stock vehicle and will set a service engine soon light (SES) or diagnostic trouble code (DTC) if unplugged or not functioning. If you are using high-flow aftermarket cats, or no cats at all, then this monitoring system is not necessary (if legal by your local laws). The post-cat sensors can be disabled using LS1 edit, HP tuners, or EFI live ECM programming software and hardware. Your retrofit company of choice can also perform this service to your stock ECM for a nominal fee. It's usually done when deleting the OEM theft-deterrent system ECM functionality.

If your ECM and harness used EGR or AIR in stock configuration and you are not reusing these, either the ECM can be reprogrammed to not set codes due to the lack of these systems. As for wiring, if you are buying a stand-alone harness, these provisions are usually not retained unless requested. They can usually be added as an option for a small fee. The majority of LS swappers do not require their usage, so removal is usually the norm.

GEN III VS. GEN IV ENGINE ELECTRONICS

While the Gen III and Gen IV LS-series engines may physically appear mostly the same, there are quite a few differences. Other than the mechanical aspects of different locations of knock sensors and the camshaft sensor itself, there are electronic differences to these sensors and others that will prevent interchangeability.

First of all, the in-between year models such as the 2005 C5 and 2005–2006 GTO and such engines with Gen IV blocks but with older 24×/1× crankshaft and camshaft signals should be regarded as electronically interchangeable with the Gen III ECMs, other than the two-wire knock sensors, as these are Gen IV–style and ECM-specific.

Yes you can take a 2006 GTO LS2 engine and put it in your 1997 Corvette without signal converters or adapters. You will need to extend the camshaft sensor wiring and relocate the Gen III knock sensors to the side of the engine. With little or no additional hassle over a normal engine swap, you can take these in-between engine designs with the 24×/1× crankshaft and camshaft signals and install them into any Gen III–based vehicle or mate them to any Gen III ECM/harness system. For reference, these Gen IV engines with Gen III electronics will have black crank and cam sensors and wiring terminals, while Gen IV engines with Gen IV electronics will have gray sensors.

Now as for the electronic variations for 100 percent Gen IV as compared to Gen III electronics. For one, *all* Gen IV engines and ECMs are equipped with electronic drive-by-wire (DBW) throttle body controls. Where only Gen III C5 Corvettes, SUVs, and 2003 and newer trucks are equipped with electronic throttle control, and some of these could even be swapped back to mechanical linkage, depending on the ECM service number (some DBW 2003–2006 ECMs do not have IAC motor drivers). An easy identifier for the pre-2003 trucks is if the vehicle had traction control, it had a electronic DBW throttle setup.

A full electronic throttle control system means that the throttle pedal and throttle body are not connected by a cable, mechanical linkage, or any physical moving. Instead, both are equipped with multiple throttle position sensors, while cable-drive systems only have the one sensor on the throttle body itself. These TP sensors read how much throttle you request and interpret that into throttle control into the electronic throttle body. There is no way around this when using the Gen IV harness and ECM. If the Gen IV engine is retrofitted with a Gen III harness and a mechanical throttle body, you do not need to worry about the electronic pedal assembly. Gen IV harnesses and ECMs require Gen IV electronic throttle controls.

In other areas we had mentioned that Gen IV 58×/4× engines differ from the 24×/1× in crankshaft and camshaft reluctor wheel count. The 58×/4× setup offers greater resolution for the Gen IV features, such as greater misfire detection and the accuracy improvement needed for AFM and VVT requirements. Gen III engine controllers cannot offer the resolution of the Gen IV architecture.

What some may not know is that the Gen III crank and cam sensors operated on a 12-volt reference signal, also known as alternator voltage, while the Gen IV is revised to a much more stable 5-volt and cleaner reference. Hence, the sensors are not intended to be interchanged between engine designs, although the gray camshaft sensor will read either 1× or 4× timing chain reluctor wheels just fine if only gray sensors are available.

The last electronic sensor difference between Gen III and Gen IV are the knock sensors. Gen III knock sensors (one wire each) are mounted in the valley plate area under the intake manifold. Gen IV engines use side-mounted knock sensors (two-wire) on the sides of the block midway front and rear and just above the oil pan mounting flange.

The knock sensors will not interchange between the two engine controllers due to wiring and functional differences. If using a Gen IV engine in a Gen III–controlled vehicle, use the Gen III knock sensors but relocate these to the outside of the engine. There are surplus mounting locations throughout the engine such as extra motor-mount bolt holes, ends of cylinder heads, unused A/C bolt hole locations, and so on. Find something away from exhaust heat and remember the wiring harness will need to be extended to match, or use a plug-and-play knock sensor extension harness.

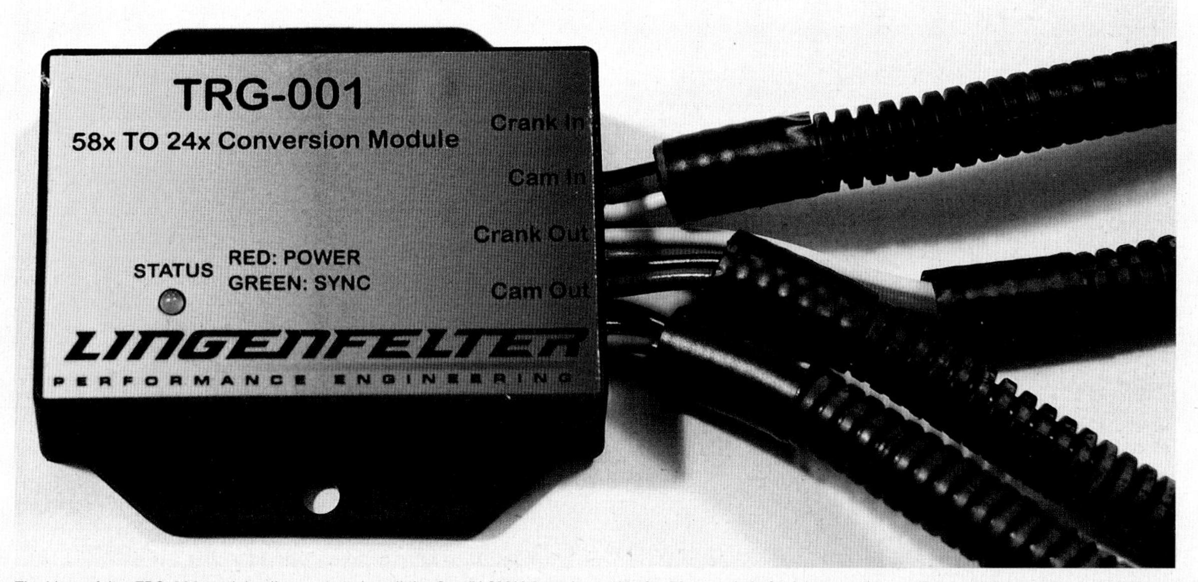

The Lingenfelter TRG-002 module allows you to install the Gen IV GM V-8 engines with the 58× crankshaft trigger or reluctor wheel and the 4× camshaft timing gear in earlier vehicles equipped with the Gen III ECM. This is far easier than having to take the engine apart to change the cam timing gear or the crankshaft reluctor wheel.

AFTERMARKET WIRING HARNESS OPTIONS

GM PERFORMANCE PARTS/ CHEVROLET PERFORMANCE

Recognizing the need for OEM Engine Control systems for the GM LS crate motor lineup, GMPP stepped up to the plate and offered a matching EFI harness configuration to supplement the needs of crate engine swaps. The GMPP harness kits come with the properly calibrated ECM, stand-alone engine harness, MAF, front O_2 sensors, and the drive-by-wire throttle pedal assembly.

Basically, the harness includes the ECM and a lot of extras that swap candidates previously had to procure individually with crate-type engine assemblies. The GM harness is the stand-alone type, where you simply connect battery, ground, and ignition and have a functional engine harness tied into your car's electrical system. There are currently five different GMPP harness offerings: LS2, LS3, LS7 (early and late), and the LS376/380 setup. All of these harness/PCM kits can be obtained for under $1,000, which is quite a steal considering the extra content included.

While these recent additions are all Gen IV–type controllers for 58×/4× crank and cam signals matching the crate motor offers, GMPP does offer a Gen III harness for use with either the Gen IV F-body PCM or a pre-calibrated GMPP PCM. The harness and matching calibration PCM would run about $2,500 from your GM dealership, which would keep most away.

SPEARTECH FUEL-INJECTION SYSTEMS

Speartech specializes in LS and other retrofits. They mainly build harnesses for swap vehicles but can modify and adapt any stock harness for use with your swap vehicle. They can build you a brand-new engine harness with features specific to stand-alone operation to your custom specs as needed.

Want to save a few bucks and have a stock LS harness that came with your engine? Speartech can adapt your stock engine harness to fit a late-model non-LS vehicle for a plug-and-play system. Send your stock harness to Speartech, and they will convert it to stand-alone by removing unnecessary wiring from the harness and adding wiring necessary for

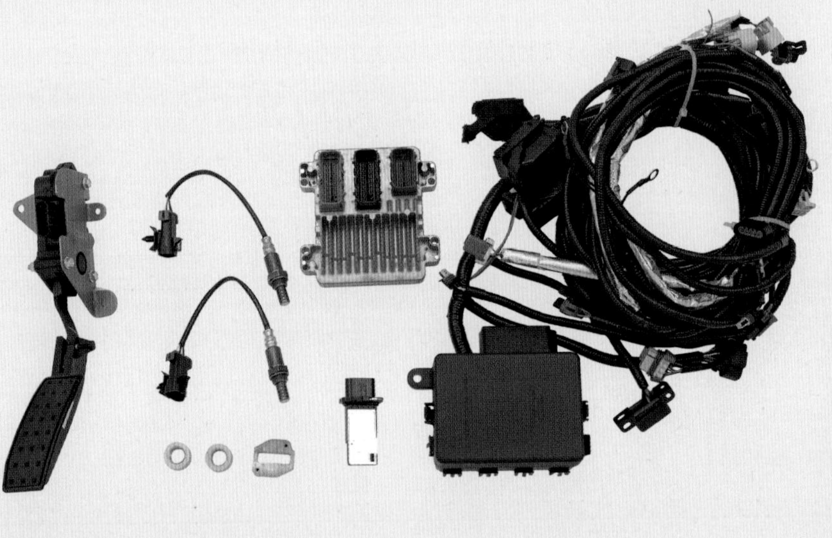

There are many options for engine harnesses and ECM packages. You have many choices, from stock harness conversions to aftermarket stand-alone harnesses, or one of the more complete setups that include an ECM, throttle pedal, and a few sensors, such as the GM Performance Parts Gen IV harness kits. *Courtesy of General Motors*

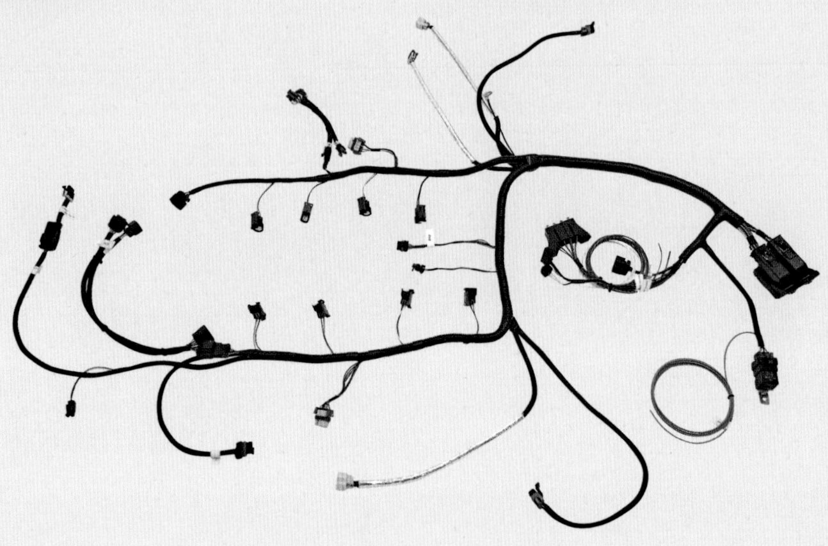

Another main option for a drama-free installation is a stand-alone harness from Speartech, a specialist in the LS swap arena. Speartech can custom build a harness or modify a stock harness for your vehicle. Shown here is the Gen IV LS3/ L99 harness kit, which controls both AFM and VVT for engines projects taken from 2010–2011 Camaros.

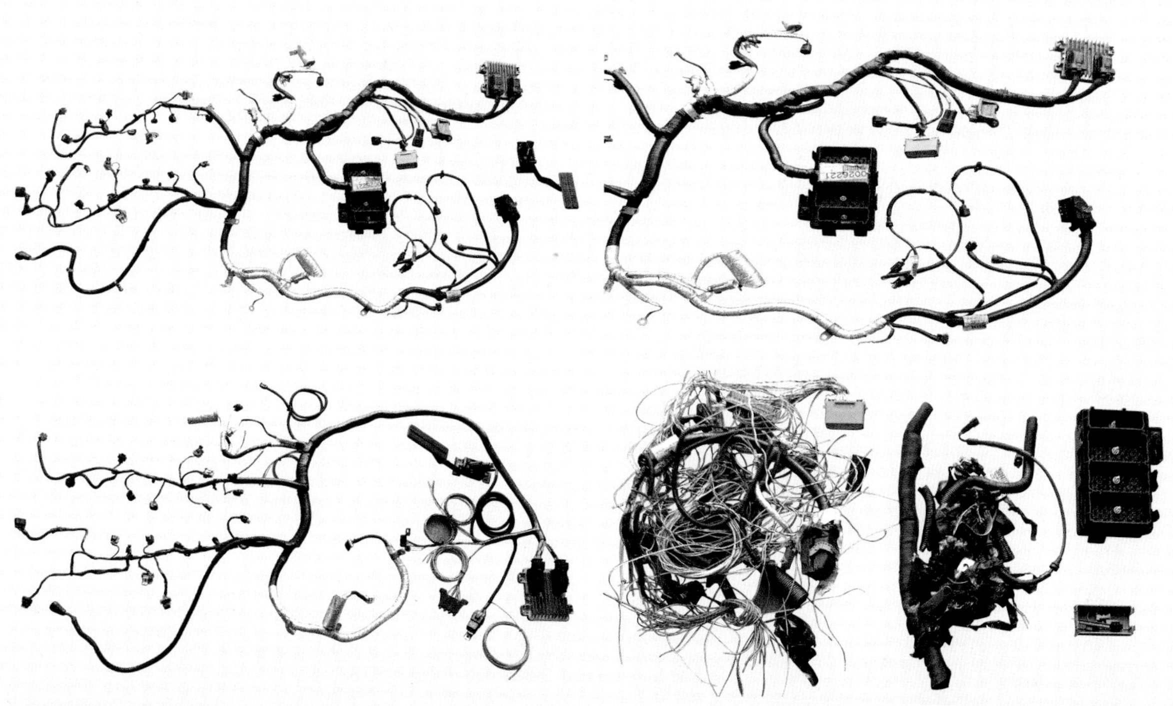

Speartech can easily modify any stock LS-based engine harness for a conversion project. This harness started off as a 2008 Corvette LS3 engine harness and is converted to a stand-alone Gen IV LSx conversion harness with some labor-intensive wiring procedures. The first two images are the stock harness, the third image is the finished harness. Note the unused wiring and connectors in the last image. *Speartech*

stand-alone operation. The converted harnesses feature some other add-ons and features similar to a brand-new harness. These may include a check engine light, a diagnostic connector, a fuel pump relay, as well as simplified ignition and battery power-up wires. When applicable, fan, speedometer, tachometer, and gauge wires are also brought out for convenience in mating the modified harness into the *new* vehicle.

Have a six-speed manual harness and want to use it with an electronically controlled automatic? No problem. Speartech will convert your harness or can supply you with a transmission harness to mate into your six-speed harness with labeled pins for convenience. They also offer complete packages, including engines, transmissions, and the works. Speartech does both Gen III and Gen IV harnesses, including compatible transmission controlling such as adding a 4L80-E in place of a 4L60-E or adding a 6L80-E with any of the Gen IV PCM options.

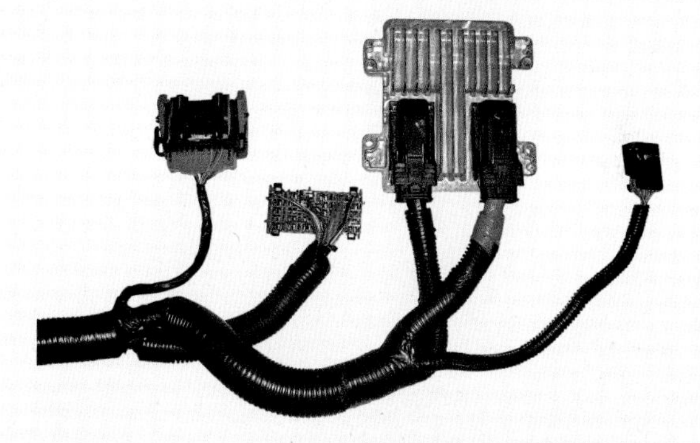

The experts at Speartech can also modify truck harnesses such as this one from a 2008 5.3-liter Silverado. Speartech removes the unnecessary wiring, adds the basic power and ignition source wires through a mini fusebox and direct throttle-by-wire signal wires. Additionally, added features such as fan controls, fuel pump relay and power wire, and a diagnostic link connector with check engine light are all implemented as well. *Speartech*

The Holley Terminator X completely changed the way people perform an LS swap, offering an extremely user-friendly interface, tuning ability, and versatility. The basic Terminator X can be used for 1999–2007 Gen III 24x engines and 2007–2015 Gen IV 58x engines with drive-by-cable throttle bodies; the harnesses are the main difference between the two choices. Holley also has the correct fuel-injector sub-harnesses to make it easy to match your donor engine.

The Terminator X Max can do everything the Terminator X can, but with added features such as electronic transmission controls and drive-by-wire (DBW). The Max does this by having two extra connections at the ECU, four in total compared with the two connections in the base Terminator X option. The Max also can be programmed for either 24x or 58x engine, but with the DBW feature it can be used with more modern LS3/LS7 crate engines that may come with a DBW throttle body out of the box.

Both the Terminator X and the Terminator X Max have an available built-in 1-bar map sensor, although it is rarely used with LS engines as the engine harness also supports the OEM map sensor. You can see the extra two harness connections and size comparing both Terminator ECUs side by side. Note that both can be used with drive-by-cable; the main engine harness for your type of engine will have TPS and IAC plug-ins for either type.

The integrated wideband oxygen sensor makes the Holley ECU options much more user-friendly than OEM. Set a target air/fuel ratio and the ECU does all the hard work by always changing the fuel correction to achieve that targeted setting. Once the ECU learns some historical air/fuel data, the learned values can be transitioned to the base fuel map table calibration.

(continued from page 56)

PCM. The cable-drive throttle body harness may be easier to find as they have been around a lot longer, but as with most choices it also depends on which harness and engine you are using.

If you have just an engine and no harness, you have many options and are not restricted to strictly matching the engine to the PCM harness. As mentioned, the cable-actuated TB is only factory supplied on certain Gen III engines, but since these harnesses are the most popular for budget builds and simplest to procure, the question is, can you use a cable-drive Gen III PCM harness with a Gen IV engine?

The short answer is yes, but with stipulations. You can put any engine with any controller if the engine is torn down and the reluctor wheel is replaced with the matching signal wheel. The earlier 2005 Gen IV engines use a 24× crankshaft signal and 1× camshaft gear. The 2006 and newer use 58× crankshafts and 4× camshaft signals.

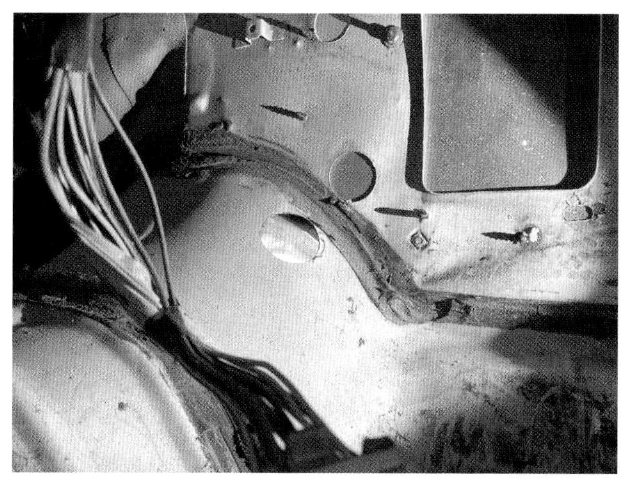

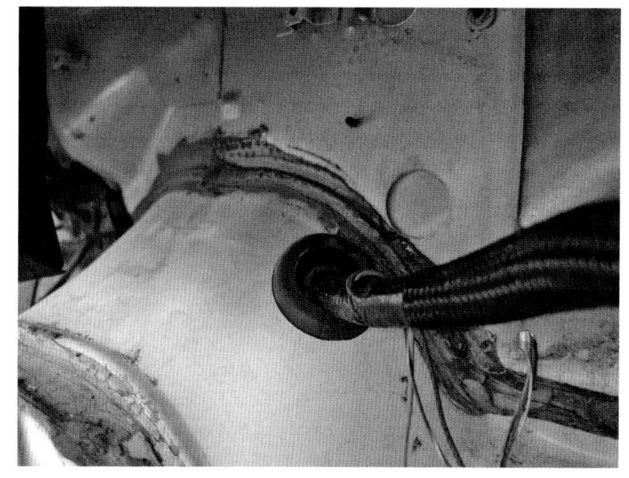

If your engine controller is going to be located under the dash and you want to hide the main branch of the wiring harness, using a hole saw near your transmission tunnel is one of the easiest solutions. This Speartech firewall grommet requires a 2 1/8-inch hole and protects the harness from abrasion through the hole you just cut into the metal tunnel.

The 24× engines are easily adapted, only requiring a camshaft-sensor extension harness or knock-sensor relocation harness, as all Gen IV engines use camshaft timing gear–mounted reluctor wheels versus Gen III, which are on the rear of the camshaft itself. The 24× and 1× Gen IV engines have exactly the same crankshaft and camshaft signal, so it's just a matter of plugging in your Gen III harness and adapting the differences. Then you can use the rest of the Gen III hardware such as the cable throttle body, intake, and injectors. Even though the Gen IV knock sensors are quite different than the Gen III sensors, the latter (if retained) can simply be relocated to the side of the engine block like the former.

What about the 58× and 4× Gen IV engines, you ask? Ideally, these should be kept with the matching PCM, but they can also be adapted for use with earlier Gen III–style harnesses (or in OEM Gen III vehicles) with a Lingenfelter 58× to 24× crankshaft/camshaft conversion box, mentioned earlier.

Aftermarket controllers such as the Holley Terminator X are a different story than OEM for throttle control. The basic Terminator X is specifically designed for cable-drive throttle-body usage, whereas the Terminator X Max can work with either drive-by-wire (DBW) or drive-by-cable (DBC) choices with a hardware and tuning change. Keep this in mind when you are sourcing your Holley systems, as there are many choices and it can be confusing.

FACTORY PCMS

1998–2002 Gen III LS-Series "Warren" Controllers (Red/Blue PCM Connectors)

As with any electronics, there is always something newer, smaller, and better every few years. Automotive computers are no exception. The earlier PCMs are larger and slower than the newest ones. The 1997–1998 C5 and 1998 F-body had the first designs of the LS1 PCMs, and then 1999–2004 performance cars received a slightly smaller, less bulky case with a little more processing speed with each subsequent model year. Almost any 1999–2002 car PCM can interchange with one another, pending similar vehicle options and correct VINs are programmed into the units. In fact, many PCMs are the same blank computer until a specific base program is flashed into it at a dealership.

The 1998 F-body and 1997–1998 C5 have unique PCMs that use different connector pinouts and cannot interchange with the 1999 and newer units. The wiring pinouts for most 1999 and newer Gen III vehicles are similar between vehicles, other than vehicle specific add-ons such as DBW controls, electric fans, transmission options, and the like.

2003–2007 Gen III LS-Series "Warren" Controllers (Green/Blue PCM Connectors)

In 2003, many of the PCMs were upgraded and deemed incompatible with prior engine harnesses due to a few small internal changes, though similar PCM external architecture remained. Two of the large changes are the PCM now controls the O_2 sensor heaters rather than the ignition switch position, and some do not have the idle air control driver function.

One of the biggest changes was that most vehicles are now drive-by-wire (DBW) systems, other than 2004 GTOs and many full-size vans, which were left cable-operated. The DBW-only PCMs are not equipped with idle air control (IAC) drivers internally as it is now controlled by the electric throttle control. If more air or idle speed is required, the PCM opens the throttle body more to increase engine airflow. If you take a DBW version PCM and reflash it with a cable-throttle body operating system, you will not have IAC motor-controlled engine speed. The PCM service number indicates compatibility.

Engine Crankshaft Reluctor Wheel and Camshaft Gear Tooth Count Information

Year	Make	Model	Engine	Crankshaft reluctor tooth count	Camshaft gear count	Cam gear bolt pattern	PCM/ECM type
2004	Buick	Rainier	5.3L LM4	24×	1× (on camshaft)	3	Warren
2005–2006	Buick	Rainier	5.3L LH6	24×	1× (on cam gear)	3	E40
2007	Buick	Rainier	5.3L LH6	58×	4× (on cam gear)	1	E67
2004–2005	Cadillac	CTS-V	5.7L LS6	24×	1× (on camshaft)	3	Warren
2006	Cadillac	CTS-V	6.0L LS2	58x	4× (on cam gear)	3	E67
2007	Cadillac	CTS-V	6.0L LS2	58×	4× (on cam gear)	1	E67
2009	Cadillac	CTS-V	6.2L LSA	58×	4× (on cam gear)	1	E67
2001–2006	Cadillac	Escalade	5.3L & 6.0L	24×	1× (on camshaft)	3	Warren
2007–2009	Cadillac	Escalade	6.2L L92	58×	4× (on cam gear)	1	E38
1998–2002	Chevrolet	Camaro	5.7L LS1	24×	1× (on camshaft)	3	Warren
2010	Chevrolet	Camaro	6.2L LS3 & L99	58×	4× (on cam gear)	1	E38
1997–2004	Chevrolet	Corvette	5.7L LS1	24×	1× (on camshaft)	3	Warren
2001–2004	Chevrolet	Z06 Corvette	5.7L LS6	24×	1× (on camshaft)	3	Warren
2005	Chevrolet	Corvette	6.0L LS2	24×	1× (on cam gear)	3	E40
2006	Chevrolet	Corvette	6.0L LS2	58×	4× (on cam gear)	3	E38
2006–2009	Chevrolet	Z06 Corvette	7.0L LS7	58×	4× (on cam gear)	3	E38
2007	Chevrolet	Corvette	6.0L LS2	58×	4× (on cam gear)	1	E38
2008–2009	Chevrolet	Corvette	6.2L LS3	58×	4× (on cam gear)	1	E38
2009	Chevrolet	ZR1 Corvette	6.2L LS9	58×	4× (on cam gear)	3	E67
2003–2007	Chevrolet	Express	4.8L, 5.3L, 6.0L, 8.1L	24×	1× (on camshaft)	3	Warren
2008–2009	Chevrolet	Express	4.8L, 5.3L, 6.0L	58×	4× (on cam gear)	1	E38
1999–2007	Chevrolet	Silverado, Avalanche*	4.8L, 5.3L, 6.0L, 8.1L	24×	1× (on camshaft)	3	Warren
2007–2009	Chevrolet	Silverado, Avalanche**	4.8L, 5.3L & 6.0L	58×	4× (on cam gear)	1	E38
2003–2004	Chevrolet	SSR	5.3L	24×	1× (on camshaft)	3	Warren
2005–2006	Chevrolet	SSR	6.0L LS2	24×	1× (on cam gear)	3	E40
2001–2006	Chevrolet	Suburban	4.8L, 5.3L, 6.0L, 8.1L	24×	1× (on camshaft)	3	Warren
2001–2006	Chevrolet	Tahoe	4.8L, 5.3L, 6.0L	24×	1× (on camshaft)	3	Warren
2007–2009	Chevrolet	Tahoe, Suburban	4.8, 5.3, 6.0 & 6.2L	58×	4× (on cam gear)	1	E38
2003–2004	Chevrolet	Trailblazer	5.3L LM4	24×	1× (on camshaft)	3	Warren
2005–2006	Chevrolet	Trailblazer	5.3L LH6	24×	1× (on cam gear)	3	E40
2007–2009	Chevrolet	Trailblazer	5.3L LH6	58×	4× (on cam gear)	1	E67

Year	Make	Model	Engine	Crankshaft reluctor tooth count	Camshaft gear count	Cam gear bolt pattern	PCM/ECM type
2006	Chevrolet	Trailblazer SS	6.0L LS2	24×	1× (on cam gear)	3	E40
2007–2009	Chevrolet	Trailblazer SS	6.0L LS2	58×	4× (on cam gear)	1	E67
2003–2004	GMC	Envoy	5.3L LM4	24×	1× (on camshaft)	3	Warren
2005–2006	GMC	Envoy	5.3L LH6	24×	1× (on cam gear)	3	E40
2007–2009	GMC	Envoy	5.3L LH6	58×	4× (on cam gear)	1	E67
2003–2007	GMC	Savana	4.8L, 5.3L, 6.0L	24×	1× (on camshaft)	3	Warren
2008–2009	GMC	Savana	4.8L, 5.3L, 6.0L	58×	4× (on cam gear)	1	E38
1999–2007	GMC	Sierra*	4.8L, 5.3L, 6.0L, 8.1L	24×	1× (on camshaft)	3	Warren
2007–2009	GMC	Sierra**	4.8L, 5.3L & 6.0L	58×	4× (on cam gear)	1	E38
2001–2006	GMC	Yukon XL	4.8L, 5.3L, 6.0L, 8.1L	24×	1× (on camshaft)	3	Warren
2001–2006	GMC	Yukon, Denali, Yukon XL	4.8L, 5.3L & 6.0L	24×	1× (on camshaft)	3	Warren
2007–2009	GMC	Yukon, Denali, Yukon XL	4.8, 5.3, 6.0 & 6.2L	58×	4× (on cam gear)	1	E38
1999–2005	Holden	Commodore	5.7L	24×	1× (on camshaft)	3	Warren
2007–2008	Holden	Commodore	6.0L L98	58×	4× (on cam gear)	1	E38
2001–2005	Holden	Monaro	5.7L	24×	1× (on camshaft)	3	Warren
2005–2006	Holden Special Vehicles (HSV)		6.0L LS2	24×	1× (on cam gear)	3	E40
2007–2009	Holden Special Vehicles (HSV)			58×	4× (on cam gear)	1	E38
2003–2007	Hummer	H2	6.0L LQ4	24×	1× (on camshaft)	3	Warren
2008–2009	Hummer	H2	6.2L L92	58×	4× (on cam gear)	1	E38
2008–2009	Hummer	H3 Alpha	5.3L	58×	4× (on cam gear)	1	E67
2003–2004	Isuzu	Ascender	5.3L	24×	1× (on camshaft)	3	Warren
2003–2004	Oldsmobile	Bravada	5.3L LM4	24×	1× (on camshaft)	3	Warren
1998–2002	Pontiac	Firebird	5.7L LS1	24×	1× (on camshaft)	3	Warren
2008–2009	Pontiac	G8	6.0L	58×	4× (on cam gear)	1	E38
2009	Pontiac	G8 GXP	6.2L LS3	58×	4× (on cam gear)	1	E38
2004	Pontiac	GTO	5.7L LS1	24×	1× (on camshaft)	3	Warren
2005–2006	Pontiac	GTO	6.0L LS2	24×	1× (on cam gear)	3	E40
2005–2006	Saab	9-7×	5.3L LH6	24×	1× (on cam gear)	3	E40
2007–2009	Saab	9-7×	5.3L LH6	58×	4× (on cam gear)	1	E67
2008–2009	Saab	9-7× Aero	6.0L LS2	58×	4× (on cam gear)	1	E67

*1999 was a transition year for the CK truck platform from the Gen I SBC engine (5.0L & 5.7L engines) to the Gen III engines and the new vehicle chassis/body.

**2007 was another transition year for the CK truck platform from the Gen III engines to the Gen IV engines and the new chassis/body.

Source: Lingenfelter Performance Engineering

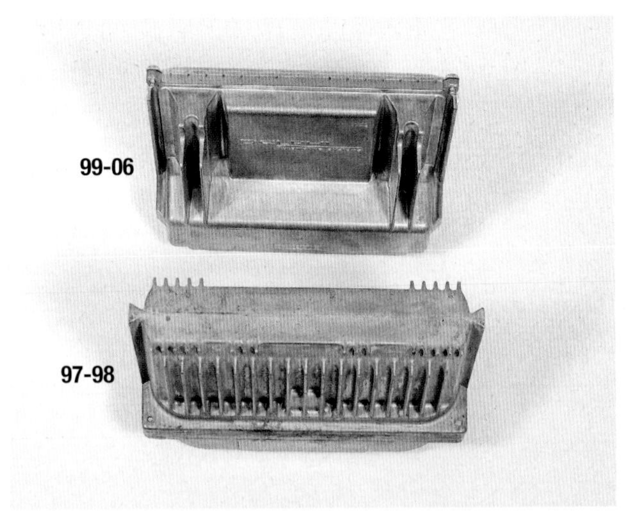

The 1998–2006 Gen III ECMs are easy to identify; they have two 80-pin connectors marked red and blue, with the later ones (2003 and newer) being green and blue, as shown. The same "red/bluc" ECMs are found in similar year trucks and SUVs. When looking for an ECM, be sure to get the correct service number if an incorrect year or make is found. You may still have to reflash the ECM at a dealership to a similar engine design as your buildup.

For the most part, green or blue Warren PCMs are found in Gen III–equipped trucks, vans, and SUVs. All are popular LS engine donors. If possible, always grab the matching harness and PCM from the same vehicle to alleviate mismatches.

In addition to parts numbers, GM also marks the PCMs with service numbers, which is another way to find compatible PCMs for your application. Such a PCM may need a GM dealer reflash to be usable behind your engine, but it is an option.

2005+ GEN IV LS-SERIES CONTROLLERS E40, E38, AND E67

The 2005 performance cars saw a dramatic decrease in PCM size that came with the Gen IV engine revisions. It is equipped with more memory and processor speed, which are necessary to comply with increasingly stringent emissions testing. The improved processor was also needed to function with newer engine controls such as active fuel management and variable valve timing. Model year 2005 also came with CAN (controller area network) communication, which ties all modules throughout the vehicle on one common communication line. These are features not necessarily required (or in some cases desirable) with an engine retrofit though.

The transfer to the newer, smaller PCM brought a few other changes. One of the main benefits is the ability to easily use the electric throttle body and throttle pedal inputs. The automatic transmission control functions are no longer integrated with the engine controller and are housed in their own separate control box. This means the engine controller manages engine-related functions only. If you use the same year 4L65-E transmission, remember to grab the transmission control unit (TCU). One of the neat things about the six-speed automatic transmissions is that the TCU is mounted internally to the transmission itself.

There are three versions of the smaller 2005 and newer PCM: the E40, E67, and E38. Each resides in a similarly sized external housing, but each is unique in operation, wiring connectors, and engine compatibility. The 2005–2006 LS2 engines (with Gen III–style 24×/1× crank/cam signals) use the matching E40 PCM for example. Refer to the PCM chart for exact vehicle donor applications.

HOLLEY EFI

Since the first edition of this book, there have been several industry-wide improvements in the LS-swap game. The Holley EFI is one of those game-changers. The Holley EFI systems offer a user-friendly hardware and software solution to aftermarket EFIs. There are several versions of the ECUs, each with different option capabilities and price points, from inexpensive to moderately priced units. We will cover each option so that you know which one is best for your application.

Much like OEM stand-alone kits, the Holley harness comes with a simple three-wire hookup (battery 12v, ignition 12v, and fuel pump/ignition relay ground) and a separate main battery harness that connects directly to the battery. Every Holley EFI harness comes prewired with the outputs required for fuel pump, electric fans, tachometer, and inputs for oil pressure and fuel pressure. Other inputs/outputs can be added, but the base ECUs are meant to control just the engine so they have fewer inputs and outputs available. More

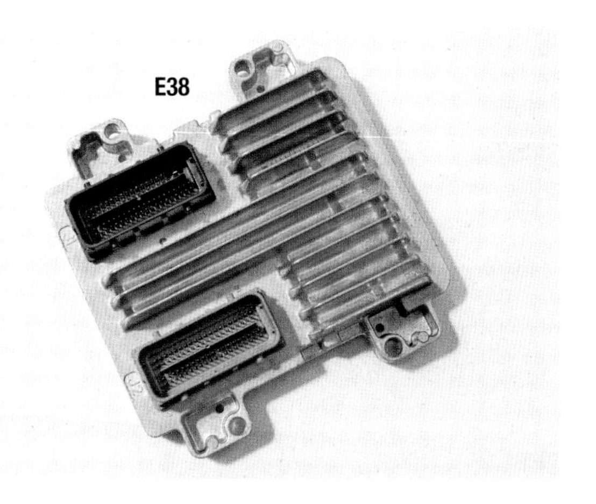

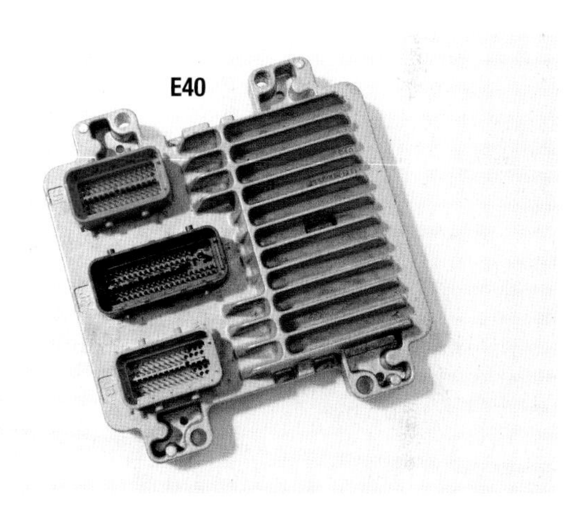

The Gen IV ECMs are Gen IV specific, although the earliest model, the E40, will control Gen III engines as they are both 24×/1× crankshaft and camshaft signals. The E38 and E67 are 58×/4× crankshaft/camshaft signal ECMs only and will only operate with the newer Gen IV engines. Gen IV ECMs are all drive-by-wire; no cable-actuated throttles were available.

You can test-fit your wiring harness on your engine while it's mounted to an engine stand to inspect wiring length and to see where items go. If something needs to be lengthened or shortened, this is the perfect opportunity to make changes. In the GMPP harness, I separated the DBW throttle position wiring from the main harness so it could be rerouted. The coolant temperature sensor wiring is also a bit longer than necessary. This excess wiring was managed by rerouting it a different way though the main harness.

elaborate ECUs (like the Holley Dominator) can control your entire vehicle's electronics much like a body control module.

Additionally, you can get a 3.5-inch base Holley screen (or larger screens from Holley) to view a live data stream, tune the ECU, or even use the screen as a gauge cluster. The screen kit simply needs an ignition source and a ground, and then it plugs into the harness using a CAN bus connector. This data stream can also be used with aftermarket gauges if desired (more on that in later chapters on gauges).

Tuning and calibration can be performed from the Holley screen, but it is easier on a laptop. If you get the non-screen Terminator X versions or the Dominator/HP ECUs, they will come with a USB adapter cable to connect to the ECU. Then you simply download the Holley software for your system and connect. With a few basic settings, you can get the engine started on a base calibration, then use the wideband air/fuel (A/F) correction to dial in the tune over a variety of rpm/cruise driving conditions.

The Terminator X ECUs have a built-in 1-bar manifold absolute pressure (MAP) sensor, but most people use the factory GM MAP sensor in the intake manifold, which already comes prewired in the Holley engine harness. The budget-oriented Term X kits utilize Bosch wideband sensors. More race-oriented Holley HPs and Dominator EFIs can use either the Bosch or the NTK widebands. One other main difference is software based: If you are using a laptop for tuning, the Term X software is specific to the Term X ECUs, while the HPs and Dominators use the more powerful Holley EFI for HPs and Dominator ECUs.

Another advantage of the Holley ECUs is the ability to mix and match crankshaft and camshaft trigger wheels. For example, if you had a 58× crankshaft in a Gen III block and wanted to use the Gen III camshaft sensor on the back of the camshaft, then you can configure this hybrid setup to work in custom ignition settings. Alternately, if you had a 24× crankshaft and wanted to use a 4× timing gear for whatever reason, you could change the settings to do so. This isn't as common as it would seem, but it does allow options for racers who may need to use something like 1× upper timing gear in an engine for a camshaft-driven mechanical fuel-pump drive.

Holley Terminator X

The Terminator X is one of the more popular units and also one of the more budget-friendly. It can operate just about every LS engine made with different kit part numbers; just choose your reluctor wheel and injector connector and you will have all the right connectors for plug-and-play install. The base Terminator X is drive-by-cable only, but can operate either 24× or 58× engines, not electronic transmissions. This means you can retrofit a 58× LS3 crate engine into an older project with a cable-drive throttle and non-electronic transmission. This is an extremely capable and budget-friendly ECU. At the same time, it has the basic minimal outputs to make it an easy engine controller for a swap.

As mentioned, the Terminator X has fewer options and does not control things like electronic transmissions, DBW systems, speedometer, or anything on the vehicle itself. It comes with the standard input/output sub-harness to add a few extra components such as electric fans, two-step rev limiter, or nitrous controls as needed. If you have a drive-by-wire throttle, and/or a 4L60/80 transmission for your LS swap, the next ECU is the desired choice.

Holley Terminator X Max

This is an upgraded version of the base Terminator X ECU. Still an affordable aftermarket ECU, it features more capabilities, offering all that its little brother can with the added features of the drive-by-wire option (or cable drive) and the ability to control electronic transmissions. There are several part numbers with this system, varying fuel injector connections, and also 24× or 58× harness versions, which is more engine specific, as the ECU itself can be programmed for either.

The DBW version comes with the harness that connects from the ECU to the DBW pedal and also to the electronic throttle body. This system is ideal for Gen IV LS3 or LS7 6.2-liter crate engines, for example, but can also be used with any of the DBW throttle-body setups. One nice feature of DBW and LS swaps is that you can easily add an aftermarket Dakota Digital cruise-control module, such as CRC-2000, to the DBW systems, and it is a mostly plug-and-play system with simple hardwiring.

Similarly, the electronically controlled transmission choices are a matter of plugging in the additional transmission sub-harness and programming the ECU for the chosen transmission option. Ideal matched transmissions are the LS 4L60/65/70/75-E and the 4L80/85-E. The six-speed automatic 6L80/90 versions require a separate, third-party stand-alone module and harness from Zero Gravity Performance to work with a Holley EFI system.

Holley HP and Dominator EFI

The Holley HP was the base-model ECU before the Terminator X systems were available. While the Term X ECUs are popular, these more robust and hard-core systems offer many more functional options. These are definitely more race oriented, but they can be used for both retrofits and race vehicles just the same as the Term X options.

The HP ECU offers a very capable system and sits in the middle ground between the Terminator X and the Dominator EFI for capabilities. The HP ECU is similar to the Holley HP version that comes with Chevrolet's 5th Gen and 6th Gen COPO Camaros. The HP can be used on either 24× or 58× engines, but not with DBW or electronically controlled transmissions. The HP uses a single Bosch or NTK wideband for a variety of fuel choices, while the Dominator can use dual widebands.

Gen IV knock sensors are on the sides of the block. Notice that the knock sensor connector is at the 12 o'clock position. This allows easier plug-in access to the sensor but can also be reset for other reasons.

One major advantage offered by the HP and Dominator EFI is that these ECUs are potted for vibration dampening and heat, meaning they can be mounted in the engine bay if needed. The HP and Dominator ECUs feature internal data logging, whereas the Term X requires a laptop or touchscreen to accomplish that. The HP is essentially a Dominator ECU with fewer features, so if you don't need everything the Dominator has but still want more than the Terminator X provides, the HP ECU may fit your requirements.

Both ECUs are intended for more hard-core race-bred applications and can do things such as nitrous control or also offering pulse width modulation (PWM) for turbocharged wastegate boost control solenoids. The Dominator ECU has many more inputs and outputs than the Terminator X and HP ECUs. As mentioned, it is the only ECU that can be optionally used with dual-wideband oxygen sensors (or a single wideband). It also can be employed with DBW throttle bodies and electronic transmissions similar to the Terminator X Max.

While the Dominator can control the 24× and 58× engines just like the HP and Terminator X systems, the generous added inputs and outputs provide another level of control of the entire vehicle by means of the Dominator, much like a body-control module would. Functions like turn signals, brake lights, headlights, and so on can be wired through the Dominator and controlled with a relay or distribution module. While the main engine harness comes prewired, most of these custom options have to be added separately, so setup calls for some wiring know-how.

ELECTRONIC TRANSMISSION PCM CONTROLS

One of the nicer features of the OEM harnesses is the ability to control the electronic automatic overdrive transmission, if so desired. If you want an electronically controlled automatic transmission behind, let's say, a carbureted SBC or BBC engine, you would need to obtain a much more expensive stand-alone transmission controller. This is a huge luxury that comes with the LS engine controllers, both from a cost and ease of integration standpoint.

The Holley EFI ECMs offer internal electronic transmission controls on certain ECUs, specifically the Terminator X Max and the Holley Dominator. Both of these can control the 4L60-E or 4L80-E options, with the correct add-on harness (P/N 558-405) from Holley. A base-model

ECU like the Terminator X does not have transmission controls or even vehicle-speed inputs, so it's best to plan ahead and get the correct ECU kit from the start if you need to control an E-version GM transmission. Holley makes it easy to do this, as they offer a variety of kit options that include the required harness options.

The 1997–2007 "Warren" PCMs have all necessary transmission-controlling functions internal to the PCM to control both 4L60-E and 4L80-E transmissions, requiring only matching GM programming to that transmission. Manual transmission PCMs do not have the programming, but it can be added. The E40, E38, and E67 use external separate transmission controller modules (TCMs) that directly communicate with the engine controller for input.

You might be wondering how you program the blank PCM for your specific application. For this example, let's

The factory coolant temp sensor is located on the front of the driver's-side cylinder head. Keep in mind that 1998 F-body harnesses use a three-pin coolant temperature sensor, while all other years use a two-pin setup. The third pin position sends direct coolant temperature values to the temperature gauge. Whereas the two-pin connectors transmit CTS data to the ECM, which then passes it on to the temperature gauge.

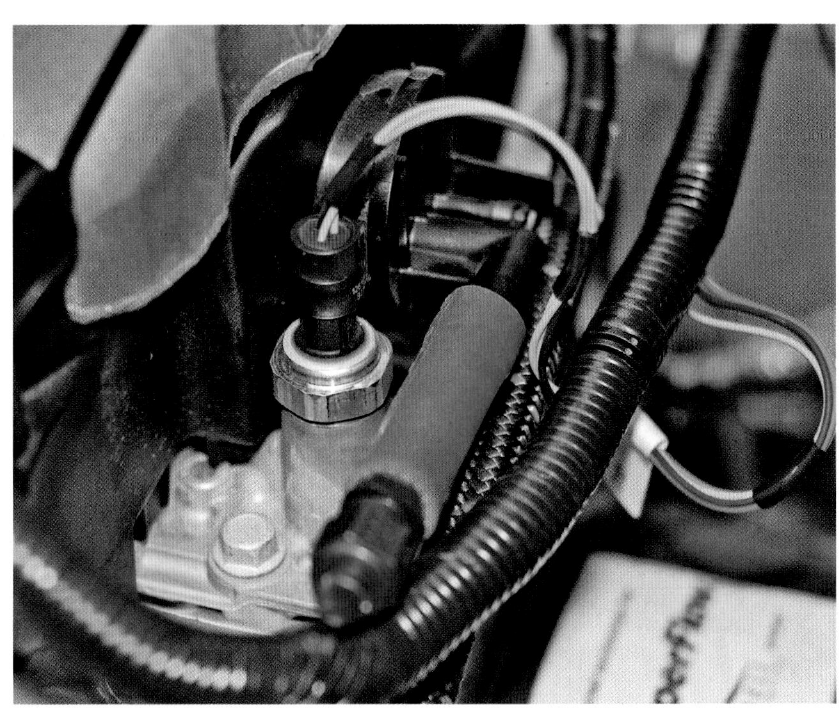

This is the oil pressure sensor and connector. In a retrofit application, the single one-wire oil pressure senders can sometimes be applied in GM-to-GM swaps, but you will rarely find the three-pin LS sensor useful other than for data-stream OBDII data usage or for a stand-alone such as a Holley dash. Many harnesses still have the connector, so if you have the sensor, you might as well use it. If not, secure the harness pigtail and either plug this oil pressure hole or use an aftermarket mechanical or electric sender and gauge.

assume you have a 2001 5.3-liter engine with a 4L60-E automatic transmission. The most obvious and easiest thing to do is find a PCM from a wrecking yard search (www.car-partinc.com is a great resource), but there are other options. If you need to purchase a new PCM from a GM dealership, you can have them flash the PCM with a similarly optioned 2001 truck (search eBay/autotrader for a donor VIN to use). Afterward, the PCM will still need a reprogram from a capable LS PCM tuner to remove the vehicle anti-theft system (VATS). This is covered in the next section.

Keep in mind that only a few GM dealerships can do an off-board GM PCM reflash without having the vehicle. If so, you may have to do this as one of the last steps by bringing your swap vehicle to a dealership, which is a hassle. In addition, the dealership cannot disable VATS or program electric fans. Talk to your preferred LS-swap specialist beforehand for PCM options; there are a plethora of routes to get the PCM setup for your particular swap project.

VEHICLE ANTI-THEFT SYSTEM

The theft-deterrent feature found in many late-model vehicles integrates the PCM with the proper signal from the body control module (BCM). There are a few variations of this basic system, but they all function similarly: If an engine start is attempted without the proper signal from the BCM, the engine will not start at all, as the PCM thinks it is being stolen. This is obviously quite a problem for the LS-swap community. Since it is unlikely that LS-swap candidates would swap the entire dash harness, steering column, and body control module, there is an extremely easy way to disable the theft deterrent by reflashing the PCM. This can usually be accomplished along with any other PCM programming required (such as deleting rear O_2 sensors, EGR, etc.) by a qualified tuner.

While the PCM is being recalibrated, the VATS function can simply be disabled by turning it off in the tuning software. With the VATS disabled, the PCM can function completely by itself when powered up, with no further input from any other modules. This adds simplicity to your swap procedures, but also allows others to start your car just as easily.

Be aware that if you leave your vehicle unattended and someone with some electrical know-how attempts to steal it, it is possible to easily start an older car with EFI. It's just as easy as with the original carbureted engine. If this scenario is likely, look into an aftermarket alarm to integrate into your vehicle to at least put up a fight.

There are many ways to integrate a custom theft deterrent, such as adding hidden switches for the fuel pump or an ECM power/ground that can interrupt the signal needed for the pump to run. The good thing about EFI swaps is that there are many ways to disable and shut them down, for example, by interrupting coil power or the fuel pump relay, fuel injectors, or the ECM ground itself.

You can even add something like an ordinary-looking under-dash OBD II connector with your own keyed wiring plug-in that loops several circuits in your OBD II "key." This is something like a $20 add-on that, when wired to your own harness, provides a simple way to kill several circuits when the car is unattended.

And don't underestimate a normal battery shut-off switch, which could simply kill the main power to the starter. You can even use one to shut down the main electronics while keeping the car alarm powered up. Since many batteries are mounted in the trunk in custom builds, a trunk-mounted kill switch hides access to powering up the vehicle versus one mounted visibly on the car's exterior.

GENERAL PCM LOCATION TACTICS

The PCM in the factory vehicle can be found near the engine either in the engine compartment or nearby. The PCM itself is sealed well from outside contamination both in the housing and for wiring connections, but as with any electronic device, care should be taken in choosing a mounting location. It's advisable to try to duplicate the OEM-mounting location in your LS-swap transplant patient, or maybe find an even better location. The PCM can be tucked behind a glove box, hidden inside an inner fender, or mounted in plain site. Mounting options are endless, provided the engine harness agrees with your location.

The Holley EFI ECMs have four integral mounting tabs on the ECM housing to ease mounting. On the Terminator X these mounting tabs do not have vibration-dampening grommets built in, but you can add isolation by using rubber-damper mounting pads. None of these ECMs require grounding, so mounting to something like a plastic glove box or an old ECM tray is an option if that is the most convenient location.

Do not rush the PCM mounting; give it some long thought as to wiring accessibility, ease of serviceability, and more obvious things such as moisture and extreme heat avoidance. As a general rule, never mount the PCM on the engine or near a heat source. Avoid mounting the PCM where it is flush against a flat panel, and leave a little air gap as with all electronic devices. Apart from these basic guidelines, there are many location options to suit just about anyone's needs.

Almost all harness options include extra main-branch wiring, allowing more mounting flexibility as compared to OEM car or truck harnesses. If you do mount under the dash, make sure to use a good grommet wherever you poke through the metal firewall. A good hidden location to add a through-hole is the top of the transmission tunnel near the passenger-side cylinder head. This location allows the main harness to be hidden and run directly into the passenger compartment.

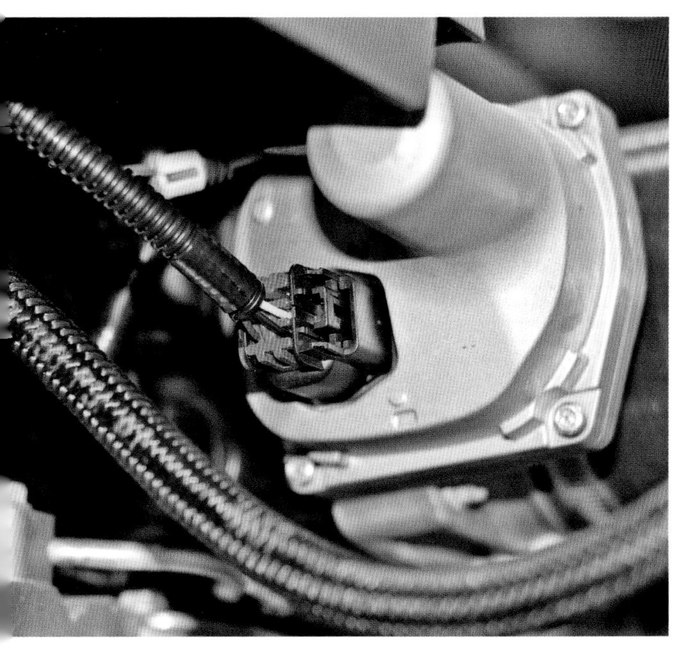

The Gen IV DBW throttle bodies have a single connector that needs to be plugged in. The connector is located on the passenger side of the throttle body and just clicks into place.

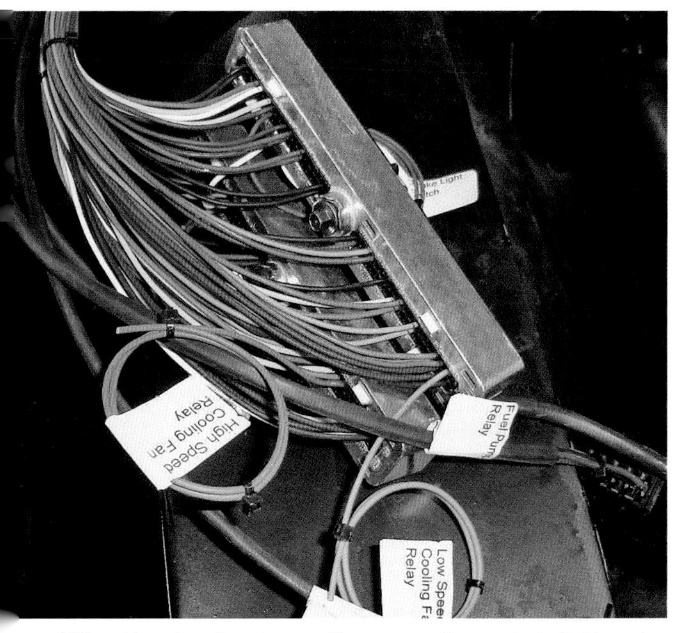

With a wiring schematic and some patience, you can trace each wire that is used in the engine harness. Or, you can buy a new harness or rework your old harness into stand-alone, which will allow the builder to mark and identify each reused wiring system.

INSTALLATION AND WIRING

Sensor Connections

Many aftermarket harnesses will have each individual component labeled for easy hookup in the proper location. If your harness is not marked, it is not hard to determine where everything goes. For instance, if you lay out your engine harness over your assembled engine in its general configuration, many connectors are obvious in how they will be connected. By process of elimination, you can get many connections plugged in and sorted out.

There may be some that do not get connected at all, such as the EGR, rear O_2 sensors, or EVAP/AIR injection solenoids, but the majority of connectors on an assembled engine will have a mating electrical component nearby to plug into. If not, connect everything obvious, and double-check the unused connectors against remaining items and compare wiring colors to the wiring harness pinouts. All wiring connectors should easily be identifiable by process of elimination. Refer to the PCM/connector pinout charts for wiring colors and sensor descriptions if all other identification procedures fail.

Keep in mind that if you are using a Gen III–style harness on a Gen IV crate engine, there will be some differences in certain connectors and sensor locations. As mentioned beforehand, the camshaft position sensor is relocated from the engine rear (Gen III) to the engine front (Gen IV), so this would require a camshaft extension harness. Additionally, the Gen IV knock sensors are relocated to the sides of the block to make room on the valley plate for active fuel management (AFM) cylinder deactivation solenoids. The Gen IV knock sensors are not compatible with Gen III engine controllers or vice versa, so you will need to use the sensors determined by your PCM harness choice. Since the older Gen III knock sensors use an M10 × 1.5 thread, relocation is accomplished easily by locating an empty bolt hole on the side of the block. Drilling and tapping the Gen IV knock sensor bolt hole is another option.

ICT Billet also makes a billet knock-sensor relocation kit that bolts to the oil pan or the side of the block. This billet solution alleviates drilling and tapping and provides a bolt-on option with versatile mounting locations. This is especially nice if your engine is already installed by the time you decide to install your knock sensors. Check out part number 551216-KN30 if this is something you need.

If using a Gen IV intake manifold, the MAP sensor is also relocated from the rear of the intake to near the throttle body itself. This would require a MAP extension harness. Last, you would require either injector wiring adapters, or splicing in matching injector pigtails to adapt your LS engine harness to the Gen IV fuel-injector connector, as these are completely different.

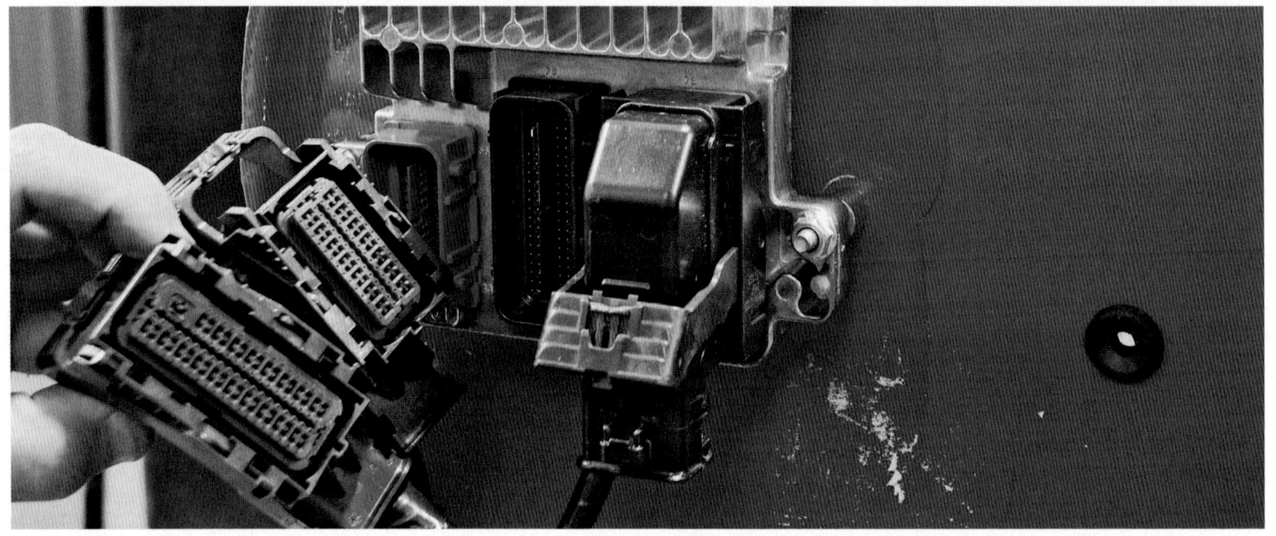

Another minor obstacle is deciding where to mount your new ECM. Likely, your swap project never had an LS engine controller, so you get to decide which location is best. On this 1969 Camaro, the E67 ECM fit perfectly in the passenger-side fenderwell and bolted to the firewall. I recommend using some spacers or rubber isolators for ECM mounting.

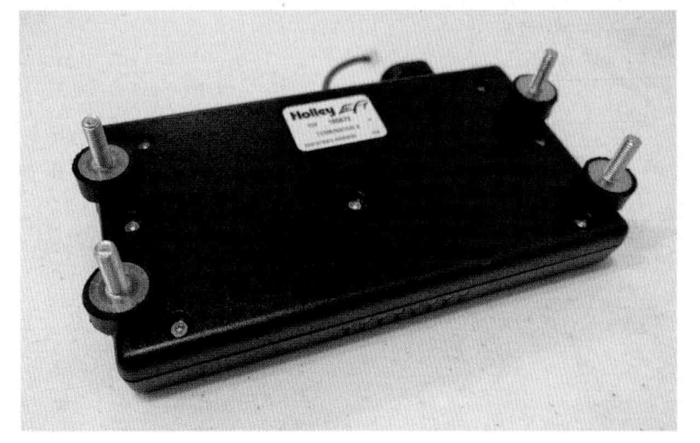

Holley ECUs come with mounting holes already equipped for direct mounting. You can add some rubber anti-vibration isolators from places like MSD or Amazon to isolate them better. These are $10 M6 x 1.0mm isolators were purchased from Amazon. Further, you can often repurpose the OEM mount from your original EFI-equipped swap project to mount your aftermarket ECU. This one is from my 1991 Camaro; drill the ECU mounting holes and remount the bracket under the dash like it was before.

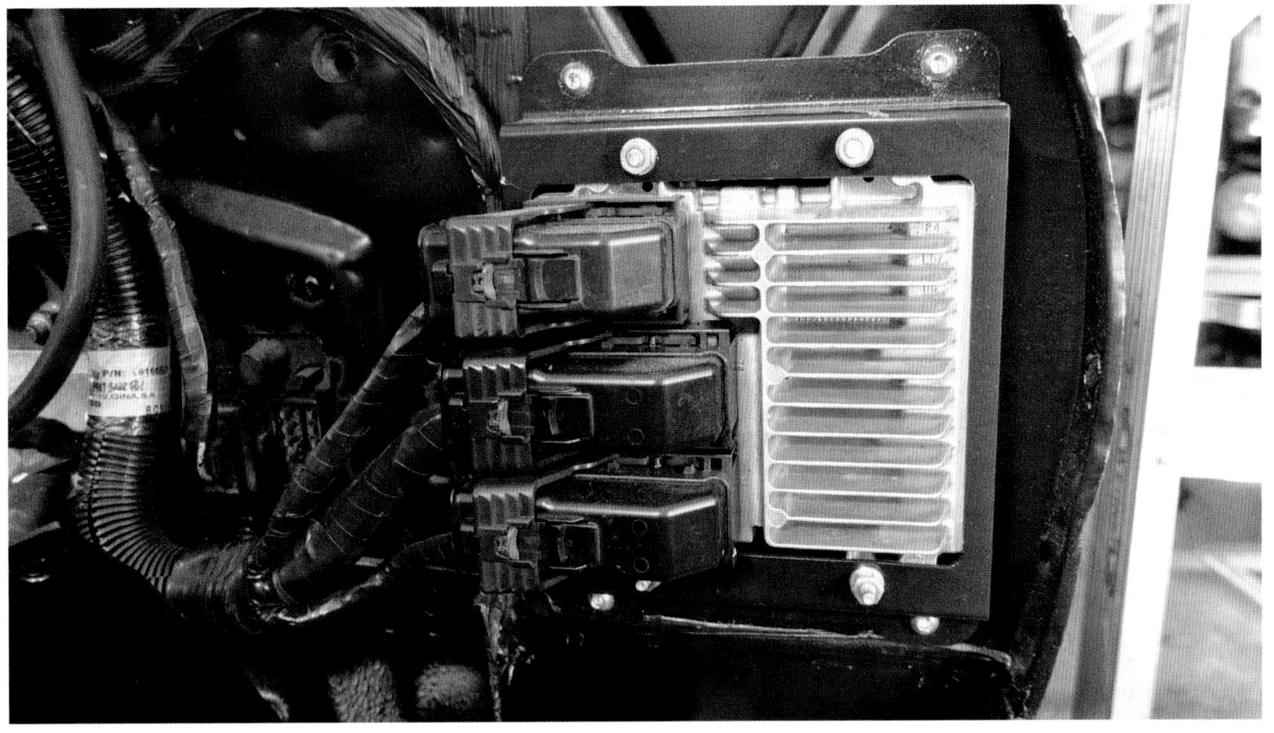

Take your time and mount your ECM in a nice, secure area, away from road debris or moisture. The GM E67 ECU in my 1967 Camaro is hidden safely behind the inner fender and inside a hollow spot on the firewall. I used Riv-Nuts and ¼-inch button-head Allen bolts to make for a clean install. This mounting bracket is from Offroad Anonymous and cost about $25 unpainted.

You can sometimes use donor original mounting solutions with your GM ECM. This one helps you mount your ECM on a flat surface and provides a secure OEM mounting point that simply clips in. Shown here is GM 15995679, which cost about $25 at the time of purchase.

Another option for GM ECM mounting comes from our friends at ICT Billet. These brackets bolt to the existing bolts on the ECM itself and provide stands to mount wherever convenient in your engine bay or under your dash. *ICT Billet*

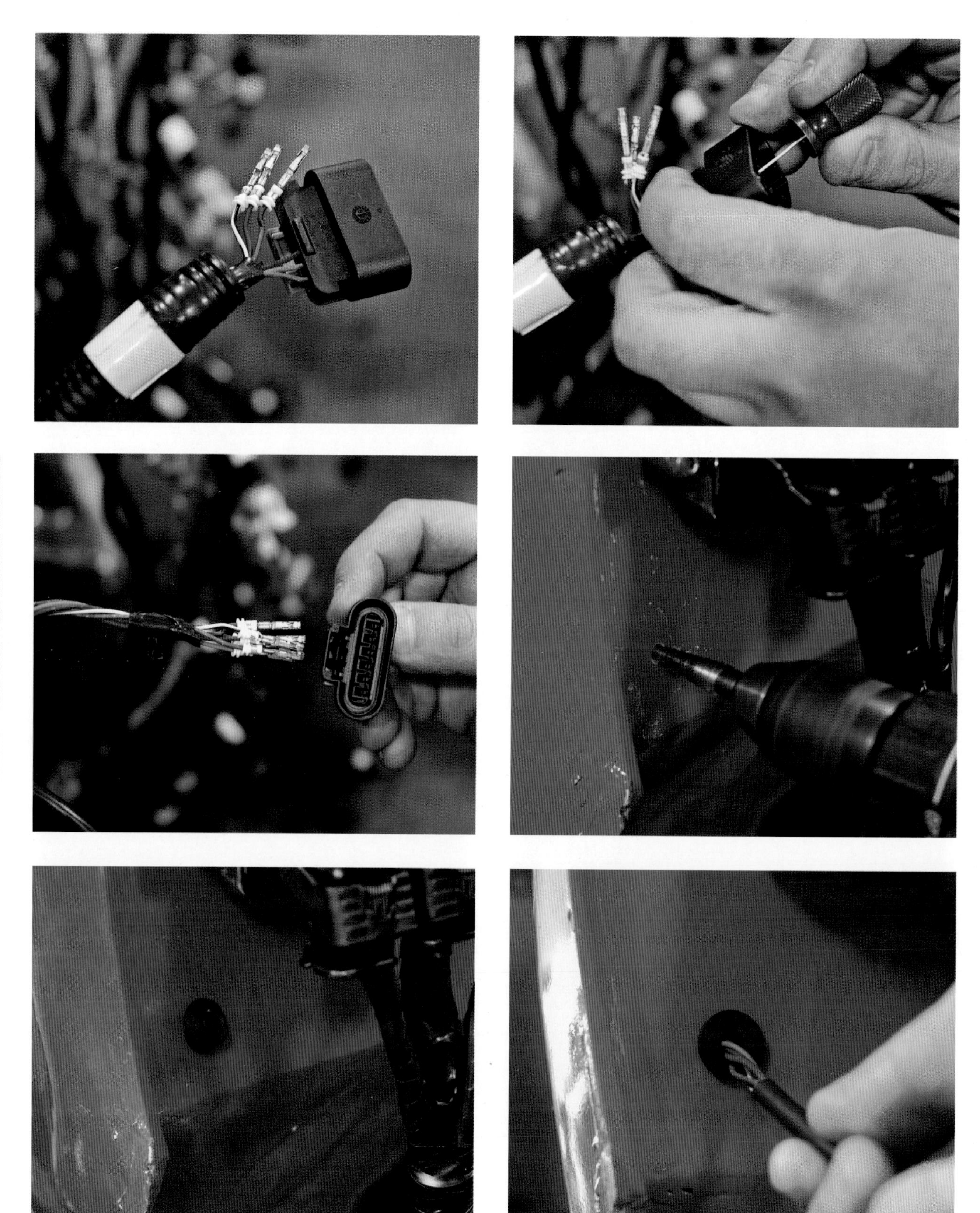

Individual wiring connectors that must go through the firewall sometimes present a problem. For instance, you don't want to cut a large 2-inch hole to allow the connector to pass through, while it uses only a ¼-inch wiring harness. The solution is to document the wire positions and colors and remove the terminals from the connector. Drill your through-hole, add a grommet, and then feed the wiring through. Once the wiring is through, reconnect the harness connector.

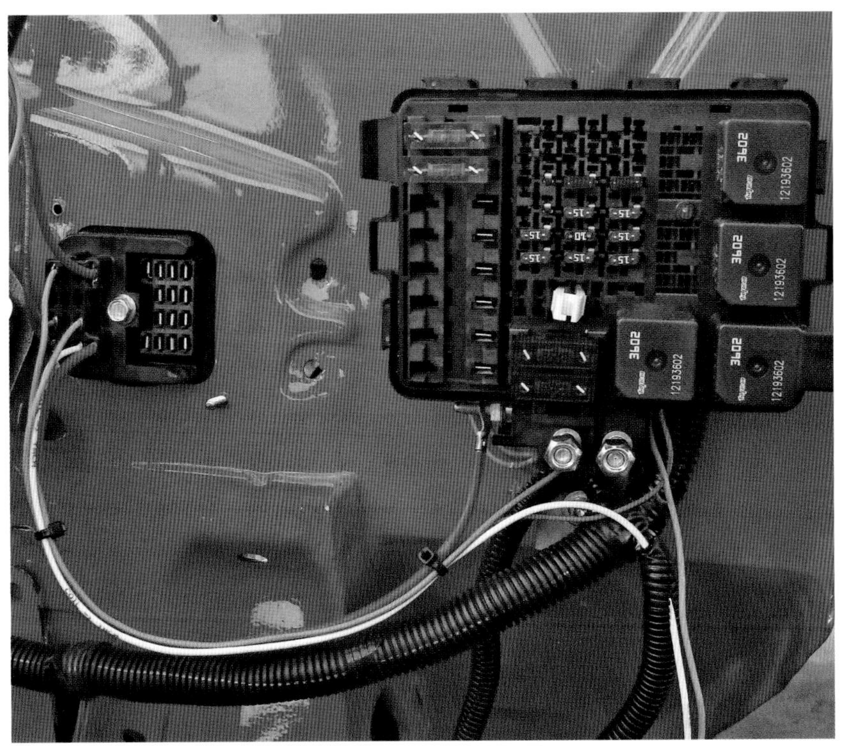

The GMPP harness like many retrofit harnesses requires two wires for power. You will need a battery (red) and an ignition source (pink). The stock bulkhead connector has what you need for connection. Usually, the required wires were used at one time with an SBC or other engine. Using factory terminals will also make for a cleaner installation.

Often you will be adding several relays to the vehicle that were never on the original setup. You will require a fuel pump relay, cooling fan relays, and sometimes a main power relay is also required. An easy mounting solution for an abundance of relays is to use aluminum strap cut to length with the edges smoothed and beveled. Make it look like it is a custom piece by using stainless bolts.

If your donor engine has some crusty old spark plug wires, these tend to tear when removed and will need to be replaced. Those shown are 10mm wires from Scott Wires, aftermarket wires are a much better solution than stock wires. And, with the 45-degree ends, they have the added perk of clearing header primary tubes much better than straight boots.

If you want an easy ignition-controller setup to use with a carburetor, look at the MSD 6LS ignition coil controller. This plugs into the stock crankshaft and camshaft signals and controls the eight stock coil packs. The 6LS is designed for LS1- and LS6-type engines with a 24× reluctor wheel, which can be identified by its black harness connector. The 6LS2 is designed specifically for the LS2/LS7 and its 58× reluctor wheel, which can be easily recognized by its gray harness connector. *MSD Ignition*

Painless Performance offers many fuse and relay panels. It is best to source what you may need specific to your application as every swap is slightly different. Organization is what keeps wiring systems clean and safe. *Tom Munsch*

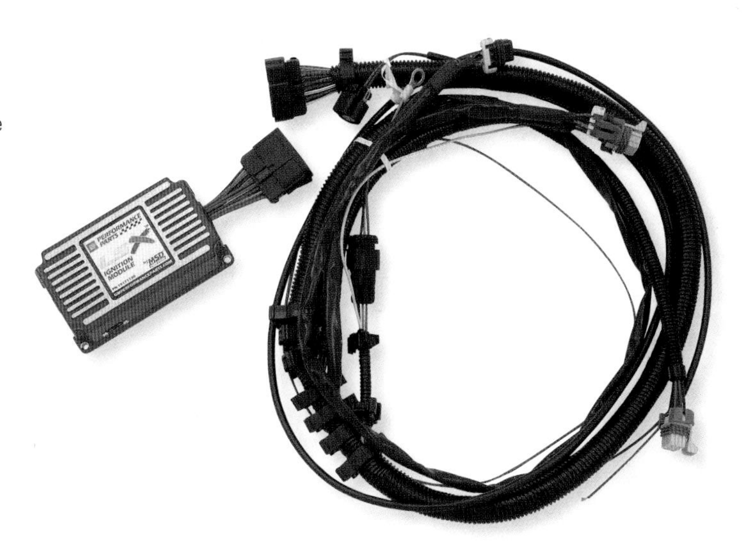

GMPP also offers a similar ignition-controller box and harness intended for their carbureted crate engine program. This controller is for LS engines that don't have distributors and use a 58× reluctor wheel only. The kit includes the ignition controller and wiring harness. Software allows custom vacuum advance curves, timing curves, programmable rev limit, and more. It is compatible with LS1/LS6 and LS2/LS7 ignition coils. *Courtesy of General Motors*

Carburetor fanatics can now rejoice at the fact that they, too, can have late-model LS-powered drivetrains without all the fancy and expensive EFI controls. The use of a carburetor simplifies the fuel system and engine fuel tuning. Fueling issues are addressed as with any other carburetor-tuning procedure. MSD provides the spark to ignite the fire on this carbed engine package. *MSD Ignition*

The MSD 6LS2 is for the 58X Gen IV–style engines when using a carburetor. Notice that the engine does not actually look any less "LSX" with the carb. The MSD 6LS2 has a sufficient wiring harness to be mounted away from the engine itself. Here, it is bolted to the inner fender of the first-generation Camaro. *Holley Performance Products*

DISTRIBUTOR INSTALLATION

GMPP offers the Wegner distributor conversion for those who want to really go old school on their LS package and ditch both the EFI system and the LS coil packs.

The distributor conversion starts out by installing a color-coded alignment dowel that matches your upper timing sprocket.

Then the mandrel for the camshaft extension stub is bolted into place; this three-bolt stub sandwiches the camshaft gear in place while providing a fixture point for the distributor and fuel pump drive extension.

Next the actual drive gear is aligned to the new extension dowel and threaded into place using an Allen cap screw. Use a bit of thread locker to keep the bolt from backing off and torque the center bolt to specifications.

Use some break-in grease on the fuel pump lobe and the distributor gear. At this point the drive portion of the distributor portion is complete. But wait, you cannot just install any GM timing cover, you will need the Wegner cover also.

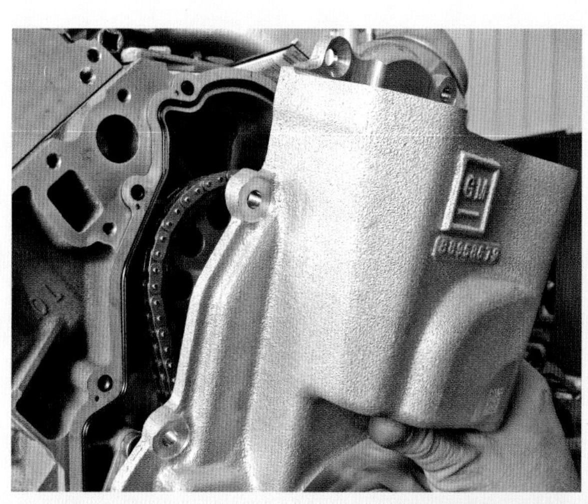

The Wegner timing cover installs much like an OEM cover, with a few small changes with bolt lengths and locations. The timing cover bolts have a 5-millimeter hex key, and the bolts are torqued to 18 foot-pounds. After installing the front cover, install the timing cover front seal with a few taps using a small hammer and seal installer.

(continued on page 80)

DISTRIBUTOR INSTALLATION *(continued from page 79)*

After the timing cover is in place, the ATI crankshaft damper is pressed into place, the extended Stewart water pump is mounted to the block, and the crankshaft pulley is bolted on. Notice that the crank pulley uses the conventional three-bolt style that SBC engines use.

Next, install the water pump pulley using the supplied Wegner accessory hardware. This particular kit uses only the alternator, so at this time the alternator brackets are bolted to the three water-pump mounting bolts and cylinder head. The Wegner accessory drive includes its own tensioner.

Once all the distributor drive components are in place, the conversion is complete. The only thing missing is the carburetor and distributor. The Wegner conversion uses a small-block Ford (gasp!) distributor that is available from a variety of sources. MSD is likely the most common source for aftermarket distributors. *MSD Ignition*

MATING THE ENGINE HARNESS TO THE CHASSIS

Many harness options only require a few wires to integrate the engine harness itself into the existing chassis wiring. For the most part there is a 12-volt battery, 12-volt ignition, and some specific outputs such as speedometer, tachometer, fuel pump, and electric fans. Wiring these properly into your swap is not hard but must be done with proper care and consideration so as not to harm electrical components (both old and new) or create unsafe wiring connections. Most harness options will have fuses protecting the components. If not, make sure you add a fusebox or inline fuse connectors to every load system.

For swaps, the wiring outputs on the GM PCM control the ground side of relays. If you are mating your own harness into the car and want to hook up fans, you cannot just connect the small fan wiring output to the fans themselves. A relay is required for all high-current outputs. The outputs that do not require relays will connect directly to the solenoid and motor via an existing GM harness connector, such as the throttle body motor, O_2 sensors, IAC motor, and EVAP solenoids. Items that require relays would be the electric fuel pump and the electric fans, as the PCM only controls the ground side of the relay.

Retrofit harnesses often have the fuel pump relay for a simple one-wire fuel pump connection. Just extend the wire to your fuel pump using wire-abrasion protection and connect to the positive terminal; ground the negative side to the chassis or battery ground. Often the fan output is just the PCM-controlled fan relay ground if you connect this wire or wires to the electric fan relay ground circuit. It is rare to have the fan relay in the harness or engine harness fusebox, although the GMPP harnesses do come equipped with a single fan power relay and wire output that connects to the electric fan itself. Remember that if the fan is on the rear of the radiator it is a puller fan. If it's on the front side, it's a pusher fan. The polarity of the fan wiring dictates the fan rotation. Fan airflow should be from the front of the car to the back. Additionally, you can use modern PWM fan-speed controls if your ECM is compatible with the system. These fans typically have three wires on a single fan—two large wires and one small ECM wire—and the power source is typically wired through a high-amp fuse for full-time power, with the small third wire serving as the PWM speed-control wire.

The 12-volt battery source can be anything directly connected to the battery full-time. You can use the battery cable connection on the starter, alternator, or the battery itself. You want to make these connections after the engine sensors, coils, injectors, and grounds are in place.

If you are not using the LS-style alternator, you must use something that has enough amperage capacity to keep up with the engine electronics, fuel pump, and electric fans. The GM alternators vary from 110-amp to 160-amp, depending on application. The factory cars these alternators are in have a ton of non-engine electronics throughout the vehicle with things like heated seats, power windows and mirrors, DVD players, GPS, and other creature comforts that your swap vehicle likely does not have. So the large-amp alternators can be overkill, but that does not hurt anything.

TACHOMETER PULL-UP RESISTOR

Some aftermarket tachometers, shift lights, or external transmission controllers may not be able to sufficiently read the tachometer output signal from the ECM. In this case, you will need to wire an ignition voltage source through a 5,000-ohm ¼-watt resistor and intersect this wire into the tachometer signal output wire. Then hook up the tachometer wire as normal. Both the Gen III and Gen IV ECM's output in "4-cyl" mode with two pulses per crankshaft revolution. Sometimes a 4,800-ohm ¼-watt resistor is needed, it just depends on what accessory reading the signal prefers. Luckily resistors are really inexpensive if you have to get a different one than what you first tried.

ECM CALIBRATION

Now the only problem with picking and choosing which vehicle your OEM ECM comes with matters if you are forced to buy a brand-new ECM from a GM dealership. Since the blank ECMs are flashed to your car and not the ECM itself, you may run into the issue that you will have to have your project complete to get a reflash. Then take the swap vehicle to the dealership to reflash using your donor VIN into the ECM.

On the other hand, if you purchase a pre-programmed ECM from either a salvage yard, or an LS-swap-intended package deal from a specialist, you save yourself some aggravation by not being required to move your vehicle for the base flash tuning. Also, before using any OEM GM calibration, you also will need to reprogram the base ECM to even get your retrofit projects engine to start, such as fuel-injector settings, VATS removal, and DBC or DBW controls.

The above considerations are all factored into using non-swap intended wiring harnesses and ECM's, if you are dealing with a retrofit company or retrofit intended kit such as the Gen IV GM Performance Parts engine controller harness and ECM for an automatic transmission. The ECM will come pre-calibrated for your GM crate-engine setup.

Using aftermarket gauges with your new LS-series engine takes a few extra steps to make functional, mainly requiring adaptation on the engine side of the installation. Aftermarket speedometers and tachometers are flexible enough to be functional with a few setup calibrations. Tachometers must be set to 4-cylinder mode for an example.

The stock oil pressure sending unit location is shown in the upper central location of this image. This is the place to tap into for your sender or oil pressure adapter. Autometer offers an M16×1.5 metric to ⅛-inch pipe adapter for this purpose. The Autometer part number is 2268.

One of the best locations for an electric bell-type sending unit is on the oil transfer plate located above the oil filter on the driver's side of the engine. The 1997–2004 Gen III versions (non-C5) of this plate have a boss that can be drilled and tapped to ⅛-inch NPT. Later non-oil cooler versions such as on Gen IV engines do not have the boss, but can similarly be drilled and tapped in the same manner—or by going with the earlier Gen III version.

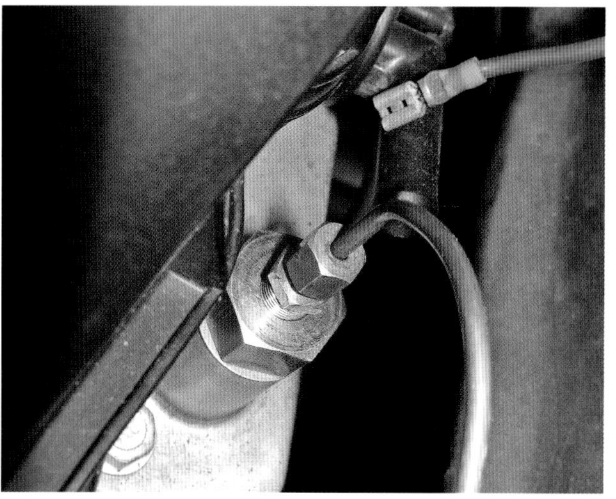

The oil pressure gauge location is ideal in one of two spots, the lower location on the side of the block offers a convenient mounting point for one of the larger bell-type oil sending units or also if using an oil temperature gauge. The upper location behind the intake manifold is the stock location for the OEM metric threaded oil pressure sender, but with a metric M16 to ⅛-inch pipe thread adapter such as the Autometer P/N 2277, you can facilitate aftermarket gauges and also adapt SBC-style OEM sending units into your LS engine.

To use an aftermarket electric temperature gauge with an LS engine, you need to either adapt the factory metric coolant hole provision by using an aftermarket metric to ⅛-inch adapter from Autometer P/N 2278 in the passenger side cylinder head or by drilling and tapping an alternative location such as in the water-pump housing.

If your donor engine is five or more years old, it is best to plan on new spark plug wires. MSD offers stock replacements in a few variations and in either black or red. The MSD cable set for car style coils is part 32819. The set for truck coils is 32829. These are a bit longer due to the higher mounting location of the truck coils.

MSD also offers upgraded ignition coils for the LS-series engines. The coil packs offer higher voltage and an auxiliary terminal connection for ignition controller boxes. Be sure to use the MSD coil identification chart on the MSD website www.msdignition.com to order the correct design coil pack, as GM has five versions.

If you don't want to have the coils on your valve covers and have a spot to remote mount them, you will need remote coil brackets and extended wiring to relocate the coil packs. ICT Billet has you covered with these coil mounting brackets. *ICT Billet*

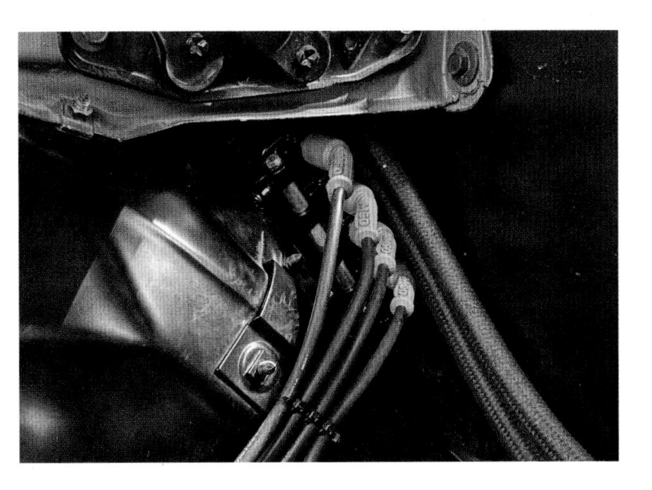

To put these Texas Speed valve covers on this 1968 Camaro, the ignition coils previously located on the valve covers had to be relocated somewhere else. I mounted them in the void between the inner fender and the firewall. This shielded the coils from road debris and moisture, hiding them from the engine bay view. This requires making custom-length spark plug wires from MSD and using a coil pack wiring extension from ICT Billet.

KATECH COIL RELOCATION KIT

The Katech coil relocation brackets are designed to fit the Katech valve covers. With some extra work, they can fit almost any aftermarket cast-aluminum valve cover though. First, set your new valve cover on the engine and mark where the bracket fits best. Usually it is centered front to rear.

Once the bracket is marked, you can drill the mounting holes. Since we are using ¼-inch × 20-thread bolts, the tap calls for drill size letter F. With these particular GMPP valve covers, it is easier just to drill through the cover and seal the bolt threads with some sealant when done.

After the hole is drilled, threads need to be cut. This is where the careful use of a ¼-inch × 20-thread tap is desired. You want the hole to tap exactly straight. Exert force toward the part when first starting the threads and continue to turn the tap clockwise until it gets about two to three threads cut. When tapping threads, it is good practice to cut one-quarter turn, back off a half turn, and repeat by using cutting oil. You will need four threaded holes on each valve cover.

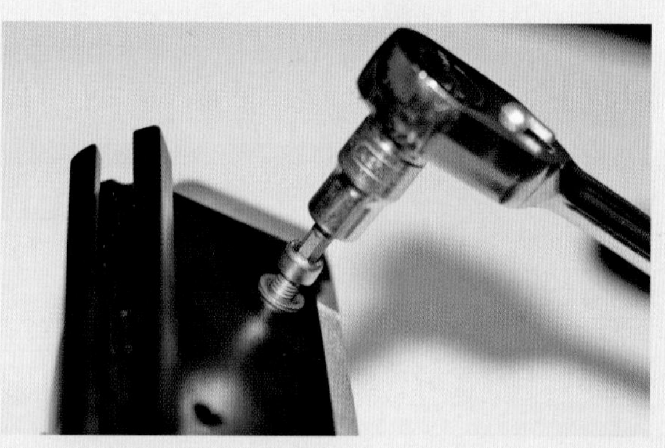

When the threads are cut, clean the valve cover thoroughly and bolt on the new coil brace using the ¼ × 20 bolts with a smidge of thread sealant. Don't go crazy on torque; just snug them up for now.

Next, when all bolts are snug, you can install a locknut to ensure that the coils will not loosen.

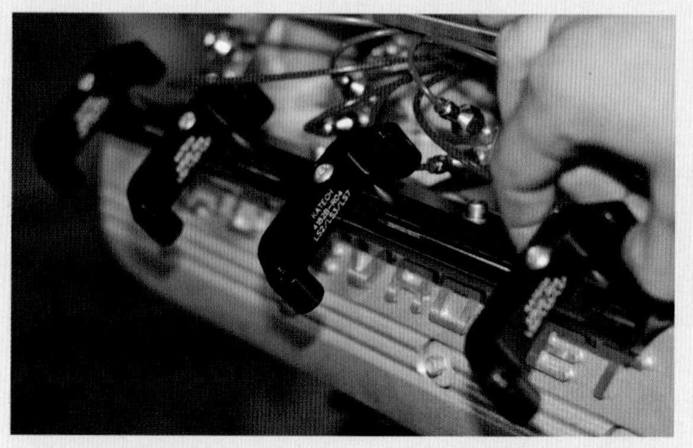

The Katech coil brackets are mounted on the lower brace via a T-shaped nut and long bolt. The bolt and T clamp down on the slider rail and lock it into position. Before mounting the coils, you need to pre-install the U-shaped brackets.

Finally, you can install the eight coil packs and tighten the mounting and slider bolts down where the coils fit best. Attach the coil pack wiring and spark plug wires and you are done. If you have aftermarket valve covers, you can actually see them with the coil pack relocation. Additionally, the valve cover bolts are accessible without removing coils as a small time-saver when working on your engine.

Chapter 4
Fuel System Design

The fuel system is an area many people do not consider when first planning the LS engine swap. It quickly becomes a major aspect of any EFI engine swap, especially if the car was originally carbureted. Often a completely new fuel system front to back is required to enable your LS engine to start. If you are working with a newer EFI non-LS vehicle, some of the costs are not as staggering. You may only need a few isolated components to adapt to the LS fuel-injector rail. In a restoration project for an older car, you often replace many fuel-system components with brand-new parts anyway, so a few extra dollars for a new EFI-specific fuel tank, lines, fuel pump, and so on is unlikely to bust the wallet.

FACTORY DESCRIPTION AND DIFFERENCES

The factory LS-based fuel systems run on much higher pressure than your standard multi-port fuel-injection setups. Many, if not all other EFI fuel systems operate at approximately 44 psi. The LS-based fuel system operates at 58 psi in stock form. What this means is that you cannot take your new LS-series engine and slap it in place of another engine without adapting the fuel system or PCM injector calibrations to match the lower pressure. Almost all EFI engines operate with an in-tank 12-volt high-pressure fuel pump to deliver the required 44 to 58 psi of regulated fuel pressure.

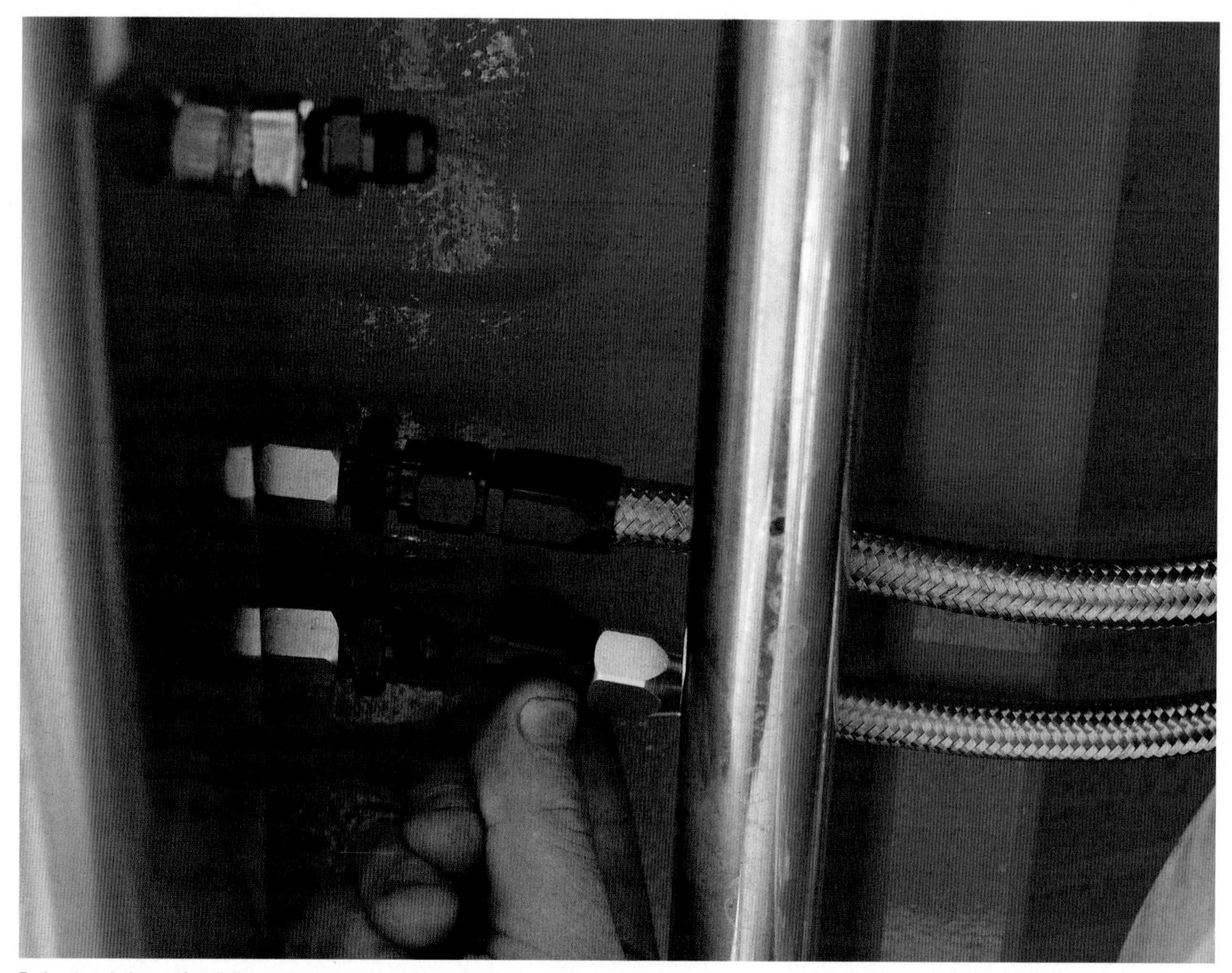

Fuel system design and installation needs to be well thought out. Don't cut any safety corners when building your fuel system, although there are correct ways to build fuel systems on a budget.

FUEL SYSTEM WARNING

Fuel vapors are highly explosive. Keep your work area well ventilated. Anytime you are working on a fuel system of any vehicle, proper safety procedures must be used. Your life depends on it. Avoid working near or around heaters, drop lights, butane torches, or conditions where there are static sparks. Use common sense: Do not smoke a cigarette while working on your car, do not cut or weld on a fuel tank or lines that have had fuel in them at any point, and keep a fire extinguisher at hand at all times throughout the project. As with any electrical work, keep the battery disconnected until the fuel system is completely sealed and any fuel vapors have had time to dissipate. It is also good practice to treat new fuel system parts as if they were flammable, even before a new fuel system is submerged in gasoline.

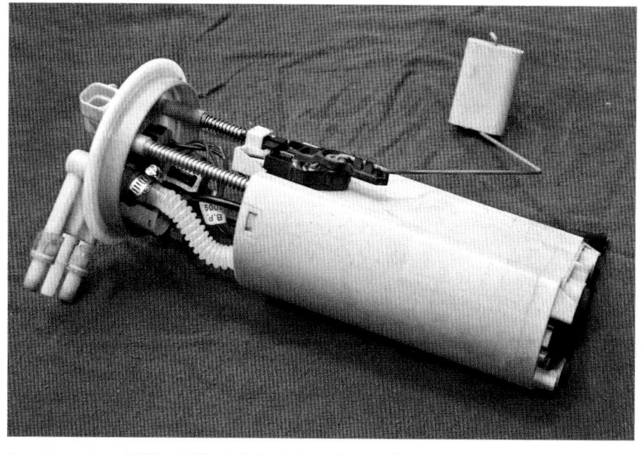

Here is a stock 1999–2002–style in-tank fuel pump/sender module. It is unlikely that you can retrofit this into a non-F-body fuel tank, although it gives you an idea of what the LS1s use. The fuel pressure regulator is built into the fuel pump module.

The Aeromotive billet drop-in fuel pump and hanger kit offers a convenient way to get an upgraded fuel system starting point, especially since many early EFI vehicles have a low-pressure fuel system. The Aeromotive 340LPH pump kit (P/N 18379) drops in the 1988–1997 GM truck stock fuel tank and can support 850 horsepower for naturally aspirated applications or 700 horsepower for forced induction. Fuel line fabrication for the -6AN fuel connections and an upgraded wiring harness are a must for these billet hanger kits.

Both carbureted engines and throttle body injection (TBI) applications use low fuel pressure. Carbureted engines often use mechanical fuel pumps to deliver the proper 6 to 7 psi of fuel pressure; whereas, TBI uses an in-tank low-pressure fuel pump to deliver about 12 psi. Neither of these systems will provide sufficient fuel to an LS-series engine with EFI. In either case, a high-pressure pump and lines must be added to keep up with the new fueling demands.

One other major change with the LS fuel system is that the fuel pressure regulator is not anywhere near the fuel injectors themselves, except for on 1997–1998 Corvettes, and 1999–2002 trucks and SUVs, with the truck version being the only vacuum-based regulator.

Standard EFI engines have a fuel-pressure regulator located directly on the fuel-injector rail, which varies fuel pressure based on engine vacuum, such as when the engine is at idle speed. The higher the vacuum, the lower the fuel pressure. Being that most of the LS fuel rails are equipped with a fuel-supply connector only, there will need to be some sort of fuel pressure regulator added into the fuel system at some point before the injector rail but after the fuel pump.

This type of fuel system is often called "returnless" since the fuel rail does not divert excess fuel back into the fuel tank by way of a return line. This is not the "returnless" designation that many newer vehicles use where the fuel pump is pulse-width modulated (PWM) to desired fuel pressure at lower engine speeds and then ramped up as engine speed is increased. The steady 58-psi setting is mechanically regulated by return spring pressure at the rear of the car near the fuel filter instead of at the fuel injector rail. This system is not affected by engine load or vacuum, unlike fuel rail–mounted pressure regulators, which lower pressure at idle and then boost fuel pressure with engine load. The 58-psi setting remains at all times the engine is running.

While known as a returnless fuel system, all of the non-PWM fuel system vehicles still require a fuel-pressure regulator; it is just in an unconventional location as compared to any other EFI engine. In 1998–2002 F-bodies, the regulator is built into the fuel pump module; the return line is connected to a T-fitting directly after the fuel filter, and this is routed back into the fuel tank where the fuel pressure regulator is located. The 1999–2003 Corvettes use a fuel filter with a regulator built in to regulate fuel pressure and divert excess back to the fuel tank. This dual-purpose external filter/regulator combination is popular in adapting retrofit LS swaps due to its convenient setup. All that is needed is the proper adapters and connectors to add it into any fuel system, but it does have its limitations.

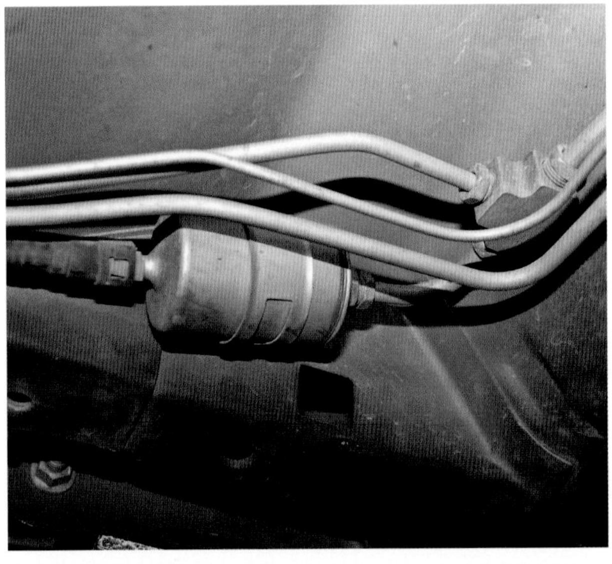

The inline F-body fuel filter does not have the fuel pressure regulator built-in like comparable year Corvettes. Rather, the T fitting after the filter provides the return route back to the in-tank fuel pressure regulator.

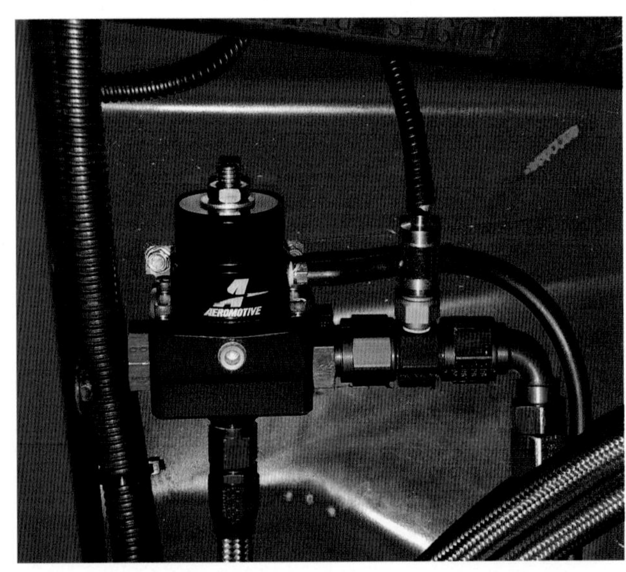

Using an external regulator is required for most retrofits. You can regulate fuel pressure to the required 58 psi by using an Aeromotive Fuel Pressure Regulator, or the Corvette C5 fuel filter/regulator.

LS engines are calibrated to operate at 58 psi, and fuel pressure regulators such as this -6AN Holley Billet FPR (P/N 12-880) come with both a low-pressure and a high-pressure spring. Remember to use the one for higher pressure and then set your base pressure to 58 psi once you've completed the fuel system.

Conveniently, Aeromotive has made two kits to help you finish your fuel system. Both include a matching fuel pressure regular, fuel filter, and -6AN fittings. One version is for OEM returnless-type stock fuel injector rails, and the other is for use with the Aeromotive billet fuel rails. Returnless is P/N 17351 and the return style is P/N 17350. The key difference between the two is the quick-connect fitting shown on the right to easily connect to the OEM LS returnless fuel rails.

FUEL TANKS AND EFI FUEL PUMPS

As we have discussed, the LS engine requires a high fuel pressure fuel system. Since most LS retrofits are intended for older non-EFI vehicles, this adds a sometimes pricey factor into the swap budget. As most 1985 and older models were carbureted, there is a huge pool of classic vehicles that fit the bill for covering the fuel system challenges when adding an EFI engine. Some vehicle choices will require complete fuel system tank-to-fuel injectors; whereas, others have some components that may be able to be adapted.

Since the main swap candidate will be the "carbureted to LSx" swap, that is a key focus here. There are many carbureted fuel systems, depending on the year of the vehicle and whether it had a six-cylinder or V-8 engine. Adding a 12-volt fuel pump can be a complicated endeavor.

Electric fuel pumps are sometimes picky; they do not often like to "pull" fuel to themselves. Rather, they are often more reliable when they are used to push fuel only. There are two ways to do this: by using a submerged fuel pump like the factory setup or by gravity. Even then, the suction line should always be generously sized so that fuel is waiting to be pressurized without the pump having to do much work to get the fuel source to itself. Sumped tanks work extremely well for this when using external fuel pumps, but not everyone desires to have a sumped tank or an external inline fuel pump.

There are several companies that manufacture fuel tanks for all of the popular swaps. Chances are if you can buy a new replacement stock tank for your application, you can likely purchase an EFI-compatible fuel tank as well. If you are not purchasing a fuel tank but instead reusing your existing tank, it will need to be adapted for EFI use. If a stock tank is the only choice, you would need a tank retrofit kit to add a drop-in fuel pump assembly. Tanks Inc. offers fabricated drop-in setups that with some legwork on the user's end can be installed into any existing fuel tank.

Aeromotive has taken fuel system modification guesswork off the table and built EFI fuel tank kits ready to drop into your vehicle using their Stealth fuel system. Starting with a new fuel tank, they modify the fuel tank to implement the Stealth internal fuel pump and Phantom in-tank baffling system, which keeps the fuel pump inlet submerged in fuel even in the most demanding conditions, such as spirited driving on a road course. This Stealth II tank (P/N 18457) is for the 1967–1968 Camaro; it requires -6AN fuel lines and upgraded wiring harness.

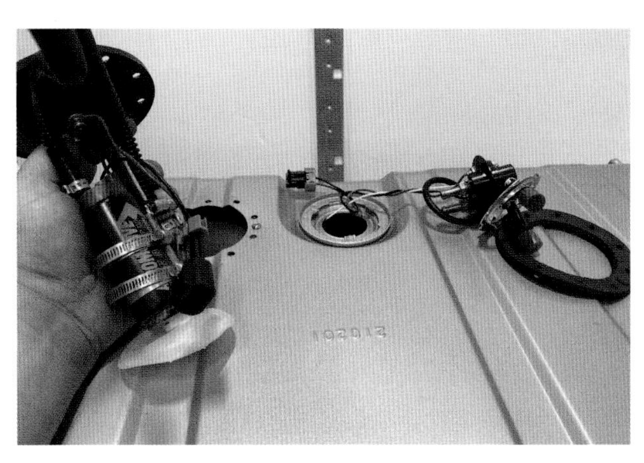

Aeromotive Stealth series tanks offer one example of an OEM tank with an in-tank fuel pump. These tanks are available for a slew of vehicle applications as a drop-in solution to your fuel system. They begin life as an OEM-style steel fuel tank and are then modified using an Aeromotive 340LPH submerged in-tank fuel pump kit. The new sending unit is relocated to the new fuel-pickup setup. There is a billet cap mounted in the OEM sending unit location, which has a -6AN outlet and a -6AN return fitting.

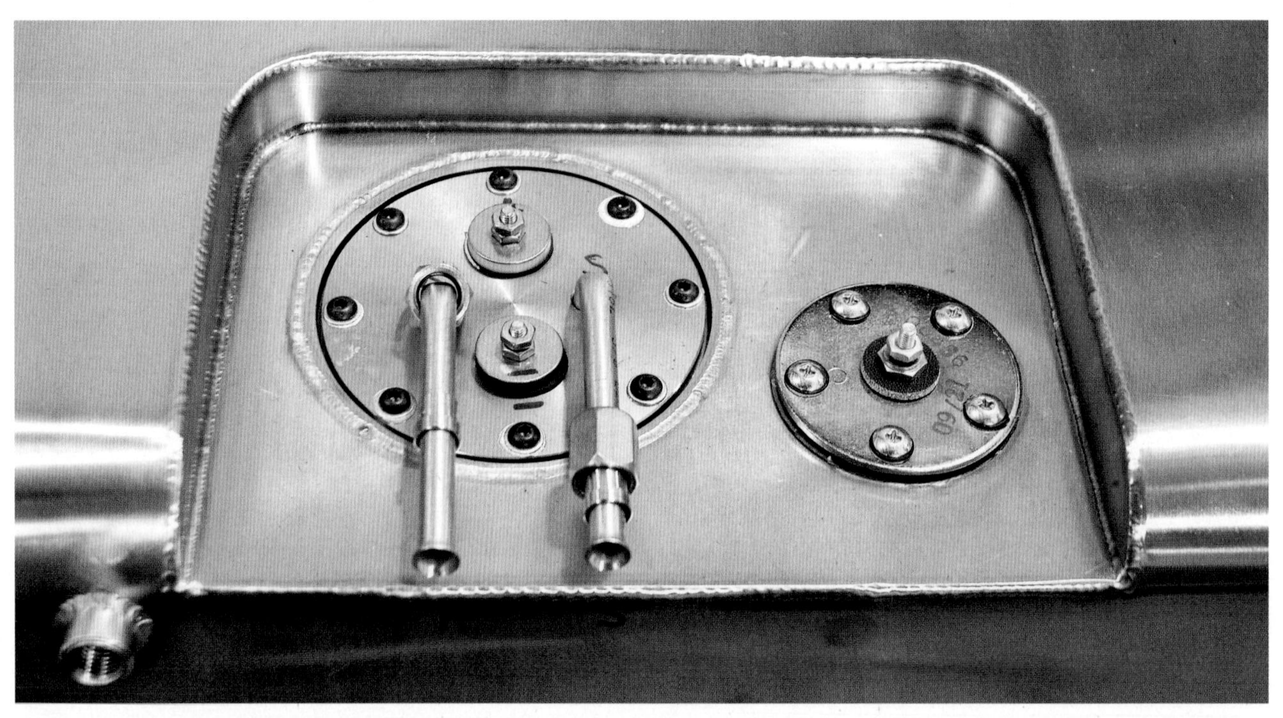

Even though the fuel pump is in-tank, much like the factory LS fuel pumps, there is usually no in-tank fuel pressure regulator, so an external regulator is needed. This is the reason for two fittings on the fuel pump module. One is the pressure side; the other is the return.

The Rick's Tank setups bolt in just like the stock fuel tank would, using matching stainless-steel fuel tank straps. These fuel tanks with in-tank fuel pumps have extra wiring required for the fuel pump power and grounding, compared to the single sending unit wire. There are now three wires: 12-volt power, ground, and the sending unit signal wire.

OEM TANKS

Certain GM chassis that were both carbureted and fuel injected can be easily retrofitted to EFI by using the later-model EFI fuel tank but with a replacement high-pressure in-tank pump. Such vehicles would include 1972–1987 GM C/K Trucks, 1982–1992 F-bodies (Camaro and Firebird), 1978–1988 G-bodies (Monte Carlo, Buick Regal, El Camino, etc.), and 1978–1990 B-bodies (Caprice, Impala, etc.). These later-model vehicle applications allow bolt-in tanks that offer EFI support. These often can be found inexpensively. Always factor in the cost of the higher pressure fuel pump, as the EFI version of many of these are the lower pressure throttle body injection.

Even some 1990s plastic tanks can be retrofitted into earlier chassis with some cutting and fabricating. These fuel tanks are found in mid-1990 Caprices and Impalas, which fit earlier muscle cars (such as Chevelles for instance) and retain the OEM-style license plate filler neck. If these donor tanks are fueling port fuel-injected engines, such as the LT1 or similar engines, they are ideal for LS engines and a good condition OEM high-pressure fuel pump can likely keep up with a stock LSx engine.

AFTERMARKET TANKS

The simplest fuel system to install, but the hardest on the wallet, is an aftermarket fuel tank that is already fitted for the EFI fuel pump. Manufacturers such as Rock Valley, Rick's Tanks, and Aeromotive offer bolt-on tanks modified for EFI usage. These can range from $450 on up to about $1,200, depending on the fuel pump used and tank material desired, as you can choose from a stock-style 12-volt pump that supports 600 horsepower to an in-tank pump that supports 1,500 horsepower or opt for a stock galvanized fuel tank on up to a polished stainless-steel tank.

Some of the nicer features of a tank such as these are their bolt-in design, not just in hardware but also in electronics. As far as the fuel pump controls, you need a 12-volt supply and a normal grounding wire for the fuel pump to operate. The fuel level sending unit will plug in and operate just like the stock tank. These tanks will have one pressure fuel feed line and one return line, which requires an external fuel pressure regulator to set the proper fuel pressure. Often an external tank vent fitting is supplied, which can be used for emissions purposes such as a charcoal canister or as EVAP fumes vented into the engine.

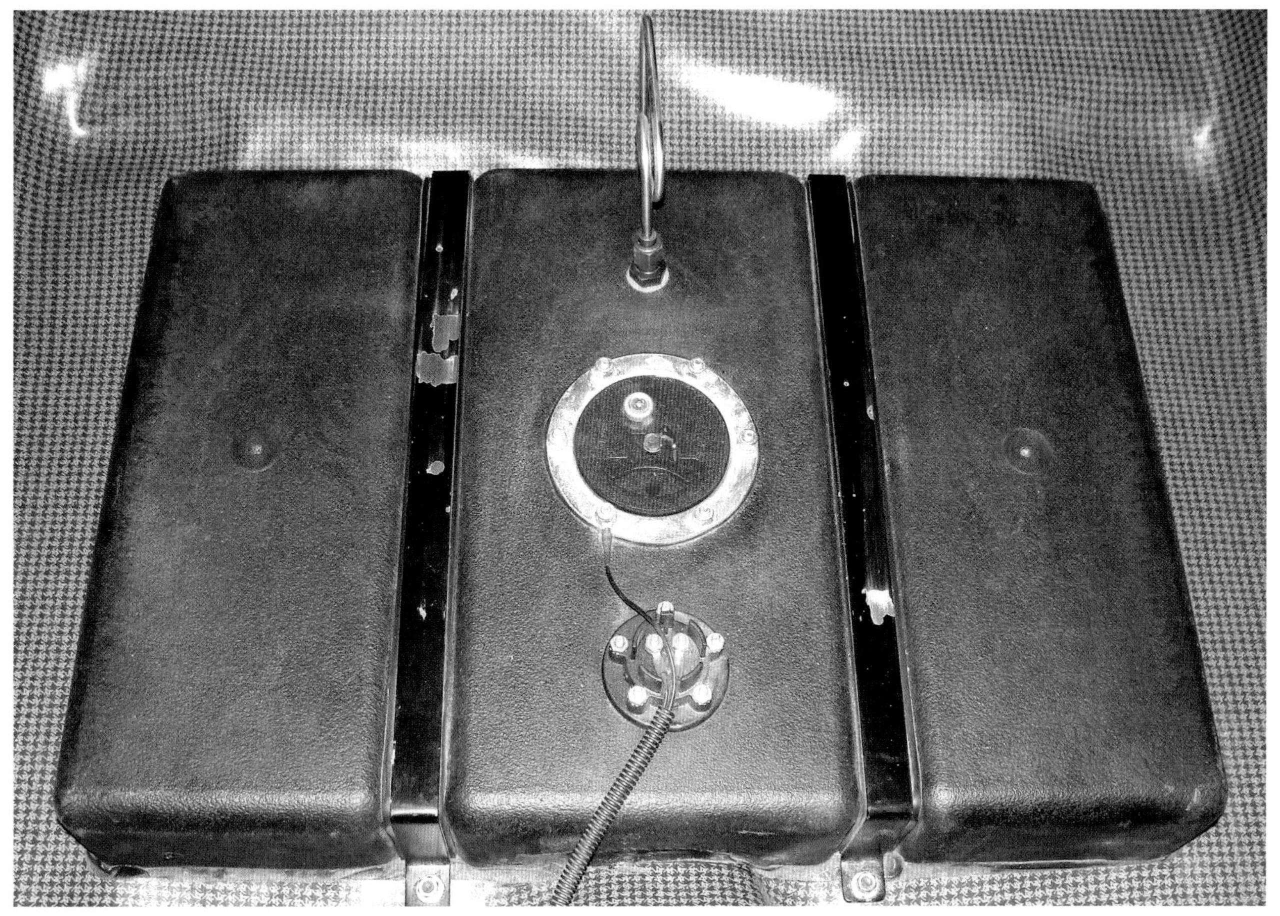

An alternative to using a sumped tank or buying an in-tank fuel tank kit is the use of a fuel cell. A fuel cell can be fabricated to fit most conventional applications and provides an easy fuel source to tie into. Fuel cells can be procured with stock fuel level sending units. About the only annoyance would be having to open your trunk to fill 'er up. Even then, there are fuel cell fill kits available that simplify that operation if you mount your cell in the stock location under the car.

I used the Aeromotive Stealth II tank in my 1st-Gen Camaro LS swap example. Obviously, this Camaro did not come with fuel injection back in the day, so my options were to use an inline fuel pump with the stock fuel tank (which I did temporarily on one LS setup) or a submerged option. I contacted Aeromotive for a solution, and the Stealth II tank assembly with the Aeromotive 340LPH fuel pump fit my requirements perfectly. My Camaro is meant to be a weekend cruiser and intended for "spirited" driving events. The Stealth II tank has features like a foam-and-rubber fuel-bucket style that keeps the fuel pump submerged while fuel is sloshing around.

EFI FUEL-PRESSURE REGULATOR

If left unchecked, an electronic fuel pump can produce pressures in excess of what the fuel injectors can handle and what the tuning calibration is set for. Since no two fuel pumps max out at the same exact pressures, some method of regulating the pressure to a constant 58 psi must be employed. The component required to handle this responsibility is the fuel pressure regulator (FPR).

Most 12-volt fuel pumps are essentially running at full throttle at all times. The FPR's job is to take the untamed mass of fuel coming in from the fuel pump and keep it at a steady calibrated pressure that feeds the engine, while diverting excess fuel flow back to the fuel tank. The FPR is bypassing and returning the most fuel at low engine speeds, while under more demanding conditions it allows additional fuel to move toward the engine as more fuel is required.

Many existing LSx swaps use the 1999–2003 Corvette-style fuel filter/regulator, which keeps fuel pressure at the constant 58 psi setting in a convenient fuel filter package. The filter/regulator assembly can be had for about $50 and works really well when matched up to OEM types of fuel pumps. Fittings from sources such as ICT Billet or Russell can adapt the C5 filter setup to -6AN lines, offering simple installation. Since the return line is built into the filter, it is often desirable to mount the fuel filter in the rear of the car near the fuel tank so that fuel return is simplified and less return line is required.

The 1999–2003 filter/regulator is GM #19239926, AC Delco #GF822, or WIX #33737 and can be cross-referenced to just about any fuel filter manufacturer. The quick-connect fuel fittings required to adapt the filter to -6AN lines are Russell part numbers:

- #640850 -6AN fuel inlet to 3/8-inch female quick connect (also used for fuel injector rail connection)
- #640860 -6AN fuel return to 5/16-inch female quick connect
- #640940 -6AN fuel outlet to 3/8-inch male quick connect

The other nice thing about the C5 fuel filter is that the replacement filters come with their own filter-mounting bracket, perfect for the LSx swapper as that is one less item required to be fabricated. Once a suitable location away from hot exhaust and moving parts is located,

A sumped fuel cell offers a good location to tap into for a fuel source when using an inline fuel pump. Gravity force-feeds the fuel pump so that the fuel pump can sufficiently perform its fuel pressurization and supply functions without working overtime to "pull" the fuel to itself.

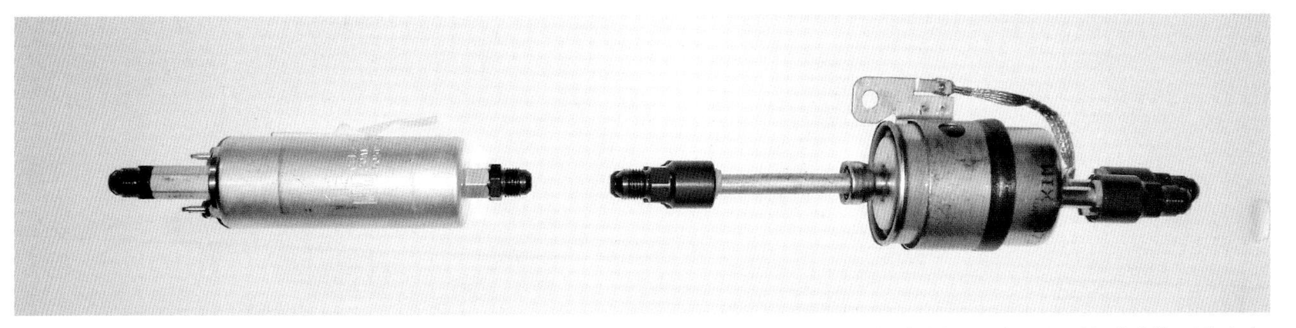

When using a sumped fuel tank or fuel cell, you can reliably use an inline fuel pump kit. Inline fuel pumps do not like pulling fuel; they are best at pushing fuel. Mount the fuel pump near the fuel tank, and ideally it is best to use at least one size larger fuel line on the suction side of the pump than the pressure side. You want gravity and fuel weight to supply the pump with fuel. *Mast Motorsports*

When using OEM fuel injector rails, you're going to need quick-connect adapters to connect your fuel feed line to the fuel rail. Many companies make a solution for this, and ICT Billet makes one of the better designed ones: they have a threaded fitting that adapts the 3/8-inch fuel rail quick connect to -6AN (P/N AN809-02B), which offers a secure, durable method for attaching your fuel system. *ICT Billet*

Not everyone needs an 800+ horsepower fuel system. If you have a stock or mildly modified engine build, you can use the C5 Corvette fuel filter/regulator to build your fuel system economically, but first you'll need to adapt the quick connects of the filter to the -6AN fittings. This ICT Billet 3/8-inch to -6AN male fitting (P/N AN817-02A) conveniently adapts the outlet of the filter to -6AN, which is the filtered fuel supply side for your fuel rails. *ICT Billet*

simply bolt the fuel filter assembly to the frame rail using the supplied bracket. Often it can be mounted in a pre-existing mounting hole left over from the old fuel-line brackets. No fuel pressure adjustments are needed; just hook it up and go as it is pre-calibrated to proper LS fuel pressure specifications.

The C5 filter does have its limitations. Because it is so easy to build power in the LS platform, this filter and regulator setup can get in over its head quickly. If your engine is stock, you can count on this setup, but if your build requires a lot more fuel, you will likely need an aftermarket return-style fuel system—not necessarily larger lines, just a better FPR that matches your fuel pump flow.

There are several adjustable aftermarket FPRs available from fuel specialists such as Aeromotive that will regulate fuel pressure as required. These are best left for larger engines, or applications such as the the power-adder crowd or big external fuel system builders who use the same brand-name fuel pumps and fuel systems. Those types of fuel systems use larger -8AN and -10AN fuel lines. A beefy FPR is needed at this level, as they are available with larger fuel-fitting diameters to match the rest of the fuel system. Often, big power cars will be set up to have a return fuel system, where the full fuel flow goes through the fuel rails and then the FPR where the extra fuel is sent back to the tank. These types of fuel systems are needed with just about anything over 600 horsepower, which is where the 3/8-inch fuel supply line and single pump simply cannot keep up.

FUEL FILTER

Protecting your new external fuel system should be high on your priority list. The use of a fuel filter on both ends of your fuel pump is highly recommended, on both the inlet and outlet. The typical pre-filter that aftermarket fuel pump manufacturers would recommend filters debris larger than 100 microns, which is a smidgen smaller than .004 of an inch, and close to the thickness of the average human hair. The outlet filter is usually a much finer 10-micron rated, which filters out debris down to about .0004 of an inch. The typical stock in-tank filter sock is 70 microns. Usually it is good practice to check or replace the fuel filter at least once a year.

If you are using one of the Aeromotive or Magnafuel fuel setups, you need to have a pre-filter and post-filter installed before and after the fuel pump. The first fuel filter separates out large debris, while the secondary filter catches anything small and any wear particles from the fuel pump itself. Keep in mind that any fuel filter used needs to adhere to EFI fuel pressure standards; you cannot use a filter intended for carbureted applications, for instance.

EFI FUEL LINES

When adding EFI to a non-EFI car, there are some special considerations. First and foremost, get rid of the OEM low-pressure line unless you intend to salvage it as a return line. The rubber connections and old steel line itself are not made to take the high pressures created by the EFI fuel pump.

In a carbureted setup with a mechanical fuel pump, these lines never see pressure at all, they are under suction instead. Many times these lines are not in suitable condition to even use with a mechanical fuel pump or sometimes cannot flow enough fuel. Take for example the 1967–1969 Camaro, the I6 and nonperformance V-8s used 5/16-inch fuel supply line; whereas, the performance SBC and BBC engines used the larger 3/8-inch lines. Stock LS fuel systems use 3/8-inch

If you use the stock EFI fuel supply lines in GM TBI, TPI, and Vortec engines from the late 1980s and the 1990s, you can also adapt these M14 and M16 O-ring fittings to -6AN easily. ICT Billet makes these fittings M14-1.5 (P/N F06ANFM1415) and M16-1.5 (P/N F06ANFM1615), which come with new fuel O-rings as well.

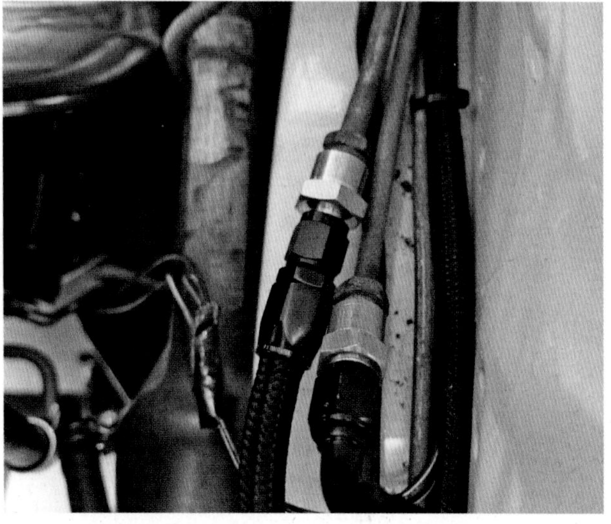

Here are the ICT Billet M14 and M16 to -6AN fuel adapter fittings in use. The OEM fuel lines in this 1991 Camaro were flushed and reused, then adapted to -6AN to connect to the Holley fuel pressure regulator. This method required only about 4 feet of flexible -6AN hose and six various -6AN hose ends to finish the fuel lines, providing a safe and leak-free fuel system.

The quick-connect fittings are convenient, but if using the Dorman fuel "repair" line, make sure to leave enough tubing so that you can replace the filter for maintenance purposes.

These are the fuel fittings needed to attach -6AN lines to the feed and return side of the C5 Corvette fuel filter. You need one 5/16-inch -6AN and one 3/8-inch -6AN adapter. I prefer the threaded fittings, but the plastic quick connect–style ones can work as well on the fuel filter.

Another option for inline fueling is the Aeromotive A1000 or Eliminator pumps. These support higher power levels than the smaller inline or in-tank fuel pumps and also would require a minimum of -8AN fuel lines. These are not recommended for the C5 fuel filter/regulator. Use a matching Aeromotive fuel-pressure regulator and fuel-injector rails. Location doesn't matter as much for ordinary engines, but it is ideal to return fuel from after the fuel injector rails when building high-horsepower fuel systems.

diameter lines and support about 600 rear-wheel horsepower maximum with a good fuel pump.

The proper installation calls for full replacement using fuel-injection–compatible tubing and hoses. Do not ever use clamp-on rubber hose without the properly rated hose, EFI-specific hose clamps, and only when connected to flared or locking barbed fittings. In fact, I discourage its use altogether, but it does have its place at times. The NHRA disallows rubber line use of lengths longer than 12 inches for racing purposes. This is an important consideration if you intend to take your newly converted car to the drag strip, but it is also a good guideline to follow for any buildup.

Suitably designed fuel systems use hard line and mounting clamps to affix to the inside of the frame rails and never near exhaust system components. Flex connections can be EFI-specific braided or push-lock–type hose. There are two major flex points to consider, from the frame rail and body to the fuel injector rail and the fuel tank to the frame rail. The entire length of the line can be rigidly mounted to the frame rail.

If full braided stainless-steel flexible hoses are used, these also need to be routed in such a way that they do not abrade electrical wiring while leaving enough flex room for moving parts. Use of rubber-insulated ADEL clamps or billet hold-downs will keep hose vibration issues on the straight lengths at bay.

There are a few different types of AN hoses such as ProFlex braided stainless with rubber liner and PowerFlex that is Teflon lined. The ProFlex hose is rated for about 500 psi, while the PowerFlex is made for nitrous, brake lines, and power steering usage and is rated for 2,500 psi. There is also Twist-Lock hose, which uses no compression fittings at all, just the matching machined barbs that lock the line into place after a push and a twist. Twist-Lock is rated for 250 psi. More recently there is the lightweight ProClassic type hose with a braided Nylon reinforcement and rated for 350 psi. The only drawback to ProClassic is that the rubber-lined hoses tend to expel minute amounts of fuel scent whereas PTFE hoses do not have this issue.

The Twist-Lock and ProClassic lines can be cut to length with a sharp razor blade or hose cutter, making installation extremely convenient. ProFlex and PowerFlex lines require a cut-off wheel, a hacksaw or chop saw to cut to length, and the fringes on the stainless braided lines tend to impale any body parts that get in the way, so use care when installing hose ends to these types of hoses.

The AN-type hoses such as these use their own specific 37-degree tapered and threaded couplers, similar to what many aircraft would use. There are no hose clamps in these systems, and the burst or leak rating is tremendous as compared to available alternatives. The AN hoses exceed the EFI fuel pressure capacities by several hundred PSI, leaving quite a safety cushion. On the other hand, rubber EFI line barely covers the pressure requirements, so why risk using it?

The minimum fuel feed line should be -6AN for a stock engine, which is comparable to a ⅜-inch fuel line and the same dimensions as many stock LS vehicles. If you are building a fuel system and using the OEM-style in-tank pump or a small inline pump, -6AN should be the minimum, but -8AN is preferred. The -8AN works well for upgraded fuel systems, and the cost difference is minimal. The -8AN also gives some room to grow in the future; whereas, -6AN is limited. If you do use an -8AN fuel feed, you will need a -6AN fuel return in keeping within the rules of EFI fuel system building.

Typical N/A EFI fuel line estimation sizing (use next size up for power-adder setups):

- Up to 600 horsepower, -6AN
- 600 to 900 horsepower, -8AN
- 900 to 1,200 horsepower, -10AN

Typically, you want a return line one size smaller than the feed for EFI systems, but in the -6AN configuration, the same -6AN return should be required. The next size down would be -4AN. Same-size feed and return lines are sometimes recommended for race-fuel systems because at lower speeds less fuel is needed at the engine, so the bulk of the unused fuel is returned to the tank.

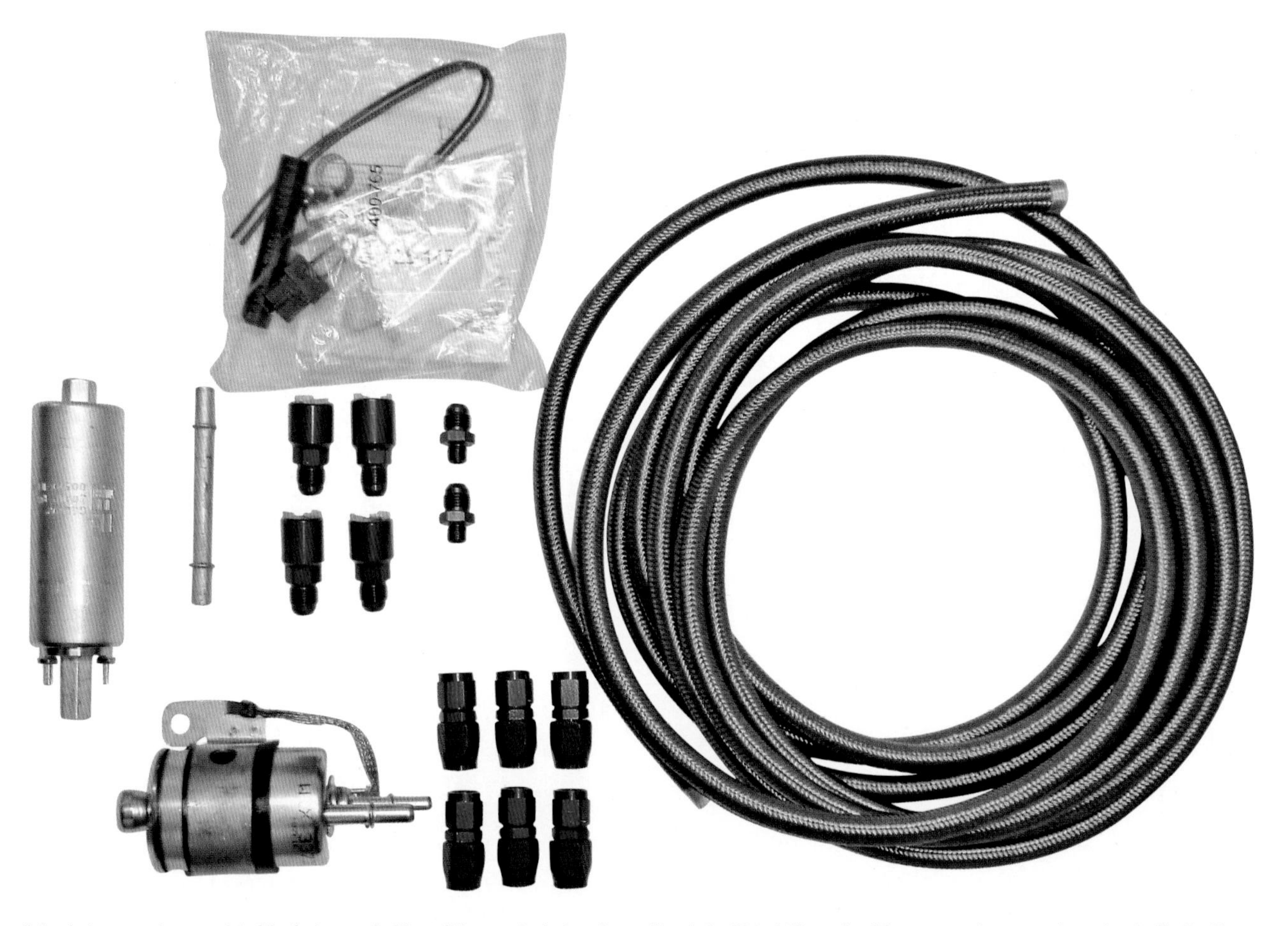

Inline fuel pump setups consist of the fuel pump, fuel lines, fittings, and adapters. As mentioned, the C5 fuel filter makes LS swaps much more easier and cost-effective than alternatives. Most kits use -6AN lines, in either braided steel or the lightweight braided black nylon. *Mast Motorsports*

EFI-to-EFI Swaps

In vehicles already equipped with EFI fuel systems, the factory fuel lines are usually sufficient unless major power upgrades are planned. Still, these fuel systems will require adapting to the LS fuel rails or need the addition of a separate fuel pressure regulator. Aeromotive makes an easy kit to adapt an external fuel pressure regulator and quick-connect AN adapter for use with the OEM fuel-pressure rail. These fittings are also available separately.

The 1991 Camaro 1LE swap I am currently building—5.7L LS6/Borg Warner T56—is one example of adapting EFI-to-EFI fuel system. This Camaro came with a 305cid tuned-port engine that had an in-tank fuel pump and a fuel-rail mounted FPR. The stock fuel pump was thirty years old and likely did not have enough capacity for the new powerplant.

If I'd done the research, I would have bought a new fuel tank and installed a new fuel-pump hanger into the tank. Instead, I decided to flush and reuse the factory Z28 fuel tank (the 1LE and 1990–1992 Z/28 had a baffled tank). Some builders use a 4th-Gen F-body fuel tank in their 3rd-Gen F-body swaps, which is a great upgrade if you have one.

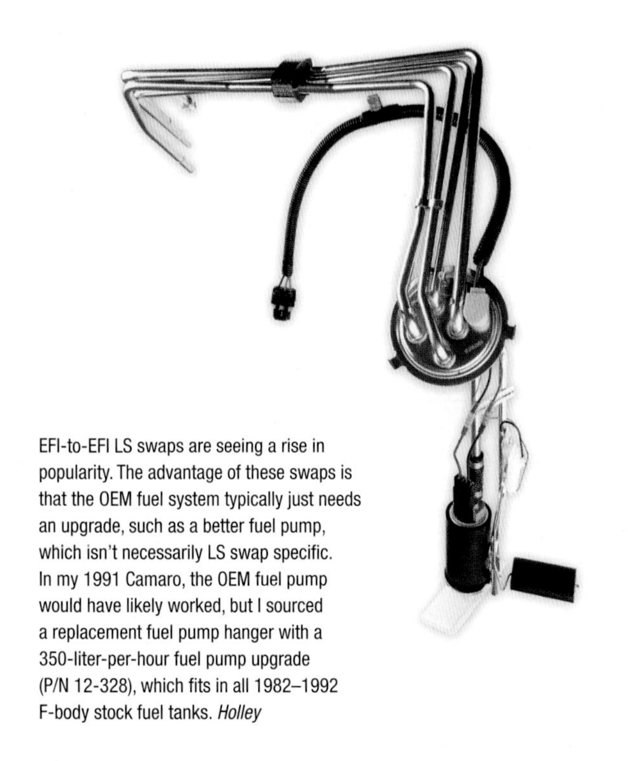

EFI-to-EFI LS swaps are seeing a rise in popularity. The advantage of these swaps is that the OEM fuel system typically just needs an upgrade, such as a better fuel pump, which isn't necessarily LS swap specific. In my 1991 Camaro, the OEM fuel pump would have likely worked, but I sourced a replacement fuel pump hanger with a 350-liter-per-hour fuel pump upgrade (P/N 12-328), which fits in all 1982–1992 F-body stock fuel tanks. *Holley*

AN LINE FABRICATION

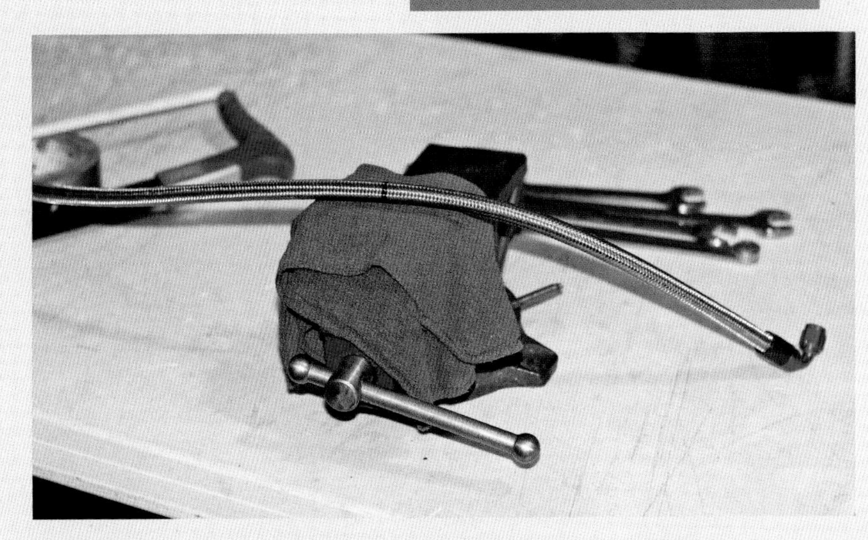

At some point during your retrofit project you may need to build a fuel, transmission cooler, or other AN line. While the instructions are not the same for high-pressure lines such as power steering or brake lines, you can get the gist of the procedures in this section. The difference is the line type and fitting design.

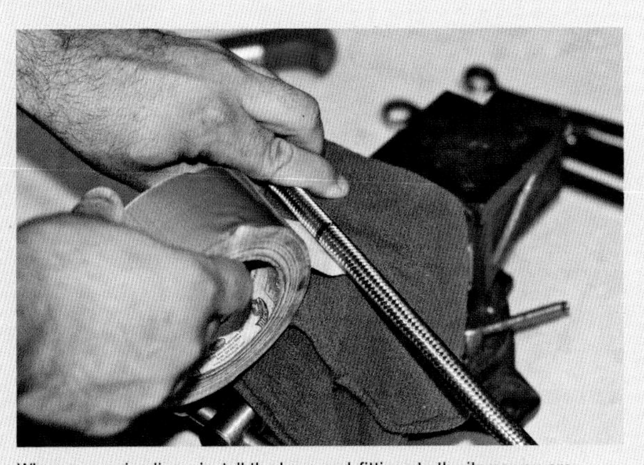

When measuring lines, install the hose end-fittings to the items you are connecting and line up the hose end to one side. Then line up the side needing to be cut to the proper location and mark the hose. The hose needs some slack; it should not be so tight that it is under strain when installed. Often an excess of ½ inch helps. Use duct tape or reinforced clear tape to keep the ends of the hose from fraying while being cut.

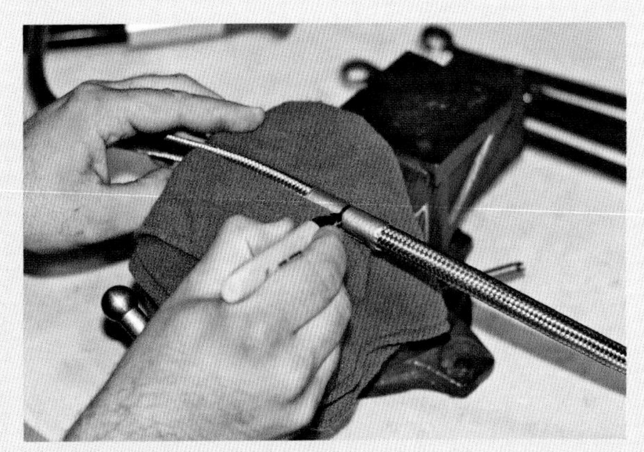

Mark the cut position on the previously applied tape section with a marker and secure the hose in a vise for cutting.

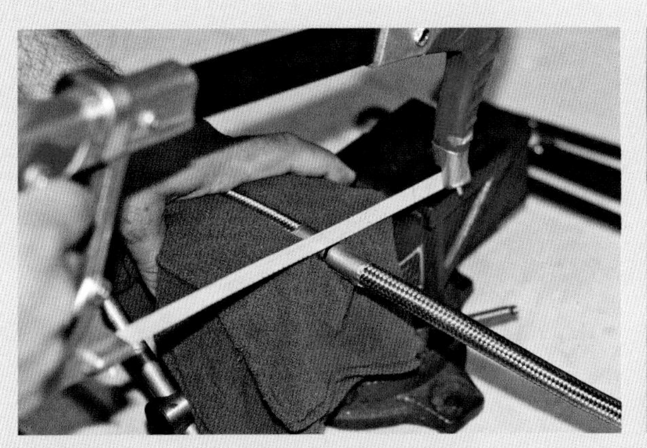

There are a few ways to cut the hose: You can use a hacksaw, a cutoff wheel, or a fine-tooth reciprocating saw. Cut the hose so that the cut is as close to 90 degrees from the hose as possible. You do not want a tapered hose-end connection. After cutting the hose end, use compressed air to flush the line of debris.

(continued on page 100)

AN LINE FABRICATION *(continued from page 99)*

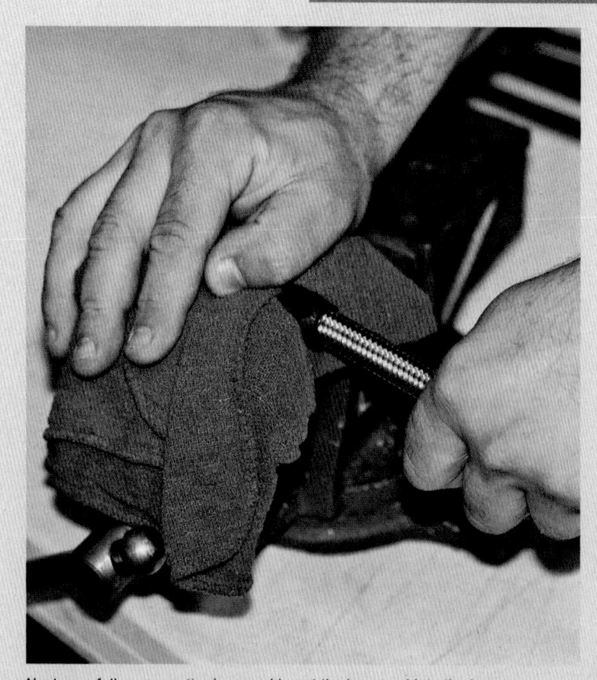

Next, carefully remove the tape and insert the hose end into the hose cap. Be careful as the frayed end of the steel braiding is sharp and will prick you if given the chance. You can twist the hose counterclockwise to help facilitate installation.

Now for the hose-end insert: This part clamps the hose itself in between the insert and the previously installed hose end cap. Using some hose lubricant on the tapered portion of the fitting that sits in the hose (use oil or assembly grease), start threading the fitting into place turning clockwise while holding the hose and hose end securely.

Sometimes the hose will push out of place. You need to make sure it doesn't push out of the fitting as you are tightening. Using AN wrenches will prevent scratches on the fitting, but steel wrenches work if you are careful. Tighten the fitting until it is close to bottoming out. You don't have to bottom it out, but it needs to be close.

The completed hose end fitting is ready for action after using compressed air one last time to make sure the lines are clear of obstructions or hose debris.

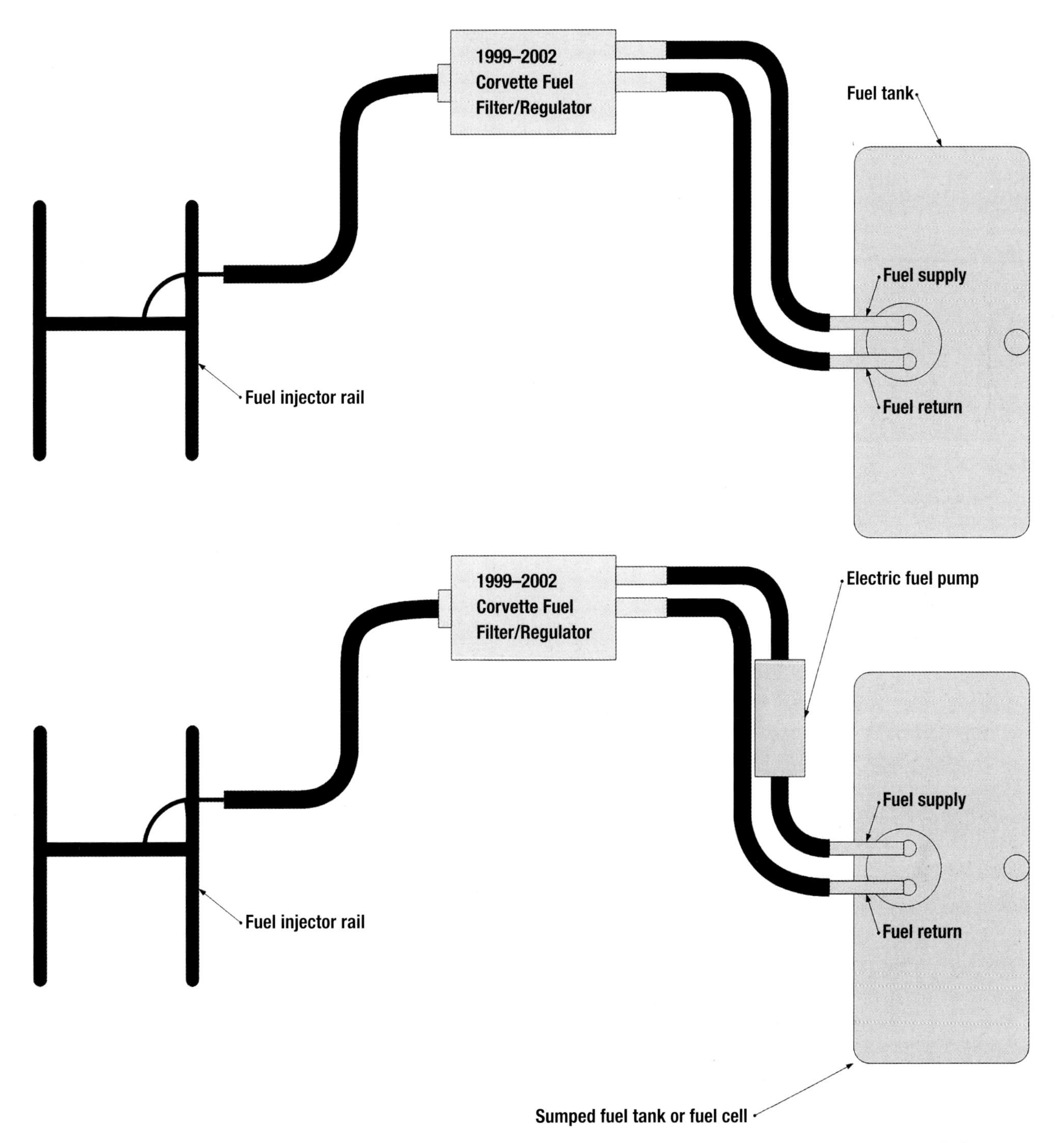

Here are some sample fuel system layouts. These are not to scale, but show the required fuel system layout necessary for functionality when using the popular C5 Corvette fuel filter/regulator.

For my 1991 Camaro, I opted to order the Holley fuel-pump hanger kit. This is a drop-in kit that comes with a new hanger, sending unit, and 350-liter-per-hour fuel pump (P/N 12-328). While you can simply plug this into the factory 3rd-Gen wiring harness, it is highly recommended to add a fuel-pump hot-wire kit with relay to avoid overloading the factory wiring. The original gray factory fuel-pump wire now becomes the trigger wire to activate the upgraded wiring harness.

While I could have added the Corvette C5 fuel filter mentioned earlier in this chapter, I decided to reuse most of the OEM fuel lines and stock filter up to the engine bay. Basically everything behind the engine is stock replacement components—new TPI fuel pump, new TPI fuel filter, OEM TPI lines, etc. The OEM steel fuel lines end at the frame rail near the brake booster and it has two metric fuel fittings: an M16 1.5mm is the feed line, and an M14 1.5mm is the return

Injector choice for larger engines is based on the horsepower level you plan to build. Stock injectors work great for stock engine power and up to about 25 percent more power than stock. You can monitor the duty cycle of the injectors with a scanner; if higher than 90 percent, you need a larger injector. There are online calculators for fuel injector calculations. Remember that most aftermarket injectors are rated for 44 psi. The LS fuel pressure of 58 psi will make the injector act larger.

In some cases, your donor engine may have contaminated or rusted fuel rails, especially on the original LS1/LS6 steel injector rails. The best thing in this case is to replace them with aftermarket billet fuel rails along with new or cleaned fuel injectors. In my LS6 5.7-liter build, my fuel rails and injectors were bad, so I replaced them with these Holley billet fuel rails (P/N 534-209) and some new 36-pound EV1 fuel injectors from FAST (P/N 303608).

Billet fuel rails can be used both for functional purpose and an improved aesthetic. When using aftermarket fuel rails, you will need to adapt the fuel supply or return to the rails differently than when using the stock fuel rails, which use the quick-connect setup.

line. ICT Billet makes M16 to -6AN (P/N F06ANFM1615) and M14 to -6AN (P/N F06ANFM1415) fuel adapters that work for 1980s and 1990s GM EFI TBI/TPI/Vortec V-6 and V-8 application LS swaps. These make a fuel system upgrade like this easy to build. Once you adapt to -6AN, you can easily plumb in an aftermarket fuel pressure regulator (FPR).

For regulating fuel pressure in this car, I used a Holley -6AN FPR (P/N 12-880), which is also a -6AN FPR. It has two -6AN inlet ports, so you can install it before or after the fuel rails. I chose to install it before the fuel rails because the Camaro will be a mild build and will not require "all the fuel." In this configuration, you connect the M16/-6AN feed to one side of the inlet FPR, and the opposite port connects to the fuel rail with a single -6AN line. The bottom port on the FPR is simply the fuel return, so it gets connected to the adapted OEM M14/-6AN return line to which we added the ICT Billet fitting before.

Because this is an adjustable FPR, once everything is connected you just need to check for any leaks, then set the fuel pressure to 58 psi (4 bar). If the regulator you are using has a low-pressure bypass spring in it, you may have to swap it to the higher spring-pressure version to get the fuel pressure range needed—some FPRs come with both springs necessary for either configuration.

FUEL INJECTORS

Many LS-series engine swaps will typically use the injectors that come with the engine, and when the engine is kept stock, these are usually sufficient. The only additional factor that requires consideration is the type of electrical plug the fuel injectors use. You don't want mismatched injectors and harness. Gen III passenger cars use the EV-1 (a.k.a., Bosch-style) connection, while Gen III trucks use the small injector connection, known as the Delphi Multec 2 marine version. Gen IV cars and trucks use LS2/LS7-style connectors, which are the EV-6 connector design. Wiring adapters and injector rail spacers are available to change between nearly all versions.

Fuel injectors can be likened to the jets in a carburetor; the more fuel needed, the larger the injectors (or jets). For stock and mildly modified engines, the stock injectors are usually sufficient. When some more involved engine modifications are performed and the stock injector is deemed inadequate, then an injector upgrade is necessary. Injector swaps can be required by a healthy heads and cam setup, or any time a boosted power-adder is bolted onto your car. Injector size is rated in pounds per hour, and with a few calculations knowing your maximum horsepower capacity, you can figure out what size injectors are needed. You want an injector that has about 20 percent more capacity than your engine actually requires. Do not just buy the largest injector possible; you want one that matches your engines needs, no less and not significantly more.

UPGRADING INJECTORS

You really only need to know three things to determine which injectors you need: approximate engine (flywheel) horsepower, fuel pressure, and estimated brake-specific fuel consumption (BSFC).

BSFC is how much fuel is burned per horsepower per hour. For BSFC we can estimate

.45-.50 for naturally aspirated engines

.55-.60 for supercharged engines

.65-.70 for turbocharged engines

From these numbers you can determine how much fuel is required. The calculation looks like this:

$$\frac{\text{(Flywheel horsepower} \times \text{BSFC)}}{\text{(\# cylinders} \times 80\% \text{ max duty cycle)}} = \text{Calculated injector flow rate}$$

This formula is based on the standard 43.5-psi (3-bar) pressure. To convert this to stock LS1 (4-bar) fuel pressure specifications the formula looks like:

Calculated injector size × ($\sqrt{}$(new FP / old FP)) = Calculated injector flow rate at new fuel pressure

So assuming a 400-flywheel-horsepower N/A engine, the example would look like this:

$$\frac{\text{(400 flywheel HP} \times .50 \text{ BSFC)}}{\text{(8 cylinders} \times 80\% \text{ max duty cycle)}} = 31.25 \text{ lb/hr @ } 43.5 \text{ psi}$$

(continued on page 106)

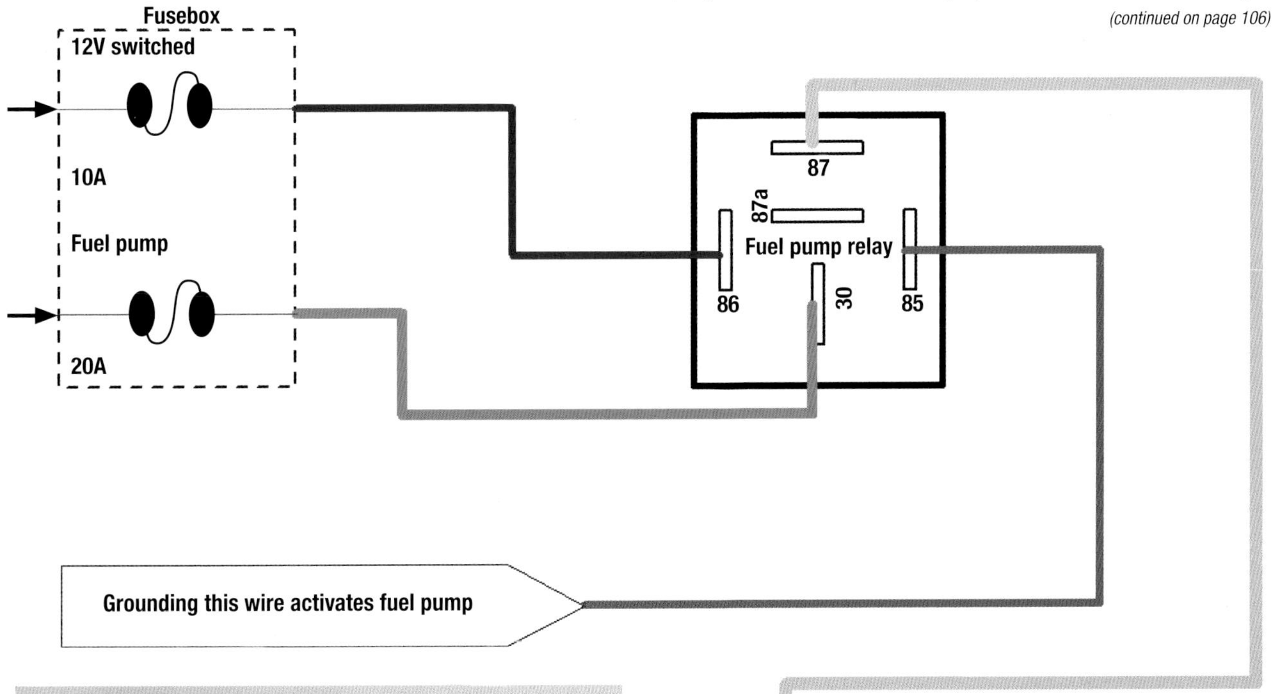

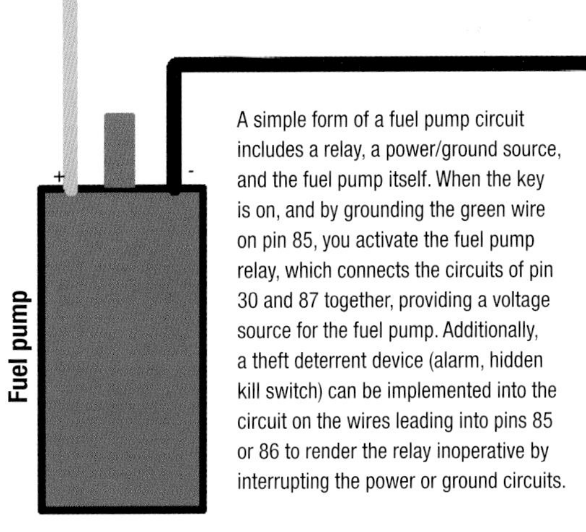

A simple form of a fuel pump circuit includes a relay, a power/ground source, and the fuel pump itself. When the key is on, and by grounding the green wire on pin 85, you activate the fuel pump relay, which connects the circuits of pin 30 and 87 together, providing a voltage source for the fuel pump. Additionally, a theft deterrent device (alarm, hidden kill switch) can be implemented into the circuit on the wires leading into pins 85 or 86 to render the relay inoperative by interrupting the power or ground circuits.

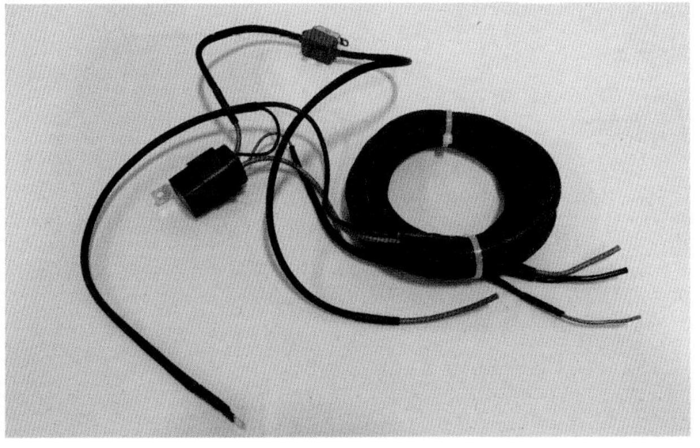

Whenever you upgrade your fuel system with a much better fuel pump, the wiring must be addressed. If you use OEM wiring and a wiring gauge that's too small you'll experience a voltage drop. The Aeromotive Stealth pumps require sufficient voltage so that the pump operates at the correct speed without straining and overheating. All aftermarket fuel pumps would benefit from a heavy-duty harness such as this one from Aeromotive (P/N 16307). It comes with 20 feet of beefy 10-gauge power and ground wires, prewired with a 30-amp circuit breaker and fuel pump relay that can easily be triggered by your old 12-volt OEM fuel pump wire.

INSTALLATION OF AN FUEL LINES

Depending on if you are using an external pump or in-tank pump determines what you need to mount on your frame rail or other support structure. In this case, we have an in-tank fuel pump with the Rick's Tanks setup. Before running any lines, you need to find a good location to mount the fuel filter and regulator that has no chance of being impacted from moving parts and is far from a heat source.

To mount the fuel filter and regulator, first mark the location for drilling, then use a properly sized drill (depends on bolt size and pitch used) to prepare the hole for threads. I used a 5/16–inch Allen bolt, so I used a 17/64-inch drill bit.

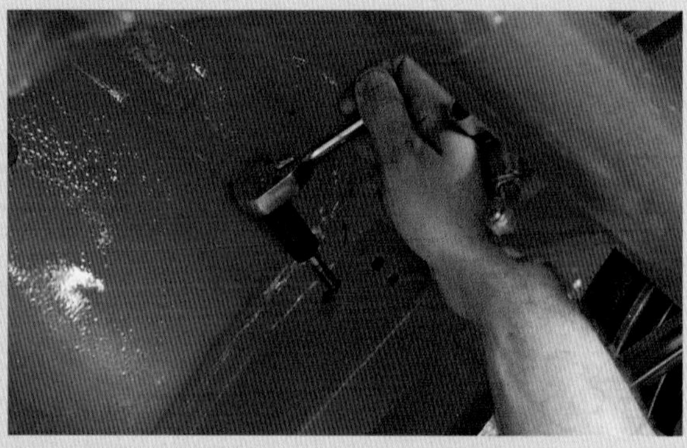

After drilling the new bolt hole location, you can start cutting threads in the frame rail using a matching tap. Make sure to keep the tap straight so that the bolt tightens down evenly. A good rule of thumb for cutting threads is that when the tap meets resistance, don't force it. Back it clean off the debris and go at it again. It is a two-step forward, one-step backward process until the length of the threads are fully cut. You can also use a Riv-Nut to add a threaded insert without tapping threads.

When the new threads are cut, place the fuel filter and regulator in place and tighten down. Don't go crazy with the torque. Something like 15 to 16 foot-pounds is adequate to hold the filter in place.

Next start routing the fuel supply and return lines to the fuel filter. The return is the smaller 5/16–inch line out the rear of the fuel filter and toward the center. Once lines are supplied and return lines are routed, you can use rubber-insulated clamps to secure them.

On the outlet side of the fuel filter and regulator, start routing the fuel line. Keep it away from moving parts and heat sources (exhaust and headers). Start at the filter, and if possible follow the OEM routing as best as possible.

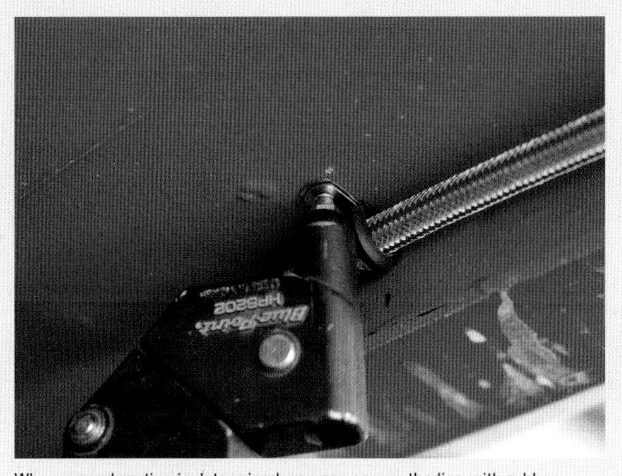

When a good routing is determined, you can secure the line with rubber-insulated clamps. I used rivets for a cleaner installation, but self-tapping bolts also work. Riv-Nuts and stainless button-head Allen bolts offer the cleanest installation.

The finished fuel line routing under the car is much like the OEM setup and is as far inboard as possible without looking tacky.

At the engine bay side of the fuel system, there are a few variations of fuel supply line connection locations and types. Depending on which fuel injector rail you are using, you will need quick connect-to-AN fitting adapter for the stock rail, or you can connect directly if using a billet fuel rail. Be sure to have ample fuel line length in the engine bay so that it can be routed and secured away from exhaust and wiring.

When using the stock injector rail and mismatched coil packs (in this instance LS1 fuel rail with truck coil packs) due to the higher mounting location of the truck and LS2/LS3/LS7 coil pack assemblies, you will be required to tweak the fuel inlet or relocate the cylinder 5 ignition coil for clearance.

(continued on page 106)

INSTALLATION OF AN FUEL LINES *(continued from page 105)*

As mentioned, you will need a quick-connect AN adapter to adapt from AN lines to the stock fuel-injector rail. In this instance, since we used a -6AN fuel line, the Russell 3/8-inch fuel line adapter. Aeromotive offers an adapter that can be used with -8AN fuel supply lines. Simply slide the fitting into place, and it will clip onto the fuel rail.

Once the adapter is in place, you can now install the matching -6AN fuel line. Notice that the fuel rail is on the driver's side of the engine. If you need to reverse it, the fuel rail can be installed "backward" 180 degrees different than stock LS applications for ease of fuel line access. In this case, you can also possibly use shorter fuel lines. Tighten the fuel fittings securely using a 5/8- and 11/16-inch AN wrench.

(continued from page 100)

If this were an LS1 fuel system, we would need to convert this to a 58-psi setting, as more fuel pressure will increase the amount of fuel the injector flows. Since many injectors are rated at the more popular 43.5 psi, another calculation must be performed to find out what they actually flow when subjected to LS1-level fuel pressures. More fuel pressure will make the actual fuel quantity increase as if it were a larger injector with less pressure.

$31.25 \times (\sqrt{(58/43.5)}) = 36.08$

or

$31.25 \times 1.1547 = 36.08$ lb/hr injector flow rate @ 58 psi

This shows that a 31.25-lb/hr rated injector will actually be equivalent to a 36.08 injector when subjected to the higher fuel pressure rating of 58 psi. From these calculations we can see that two different fuel pressures yield different fuel injectors at the same engine horsepower rating.

To determine the injector requirement for an LS1 58-psi rating, we can simply divide the 31.25 requirement by the same 1.1547 difference.

$31.25/1.1547 = 27.06$ lb/hr injector flow rate @ 58 psi

The stock 26–27 lb/hr injector would be the ideal size for a 400-horsepower LS1 at 58-psi LS1 fuel pressure levels. This would explain why GM went with a 26–27-pound per hour injector for the 1999–2000 LS1s but later went to the 28.8-pound per hour injector as further breather room was allowed for the 385-horsepower and 405-horsepower LS6 Corvette Z06.

Keeping all other variables the same but lowering the fuel pressure, you must have a higher capacity injector rating to keep up with the engine's power-production capability. Raising your fuel pressure and keeping all other variables the same, you can lower your pound-per-hour required injector rating.

The differing fuel pressure ratings only slightly complicate the injector sizing math. With these handy calculations and estimations you can figure out your desired fuel-injector flow rating size for any type of engine setup and power-adder. LS fuel systems can operate between 58 and 62 psi. The ECM has the capability to adjust fuel trims based on input from the oxygen sensors in the exhaust, so long as the base calibration matches the actual fuel pressure and injectors you are using.

As mentioned, the original Gen III LS1 came with 26–28-pound injectors. The Gen III trucks with the small Multec 2 connector are 22–25-pound rated. These all have a small window for engine upgrades and, being older and often having been sourced from non-running donors, including them may lead to problems. Commonly, injectors that sit for long periods of time with old fuel present end up plugged from fuel gunk or rust accumulation from steel fuel rails. Either of these scenarios will require that the injectors be sent for cleaning, or possibly replaced in the worst cases.

The original LS1 fuel rails are steel and will rust if water intrusion occurs. If this is the case, the best option is to get billet fuel rails to replace the old steel rails. There are a number of sources, such as Aeromotive and Holley. I am using Holley Billet Rails (P/N 534-209) on my 1991 Camaro LS swap with the LS6 intake, because my fuel rail was packed with rust that also plugged the injectors. Fortunately, I discovered this situation early in the build and ordered the Holley fuel system parts and new fuel injectors while working on other swap projects.

OEM Gen IV engines have the Bosch EV-6 connectors, so this makes injector swaps a little easier on the connector side. Some of the heights are different and some use different-size O-rings where they seal into the intake manifold. Gen IV car intakes typically use large thick orange O-rings, whereas truck Gen IV intakes use a tiny, truck-specific O-ring. Stock Gen IV injector sizes range from 34 pounds on the taller (truck-dimension) LS2 injectors to 57 pounds on the supercharged 6.2-liter LSA and LS9.

The most common Gen IV car-style injector is the 42-pound one found on LS3/L99/L76/LS7 engines. Gen IV truck engines without FlexFuel come with 30-pound injectors, whereas FlexFuel injectors can be from 36 pounds on the 5.3-liter up to 54 pounds on the 6.0-liter trucks; check the part number if you are unsure. There are many injector-part interchanges available online if you need to find the injector that works for you. Alternately, you can do a calibration read on your stock ECU to determine the stock injector sizing. If you send the injectors to be cleaned, you can often get data related to the actual injector flow.

If you are using new injectors, make sure your ECU injector calibration matches your new injector size. In some software, such as that used in the Holley Terminator X, you can choose your injector size from a drop-down menu. In OEM calibrations, you need to enter the injector data into the correct injector-size table matched to the fuel pressure.

There are three main fuel injector external dimension heights for the LS engine family LS1/LS6, LS2, and LS3/LS7, plus the multiple truck variant engines. Mixing and matching different fuel rails and injectors between engine families is possible by using injector spacers on short injectors, or fuel rail spacers and longer bolts on long injectors. ICT Billet has all of these combinations listed on their website along with the different wiring connector adapters available for these injectors. *ICT Billet*

Chapter 5
Induction and Exhaust Systems

Choosing a correctly matched intake and exhaust setup for your transplant is something many people tend to overlook. Much like any recipe, if you leave out a few key ingredients or, worse, use the wrong ingredients, the end result predictably is unsatisfying.

Like other LS swap parts, there are numerous options for both the intake and the exhaust. Many choices are better than others, and the price often tends to reflect this. As the LS is swapped into ever more vehicles, many induction and full exhaust systems are not available in kit form for specific non-common applications, whereas common swaps can be sourced very easily.

The induction assembly consists of the air-cleaner assembly that connects the air filter to the throttle body and includes a mass airflow sensor in the middle. Building an induction system can be a simple process and one of the easiest installations on your LS swap. Some OEM induction setups can be adapted to work on your project, while on some setups it might be best to build a custom setup to your liking.

Other than the obvious stock exhaust manifold choices, there are a slew of specific header applications for LS swap vehicles; often, it's just a matter of choosing which company to purchase from. In this section we will cover manifold and

A nice induction is one of the last finishing touches in the engine compartment. It provides a source of fresh air for the engine to consume. Do a fair amount of research and time how you add this component. Rotofab makes this induction kit with a sealed airbox for 1st-Generation Camaro/Firebird LS swaps—it fits in right at home in the engine bay. Due to varieties of engine bays and dimensional constraints, the induction can be an easy bolt-on or involve fabrication.

There are three styles of intakes for the tall cathedral port intake runners. All Gen III are cathedral heads, along with the Gen IV LS2 and all 4.8- and 5.3-liter Vortec truck engines. Here, we can see the difference between the more free-flowing LS6 intake (left) and the LS1 intake (right). The LS6 intake has the flat bottom, while the older LS1 intake has a bulge and tapered floor. A further benefit of the LS6 (besides a power increase) is that the LS6 intakes do not have EGR provisions.

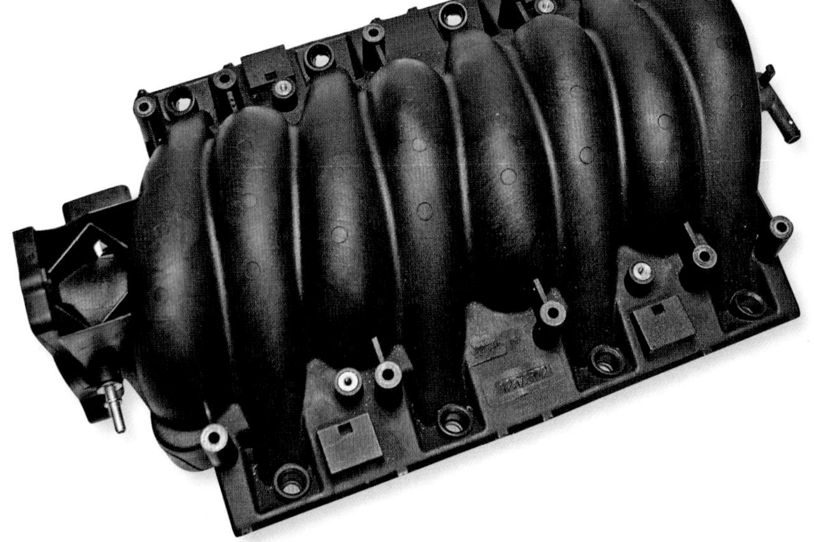
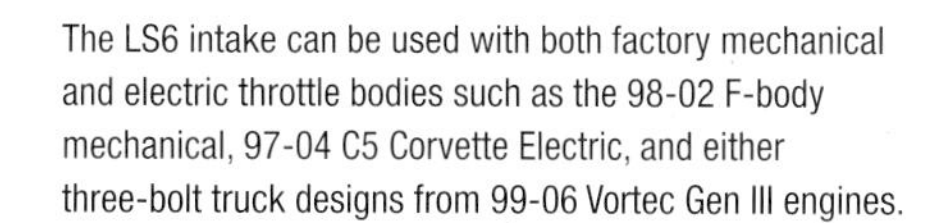
The LS6 intake can be used with both factory mechanical and electric throttle bodies such as the 98-02 F-body mechanical, 97-04 C5 Corvette Electric, and either three-bolt truck designs from 99-06 Vortec Gen III engines.

header choices and which interference and incompatibility concerns may arise.

During the LS retrofit process, there are a few things that can be upgraded that allow easier installation or increased performance. Exhaust is one such area that is beneficial, as the difference between restrictive factory manifolds versus just about any long tube header can be anywhere from 20 to 25 rear-wheel horsepower. Furthermore some cars even pick up better fuel mileage with the addition of headers over the restrictive stock manifolds.

INTAKE MANIFOLD IDENTIFICATION AND DIFFERENCES

For OEM intake manifolds the choices are limited to the production parts. Basically, you have two options: car style or truck style, in both cathedral- or square-port style. The original Gen III LS1 intake found on 1997–2000 LS1 engines would be the first LS-based intake manifold and also has the least power potential. Think back to when the first LS1s were being modified; it was four long years before anything better was available for intake manifolds. People made due, and the power potential of the time showed, as parts for all areas of the engine were still lacking. With this intake and an N/A 346-cubic-inch LS1, you're going to be limited to roughly making about 20 horsepower less than the LS6 intake we discuss next.

When the LS1 was minutely revised in 2001, it received the LS6 intake that made its debut most notably with the 2001 Z06 Corvette 385-horsepower LS6 engine. LS6 intake manifolds are found on all 2001–2004 LS1 and LS6 engines, as delivered. For cathedral heads on the stock 4.8-liter to 6.0-liter engines, this was a good economical upgrade over the LS1 manifold, and they are plentiful in the used LS marketplace as people outgrow them and move on. As swaps have become more popular, this intake choice is not as budget-friendly as in years past. In some cases, the original lower performing 1997–2000 LS1 intake may be a better swap choice just due to availability and budget.

For N/A 346-cubic-inch engines, the general consensus is that you can make 450 to 460 rear-wheel horsepower using the LS6 intake. While the intake is starting to hold the engine back in the 400-horsepower range, sometimes the cost to upgrade does not justify the extra power that could be had for a majority of users. With power-adders, intake selection matters a little less; there are LS6 intakes supporting well over 1,000 rear-wheel horsepower on boosted setups.

With the Gen IV engine debut in 2005, the LS6 intake manifold was shelved and replaced with the LS2 intake manifold, and remembering our engine discussions, this too is a cathedral port intake. The big news with the LS2 intake is the addition of the 90-millimeter throttle body opening, although surprisingly power capacity still seemed to match the LS6 75-millimeter intake manifold. The manifold absolute pressure (MAP) sensor is relocated to the front, so a MAP extension harness would be required for retrofitting into an F-body or C5 engine harness. The LS2 and all Gen IV intake manifolds are exclusively throttle-by-wire and in OEM form do not support a cable-driven throttle body. For LS retrofit with DBC vehicles, a 90-millimeter cable-driven throttle body and throttle cable bracket must be purchased. Luckily, the aftermarket is on top of things, and parts are available to make this a painless bolt-on ordeal. The LS2 intake is factory on 2005–2007 LS2 vehicles, mainly C6 Corvettes and GTOs. Due to the limited years and vehicles, this intake is harder to find without buying a complete LS2 engine, or having some luck on some of the online marketplaces.

LS2 intakes are equipped on Gen IV LS2s, most commonly found in 2005–2006 GTOs and 2005–2007 C6s. This intake fits all cathedral port heads, but uses a different fuel rail, injector, 90-millimeter four-bolt throttle-body bolt pattern, and MAP sensor location than the previous year's LS6 intake. If using this intake with a Gen III ECM and a mechanical throttle body, you will need one of the aftermarket throttle cable brackets as Gen IV are drive-by-wire. *Courtesy of General Motors*

LS3 intakes are for the rectangle design heads, such as those found on the LS3, L92, and L99 engines. The LS3 intake uses the same MAP location (bolt-on instead of clipped-on) and same four-bolt TB pattern as the LS2 and LS7. The LS3 MAP sensor has a different resolution than LS2 and earlier engines, requiring custom ECM tuning, retrofitting an older MAP sensor, or, of course, an LS3 ECM/harness. LS3s and LS2s share the same fuel injector rail. *Courtesy of General Motors*

The LS2 and LS3 intake uses its own fuel-injector rail, as the older LS1/LS6 injector rail has a different design. The LS2 and all Gen IV factory injectors are also shorter and use a thicker lower O-ring to seal the intake manifold. When using the more common tall Bosch-style injectors on Gen IV intakes, the lower injector O-ring must be replaced with an LS2 injector O-ring and a spacer and longer bolt kit installed under the fuel rail mounting pedestals. If using the shorter LS3-style EV6 injectors, you must have injector spacers to take up the difference. ICT Billet offers an injector hardware kit with these items to space the injector rail or injectors as required.

The intake manifold on the LS7 is specific to LS7-style heads and intake ports due to the ports' higher location and the fact that recent LSX heads have migrated to rectangle intake ports. The LS7 intake manifold is physically wider than any other OEM composite intake manifold. The LS7 intake uses the same 90-millimeter electric throttle body as its LS2 little brother. This intake has some untapped power potential gained by porting the intake runners and opening up the throttle body inlet opening.

With the 2008 Corvette upgrades, the highly anticipated LS3 engine arrived. The LS3 uses rectangle port heads much like the LS7, carried over from the L92 truck engines. The LS3 and L76 intake manifold ports only physically line up to the LS3/L76/L92 rectangle intake port cylinder heads. Like the LS2 and LS7, it predictably uses the same 90-millimeter electronic throttle body as well. If using these heads, the same retrofit guidelines apply to this intake as the LS2 intake manifold.

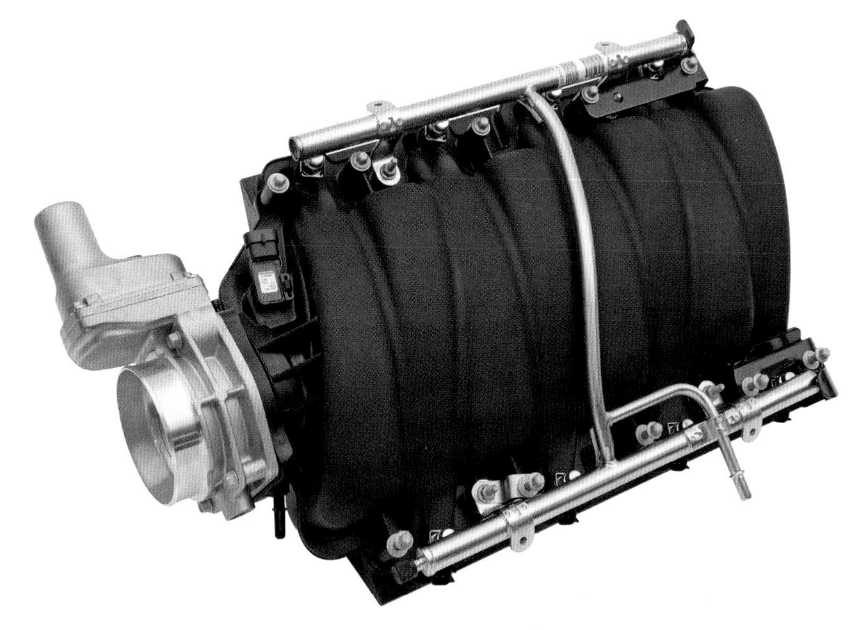

The LS7 intake fits only LS7-style heads and has a wider intake bolt pattern so as to not mis-install it on other incompatible engines. LS7 intakes use their own fuel-injector rail, but share 90-millimeter TB designs with LS2 and LS3 engines. The early LS7 engines share MAP sensor calibrations with all Gen III engines, so with the throttle cable adapter bracket, injector adapters, and MAP sensor extension, it will easily operate with Gen III ECMs, in addition to Gen IV ECMs and harnesses. *Courtesy of General Motors*

Earlier we mentioned the difference between the two newer Gen IV intake ports, the LS3 and LS7. Before these engines were available, there were speculations that they had identical intake ports, but as you can see comparing them side by side, they vary quite a bit from each other. The LS3 cylinder head is on the left; the LS7 cylinder head is on the right.

Being that the LS9 uses a 2.3-liter blower on top of the engine in lieu of an intake manifold, not much can be said about that other than with the ZR1 engine being loosely based off the 6.2-liter as far as displacement and measurements; in the future we may see blower setups that easily retrofit to other GM engines using the 6.2-liter heads. The Caddy LSA 6.2-liter engine is the cousin to the LS9 engine and is also supercharged, but with a 1.9-liter roots-type blower netting about 80-horsepower less in stock form.

Fuel Injector Rail Requirements

LS1/LS6 intake manifold	Uses LS1/LS6 fuel rail or aftermarket billet rails
LS2 intake manifold	Uses LS2 or LS3/L99 or aftermarket billet rails
LS3/L99 intake manifold	Uses LS3/L99 or LS2 or aftermarket billet rails
LS7 intake manifold	Uses LS7 ruel rail only or aftermarket billet rails
FAST 102-millimeter intake	All use LS3/L99 ruel rail*

*Requires injector nozzle spacers for LS3/L99 and LS7 injectors.
LS1 + replacement Bosch-style injectors and LS2 + replacement injectors require fuel rail spacers in lieu of injector nozzle spacers.

Using a 90-millimeter aftermarket mechanical throttle body on the 102-millimeter intake manifold causes a sealing overlap on some designs. The 90-millimeter stock throttle bodies do not have this issue, but use caution if you have a cable-drive throttle body on the FAST 102-millimeter intake, as you will likely have a hidden vacuum leak.

Aftermarket intakes such as the FAST 102 on this LS7 427ci engine are not specifically required for LS swaps, but you can see complementary power gains when combining an aftermarket intake with other modifications, such as a performance camshaft grind, exhaust, and ported cylinder heads.

THROTTLE BODY AND CABLE MOUNTING

Remember that your throttle-body diameter needs to match the inlet bore diameter of your intake manifold. You accomplish nothing by running a larger TB without having the matching intake. For instance, if you have the stock LS6 75-millimeter inlet intake, your best choice is the stock 75-millimeter TB sizing. If you have a 102-millimeter inlet intake manifold, then of course a 102-millimeter TB would be ideal, although the stock 90-millimeter DBW electric throttle body also would work on the 102-millimeter FAST intake manifold.

There are performance modifications to the stock throttle bodies that actually work if you are keeping the stock components. One of the big things that will make a difference in any setup using the factory intake manifolds is to port the throttle body. They come with a huge airflow-impeding ridge in front of the throttle blade that can be removed and blended to benefit engine airflow into the intake manifold. This applies to the OEM LS1, LS6, LS2, LS3, and LS7 throttle bodies, and gains of 5 to 10 horsepower at the wheels are commonplace with quality ported throttle bodies.

With the LS2, LS3, and LS7 with 90-millimeter openings, an aftermarket 90-millimeter cable-driven throttle body may be used, or the factory electronic 90-millimeter LS2-LS7 TB can be used in vehicles that are factory drive-by-wire setups. With the C5 Corvette's engine harnesses, an adapter wiring harness must be used when using the Gen IV 90-millimeter throttle body for proper function. There is a ton of variety for aftermarket cable drive 90-millimeter throttle bodies, but just remember that when coupled to an OEM Gen IV intake manifold, you will need a throttle cable bracket solution, such as the Gen IV intake throttle cable brackets available from Scoggin-Dickey Parts Center.

Motion Raceworks has several innovative throttle cable brackets and cable solutions for various intake and throttle body choices. They make throttle cable brackets for both the OEM and aftermarket cables.

The aluminum spacer with the throttle mount sandwiches in between the mechanical throttle body and the Gen IV intake manifold, LS7, shown here. This spacer provides a mounting location for the Gen III F-body LS1 throttle bracket.

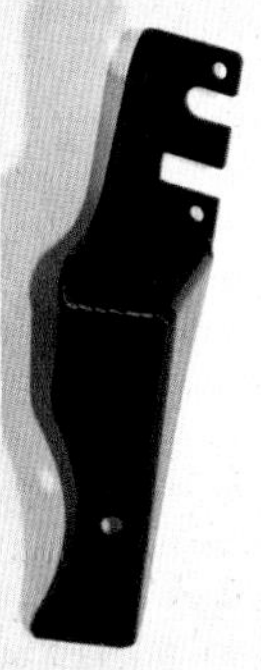

With Gen III engines, the throttle cable and mounting bracket provisions are built into most intake manifolds (1997–2000 C5 LS1 intakes are the exception). When using an OEM–Gen IV intake manifold, they present a small problem. As these are all drive-by-wire from the factory, there is no throttle bracket locations. To solve this with a cable-drive throttle, you need a throttle bracket or throttle bracket mounting spacer to mount your throttle cable to.

Adding a throttle cable to a chassis that was originally a mechanical linkage throttle setup can sometimes be intimidating. You can fabricate the required mounting brackets and retain the stock throttle pedal and lever if desired. You will need the Lokar throttle cable and some minor fabrication time to accomplish this. In first-generation Camaros you can also replace the pedal assembly with a I-6 cylinder gas pedal to use an OEM pedal design. *Tom Munsch*

Many builders who want carbed LS engines will need to use an aluminum intake manifold. GM Performance Parts offers few varieties for the varying engine designations along with such manufacturers as Edelbrock and Performance Inductions.

If you want to use a stock-style throttle body with a stand-alone EFI harness and a carbureted intake manifold, look to Mast Motorsports for this trick adapter. You can also use this adapter with the stock ECMs, although you would likely need to have an aftermarket ECU, or have your ECM custom tuned for speed-density mode, due to the lack of MAF provisions.

HOW THE PCV SYSTEM WORKS

A closed crankcase ventilation system must be used in order to provide a more complete scavenging of crankcase vapors. Filtered air from the air-induction system (air cleaner) duct is supplied to the crankcase, mixed with blow-by vapors, and passes through a crank-case ventilation metering device before entering the intake manifold. The primary component in the positive crankcase ventilation (PCV) system is the PCV flow-metering orifice. Vacuum changes within the intake manifold result in flow variations of the blow-by vapors. If abnormal operating conditions occur, the design of the PCV system permits excessive amounts of blow-by vapors to back flow through the crankcase vent tube and into the engine-induction system (air cleaner) to be consumed during normal combustion. This engine ventilation system design minimizes oil consumption and significantly reduces the potential for oil ingestion during vehicle limit handling maneuvers.

HOW TO SET UP YOUR PVC SYSTEM

- There are three ports on the LS long block that make up the PCV system. There are two foul side ports. Both of these ports should be connected to the intake manifold and be exposed to vacuum at idle.
- The two ports are (1) front port on the valley cover and (2) left rear (driver) valve cover. These two silver tubes may look simple, but they should not be modified. Both of the tubes have a small orifice within them that is used in place of a PCV valve of early designs.
- There is one fresh air port on the front of the right (passenger) valve cover. Again, this is a silver tube that faces forward on the valve cover. This port should be connected to filtered clean air. This is typically within the engine's air cleaner system or can be a separate air filter if using a carburetor. If you are planning on an electronic fuel-injection system that uses a mass airflow meter (MAF) then the fresh air to the PCV should be installed between the MAF and engine's throttle body. The engine burns the air that enters the PCV system so, if the fresh air port is prior to the MAF, then this air will enter the engine without being measured by the MAF.

Source: GMPP Crate Engine Instructions

DBW ELECTRIC THROTTLE PEDAL

Using a Gen IV ECM or C5-style throttle control has some other specific requirements due to the electric throttle pedal. Mounting the throttle pedal does require some specific guidelines as lined out in the *GM Performance Parts* manual for Gen IV–style controllers. There are some certain measurements that must be kept so that the throttle pedal is located correctly in relation to the brake pedal.

GEN III THROTTLE ACTUATOR CONTROL MODULE

If using the Gen III–style ECM that uses the electric throttle control, you will need the matching TAC module. The throttle actuator control (TAC) module is an interface that interprets and controls the throttle body via inputs from the throttle pedal. You will need a TAC module that matches the ECM style and the throttle body, which varies each few years.

The TAC module has two connections, one for the electric throttle pedal harness, and the other side plugs into the main engine harness for the throttle body and ECM integration. There are also two differing styles: the nine-pin connector for 1997–2002 ECMs with the blue and red bulk connectors and the later six-pin connectors for 2003–2005 ECMs with the blue and green bulk connectors.

TAC modules have to match up to the specific year range of both the ECM and throttle body. The 2000–2002 TAC module works with the 2000–2002 GM truck and SUV throttle body and the 1997–2004 C5 Corvette throttle body, and it requires the 1999–2002 PCM (blue and red connectors). The 2003–2005 TAC module works with the 2003–2005 truck TB and the 2003–2005 PCM (blue and green connectors).

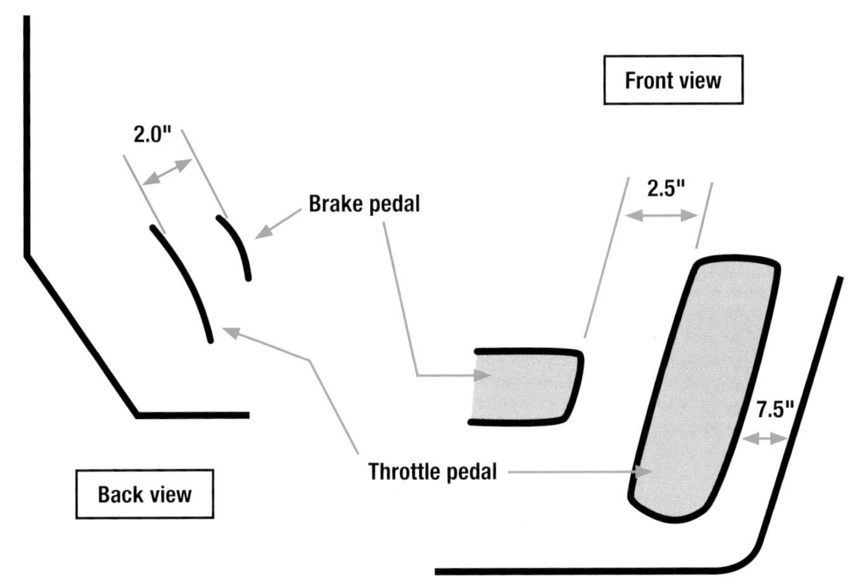

Setting up the electronic throttle pedal is one of the more involved aspects of the drive-by-wire system. In almost all cases, the car getting the new electronic throttle pedal originally had either a mechanical linkage throttle or cable throttle from the factory. The GM Performance Parts manual outlines the correct measurements and dimensions for properly installing the throttle pedal assembly. *Courtesy of General Motors*

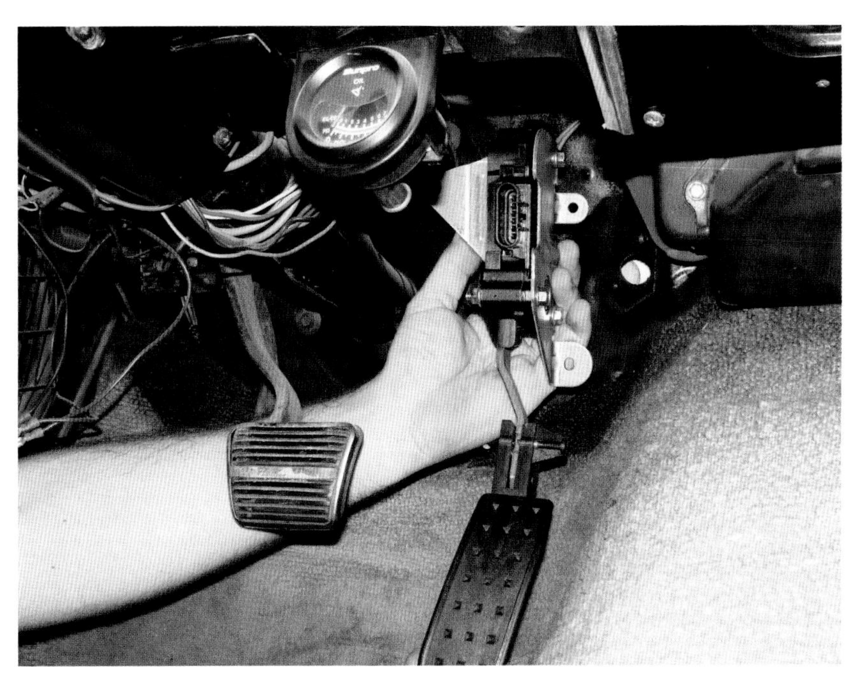

Mocking up the electronic throttle pedal requires taking some careful measurements so that the pedal is in the right placement as compared to the brake pedal and transmission tunnel. The pedal assembly that comes with the GMPP harness kit will require some spacing or extra brackets in this first Gen Camaro application. Gen III drive-by-wire vehicles require matching throttle pedal, TAC Module, and throttle body from the same type and year vehicle (recommended), whereas the Gen IV throttle pedal selection must match the base calibration in the ECM. (e.g., a Trailblazer SS base calibration *requires* a Trailblazer SS throttle pedal)

To find a donor TAC module, you need to know where they are located. On C5 applications, the TAC module is located directly in front of the ECM located behind the right front tire under the splash shield. This would be directly below the battery, but you need to remove the wheelwell access panel to remove it. On trucks and SUVs, the module is located beside the factory brake booster and master cylinder location bolted onto the firewall. If you are making your own harnesses, make sure to grab any wiring possible such as the wiring from the TAC module to the accelerator pedal. Note that the TAC module is also a key component of the cruise control system and allows easy cruise adaptation.

INDUCTION

The induction is simply the air inlet in front of the throttle body. In the easiest form, it consists of a tube, connections, mass airflow sensor, and an air filter. Building an induction is not hard, and other than tight engine compartments will likely be something that takes less than an hour's time to design. Imagination and creativity are required to clear any pulleys, cooling fans, or radiator hoses. If you have fabrication skills, this is your moment to shine.

The diameter of the piping necessary depends on the throttle body sizing and MAF diameter. For the stock LS1 F-body, C5 and Gen III truck throttle bodies, and MAF combination (if required), all that is needed is 3.5-inch piping and couplers, such as a donor 90-degree LT1-style elbow. For Gen IV engines with the larger throttle body, a 4-inch induction is recommended.

In either situation, you will likely require a 90-degree bend away from the throttle body to point the induction toward the front of an inner fender with the straight section traversing the engine bay to the filter location. For maximum power results, an outside (cooler) air source would be necessary for a typical and functional swap; outside air is not required, but engine performance is improved if you can provide it. The filter can be situated anywhere inside the engine bay away from a direct heat source; typically to the left or right of the radiator's airflow is best, although the ideal location may have something occupying that space, such as a battery or washer-fluid bottle.

The GM Gen III engine style 75-millimeter and 85-millimeter MAFs are less critical for mounting locations and can be mounted anywhere without having ill results. A good rule of thumb for MAF placement is to try to keep it away from bends so that it can accurately measure airflow. The GM Gen IV MAF sensor stalk is recommended to be in the middle of a minimum 6-inch-length straight section and a minimum 10-inch location away from the throttle body, using a 4-inch-diameter tube. Due to the sensitivity of the GM MAF sensor stalk, the MAF placement is more critical than the other MAF sensors and it absolutely cannot be located on a bend.

It's not likely you'll find an induction from a donor vehicle that fits perfectly. Most people will build their own induction for a given vehicle application. Using mandrel-bent tubing is possible.

One solution I have been using for LS swaps is the Airaid Master U-Build-It kit. This kit comes in several part numbers with a plastic Rotomolded length of 3.5-inch or 4-inch tubing with 30-, 45-, 60-, and 90-degree bends that are cut at premarked increments and then clamped together in the configuration your application requires. The kit comes with couplers, clamps, a reducer for the throttle body, an air filter, and some bend-to-fit brackets for support.

Several companies offer vehicle-specific swap-induction kits. One such company that makes a very nice solution for the 1st-Gen Camaro is Rotofab. In addition to the complete hardware required for a 4-inch-diameter induction kit, the setup also includes an air box that seals off underhood, allowing fresher outside air. The Rotofab kit can be ordered with or without an MAF-mounting location as well.

STOCK MANIFOLDS AND CATALYTIC CONVERTERS

OEM manifolds are inexpensive, compact, and offer long life with nearly a zero failure rate (other than typical broken exhaust manifold bolts on used donor engines), making them perfect for stock engines. The plethora of different stock LS applications provides a variety of alternatives to buying the sometimes pricey retrofit headers or custom-building exhaust components. There are a few that are of tubular-steel construction, such as the pre-2000 style of C5 and F-body LS1 manifolds and the LS7/LS9 "shorty" header style manifold. Otherwise, pretty much everything else uses cast-iron manifolds.

Many older swap candidates can use various OEM LS manifolds with or without cats, as a functional alternative to headers. There are no set guidelines, although due to the

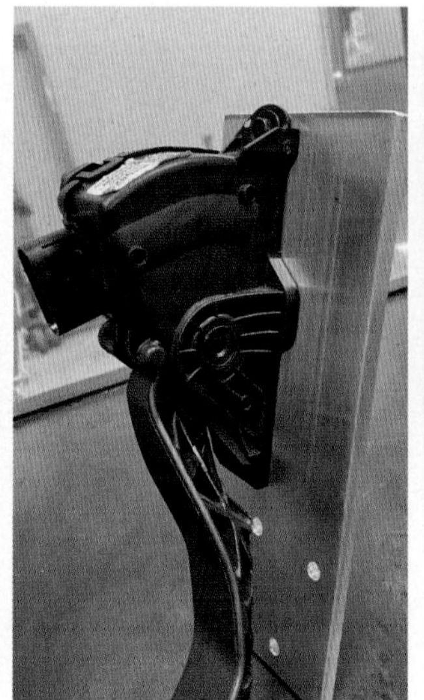

DBW can present some hurdles in mounting the throttle pedal itself. It may not bolt up to the firewall directly, which will require sourcing an adapter. One of the simplest ways to do this is by using flat aluminum plate such as this 4-by-10-inch, $3/8$-inch-thick plate I used on a 1998 Chevy 1500 conversion. I drilled three mounting holes for the original throttle pedal location and two holes for the OEM 2007-2011 Silverado DBW truck pedal. Trim the excess material off and then paint it black so that it is hidden from view.

For some swaps, you can buy premade DBW pedal brackets. This is the Detroit Speed 1st-Gen (1967–1969) Camaro/Firebird pedal mount I used in mine. It simply bolts to the existing steering column plate bolt locations and adapts the car to use the OEM C6 Corvette DBW pedal. The new pedal feels right at home in the 1967 Camaro. And, as an added bonus, the original stainless-steel pedal trim still fits.

variety of setups, there is likely a manifold that fits almost any LS swap. Car manifolds would normally be best suited to cars and trucks to trucks, but there are instances where they can be interchanged since they all share the same bolt pattern. The most common manifolds to be used tend to be the F-body and the truck manifolds, although other designs may better suit your chassis requirements.

One such manifold combination that seems to be versatile in GM muscle cars is the 2010–2015 Camaro LS3/L99 manifold and catalytic converter assembly. These are tucked close to the engine and offer a rear exit design that mimics many factory SBC manifold designs. The catalytic converters exit to a 2½-inch exhaust pipe offering simple hookup with a two-bolt flange to existing exhaust systems or as a great starting point to a new exhaust system. This manifold and catalytic converter setup is the same setup used on the emissions-legal GM Performance Parts E-rod crate engine.

These 5th-Gen Camaro manifolds (and those found on Pontiac G8 GTs) do fit a variety of LS swap projects, namely other generations of Camaro, 1955–1957 Chevys, most trucks, Jeep LS swaps, and almost any 1960s–1990s GM rear-wheel drive project. These are my go-to manifold choice as they are plentiful and inexpensive in the used market.

As a step further on the 5th-Gen Camaro manifolds, if you can find 2014–2015 Camaro 7.0-liter Z/28 tubular headers featured, these offer a mix of a manifold and a header in one that would work great in a tight engine bay with limited clearance. For instance, when we used these headers in a 2008 Saturn Sky with an LS3 swap, the headers fit with minimal clearance issues, the cast-iron 5th-Gen manifolds would not fit, but the Z/28 tubular shorty headers fit this odd application very well.

If you are picking up salvage yard manifolds and have access to the head pipes (the flange that bolts the exhaust to the manifold), it is a good idea to cut these off the donor vehicle and set them aside for when your new exhaust system is built. Many exhaust shops do not have the capability to build new flanges, and the OEM flanges give the exhaust shop a decent starting point. An added perk is that most of these head pipes have the oxygen sensor locations already placed and indexed in a location that works great with the wiring harness.

The factory catalytic converters (cats) operate basically as a post-combustion incinerator. Anything that is not consumed in the combustion cylinder is put through the "cats" and burned up. Much more than that actually happens at the molecular level, changing the physical composition of the spent combustion.

Since the main idea of this manual is taking newer engines and putting them in older cars, you may wonder why emissions are discussed. In short, if you add everything from the donor engine's model year into an older car and keep the emissions equipment functional, it is usually legal to do so. Most 25-plus-year-old vehicles do not adhere to the emissions requirements of the more modern models, but most of the components are easily reused and nonintrusive to the complete engine package. These laws do vary by state, so always double-check locally if this a concern.

The 1997–2004 Corvettes and 1998–2002 F-bodies (and some CARB-certified trucks) use air injection in the exhaust manifolds. With the use of an electric motor, fresh air is pumped into the exhaust during cold starts to help heat up the exhaust quicker, much like using a bellows to make a fire burn hotter. The motor for the air injection is controlled by the PCM through a relay, and often is deleted during a retrofit. If you are using exhaust manifolds that are equipped with these ports and do not wish to use that system, you can purchase or make block-off plates and remove the bolt-on hardware. PCM reprogramming is necessary to delete its electronic presence fully at that point.

In addition to the AIR injection, the 1998–2000 F-body LS1 engines and 1999–2002 Gen III–powered trucks used exhaust gas recirculation (EGR) systems. EGR is mainly a pollution device where metered exhaust gas is recirculated into the intake manifold and recombusted along with the fresh air and fuel charge. This has the effect of cooling the combustion temperatures down due to diluting the fresh air and fuel charge, allowing slightly more ignition timing. Most LS engine swaps will not use this system. If you wish to use both, the best way is to use the early and matching OEM manifolds.

The EGR system can also be plugged off like the AIR system at the intake manifold if disabled. Many people use a rubber expansion freeze plug, and it "works" but is less than appealing visually. GM made this job really easy, providing

Intake tubing configurations can be easy if you have generous space to work with. This 3rd-Generation Firebird has a simple yet effective induction. The owner used 4-inch-diameter 90- and 45-degree silicone elbows from Vibrant Performance, with an MAF housing from Airaid in the middle and an air filter adapted to the end, which then becomes a location to add a support bracket as well.

Induction setups are often the last, and the simplest, installation to perform along with your LS swap. The induction consists of an air filter, mass airflow sensor, and a tube between the air filter to MAF and from the MAF to the throttle body. The variations between setups are due to chassis interference points such as battery, radiator, fans, or actual inner fender interference. Note that most setups will require a 90-degree bend at the throttle body to locate the filter in a good inlet location. The widely available LT1 intake elbow (3.5-inch outlet) fits the bill for stock 75-millimeter throttle bodies. If using a 90-millimeter or larger throttle body (4-inch outlet), you will likely require a 4-inch 90-degree elbow, or use a 4-inch pipe with 4-inch silicone couplers.

There are several MAF varieties to select from. While the larger ones will interchange easier for the larger inside diameter with ECM recalibrations and wiring, you might be best off to stick to the one that matches your ECM for a fresh startup. The MAF reads inlet air, and if skewed numbers are present, the fueling will not match what the engine requires. From left to right: Stock 75-millimeter (1997–2000 C5, 1998–2002 F-body), Stock 85-millimeter (99+ Silverado/Sierra/SUV), aftermarket 100-millimeter for 2006 and newer Z06, and 2010 Camaro MAF setups. *GMPP engine harness MAF*

a bolt-in EGR plug that is a perfect match to the OEM EGR pipe, only it is plugged rather than being connected to the EGR tube. This GM part is hard to find, but there are several similar solutions from aftermarket companies that make billet plugs to block off the EGR intake port in the same manner. The plug can be used on all passenger car LS1 intake manifolds, and all truck EGR-equipped intake manifolds. If EGR is not retained, such as when using headers, you have to either plug the intake, or get an intake not equipped for EGR. This means either another stock intake or going to the aftermarket.

C5 and C6 Corvettes never used EGR systems, and 2001–2002 F-bodies did not require the system either, due to the camshaft overlap characteristics providing an artificial EGR system internally. The 2003 and newer LS-based trucks also lack external EGR systems.

If you can keep it emissions legal, great; but remember, a good running LS1 without cats is likely to be much less polluting than any carburetor-equipped stock 1970s-era car with functioning catalytic converters. The exhaust is much improved when the later-model high-efficiency cats are retained. The other aspect is that most older stock vehicles only used one restrictive cat, which was not very good for performance. LS engines use excellent, less-restrictive cats, and always in pairs, usually one per cylinder bank. A perk of having cats is that any excess exhaust smell is greatly diminished.

Catalytic converters by function operate at a high temperature. It is important to protect the underside of the vehicle with heat shields and to route or relocate flammable items away from the cats. Most non-catted OEM cars do not have this type of heat protection, so either adding heat shielding to the floor or using cats with integrated heat shields is highly recommended. Aftermarket heat shields can be had for heat protection from 900 degrees F to more than 2,000 degrees F continuous temperature. Heatshield Products offers numerous shielding products for such applications (www.heatshieldproducts.com).

As shown, air-induction systems are the simplest installation once the design is figured out. There is no rocket science to the induction, although it does help a ton to have the MAF located in a straight section of pipe with 4 inches of straight section upstream from the sensor. *Tom Munsch*

Use of the MAF stalk-type sensor that is required of the GMPP engine harness requires either welding on the supplied MAF tab to an induction tube or purchasing an aftermarket 100-millimeter MAF sensor or housing. The stalk sensor is sensitive to airflow paths and should not be located anywhere near a bend in the induction. This type of MAF housing is available from Texas-Speed and Performance and makes easier work of MAF mounting for those without access to a welder.

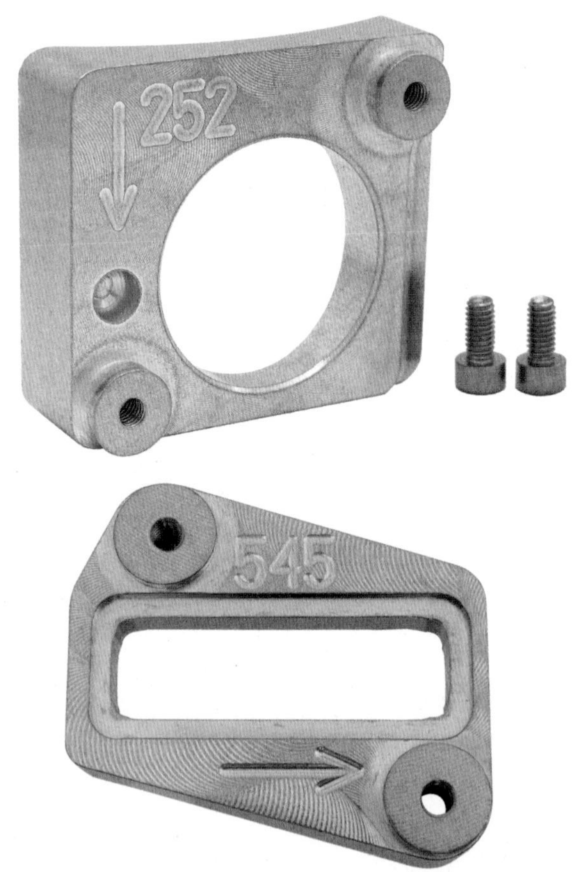

For custom air induction fabrication, ICT Billet has these convenient weld-in plates for your MAF sensor location if you're using GM OEM ECMs. For those vehicles with MAF sensors, make sure to put these in a straight section of induction piping to avoid skewing the sensor data readings. *ICT Billet*

In almost any case of an LS retrofit, the new induction will have to have a 90 degree elbow implemented to direct air into the throttle body from either side of the engine bay. Newer LS-equipped vehicles usually have a straight shot over the radiator for the induction system, but this is not feasible on retrofits for the most part without fancy fabrication. Several silicone tubing companies make an elbow that works great for such a 90-degree bend.

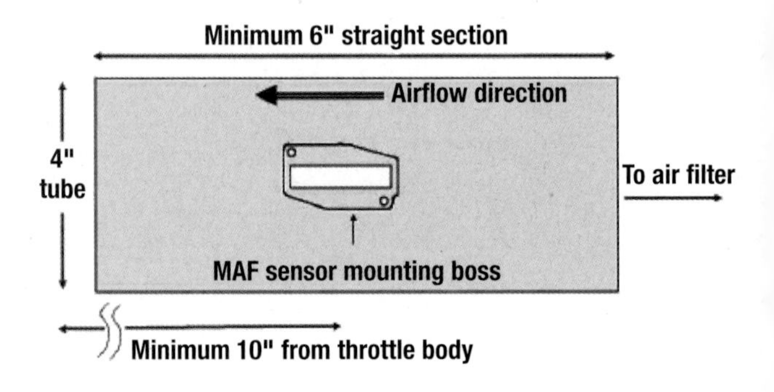

Use of the GMPP harness kit requires using the Delphi MAF stalk insert. The GMPP harness kit will come with the mounting tab and the calibrated sensor for your engine. You need to mount the sensor as shown in the GM diagram for proper and accurate airflow measuring as this sensor design is picky on where it operates best in the inlet air stream. *Courtesy of General Motors*

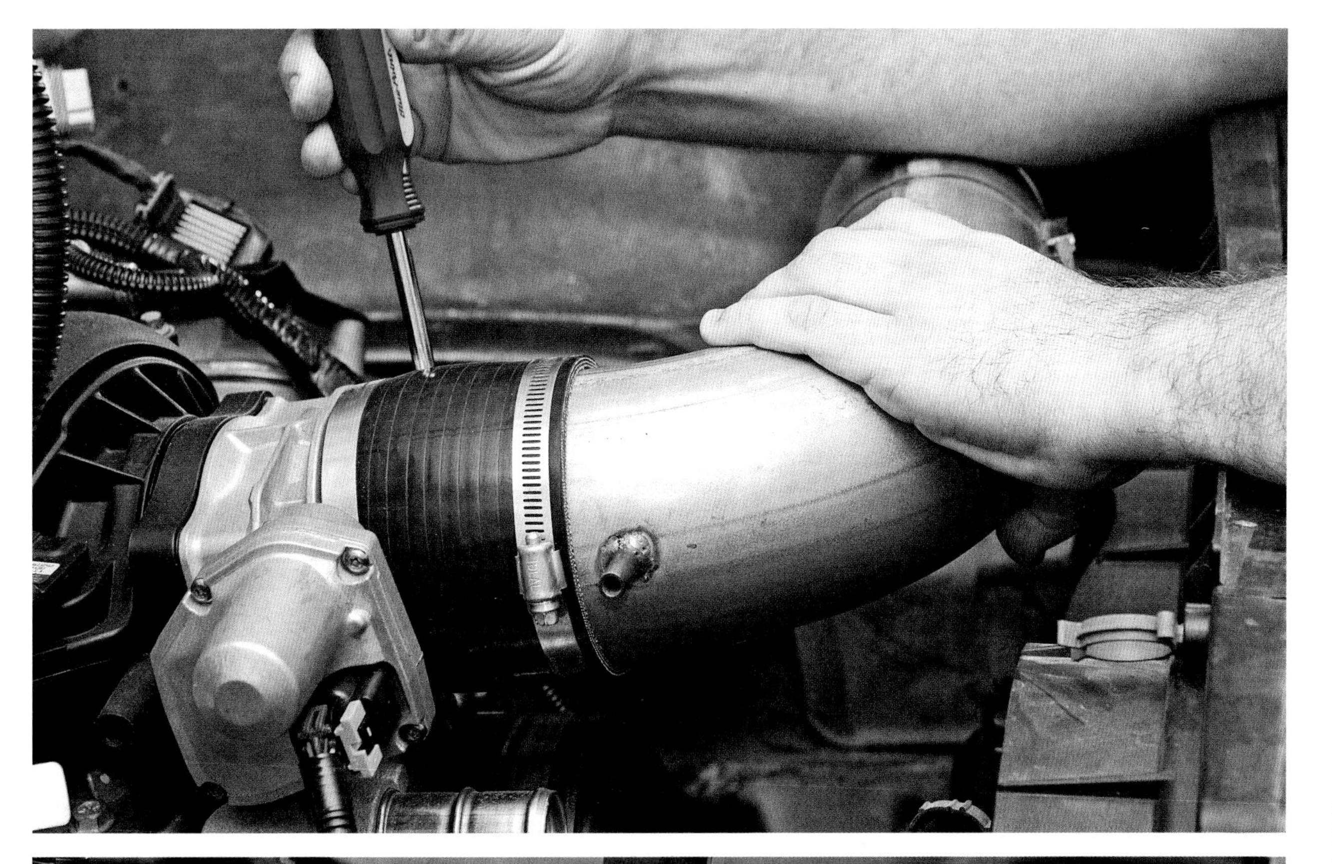

Depending on radiator hose interference, accessory kit component location, and battery location, you may need to angle the induction to the passenger side or to the drivers side. It is a trial-and-error fabrication routine, and you may decide later on to change it even further when more items are on the engine. Usually you want to save the induction fabrication for last.

RETROFIT HEADERS

There are applications that can use stock manifolds and work perfectly well, and some build goals are much better achieved with headers. In some cases, it is easier to buy LS swap headers for your project than trying to find a manifold that fits and then fabricating an exhaust from scratch. As stated earlier, LS engines using headers gain about 20–25 horsepower or more compared to manifolds, depending on the engine displacement. Other internal engine modifications can complement the gains, such as ported cylinder heads or a high duration camshaft can further impact this gain.

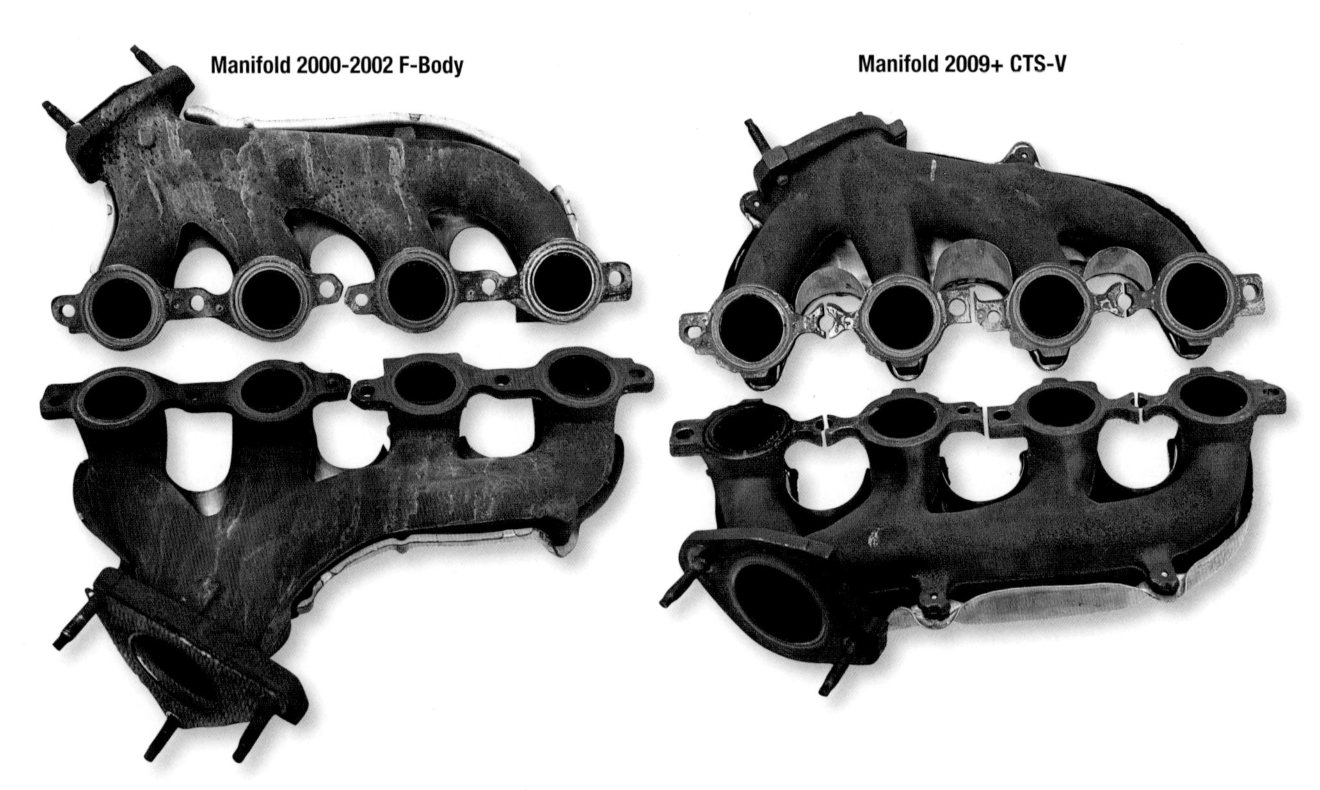

Often a set of factory manifolds are all that is needed to get your retrofit going. Although manifolds hold back some power as compared to headers, from a budget standpoint, they cannot be beaten. The issues that arise will be dimension conflicts with chassis and frame components. *Courtesy of General Motors*

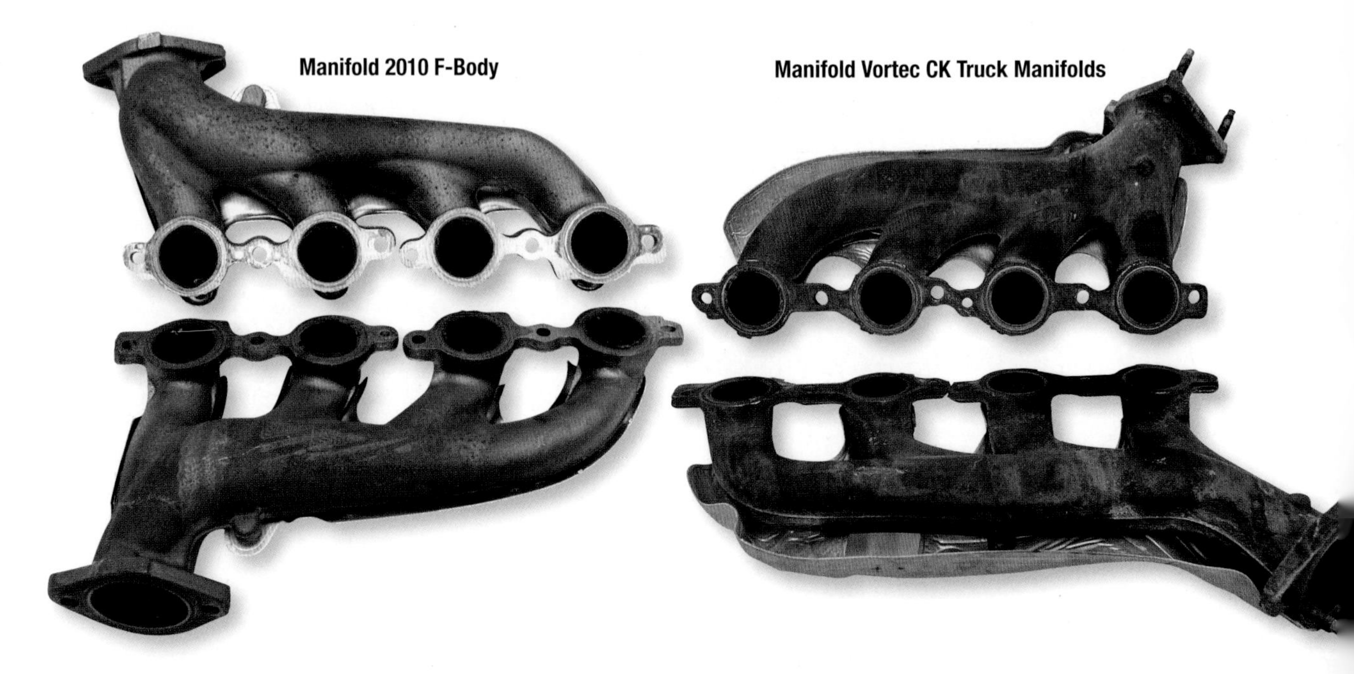

Camaro Z/28s from 2014–2015 had the amazing 505HP LS7 427ci engine, but they also came with these swap-friendly tubular tri-Y exhaust headers (P/N 12652399 and P/N 12652400). They have really good fitment and exit the engine bay in a convenient location. For our Saturn Sky LS3 swap application, these headers fit perfectly with very minor trimming needed. We also wrapped these with header wrap to cut down on underhood heat with the small engine compartment and nearby wiring harnesses, and afterward added the OEM header heat shields as extra protection.

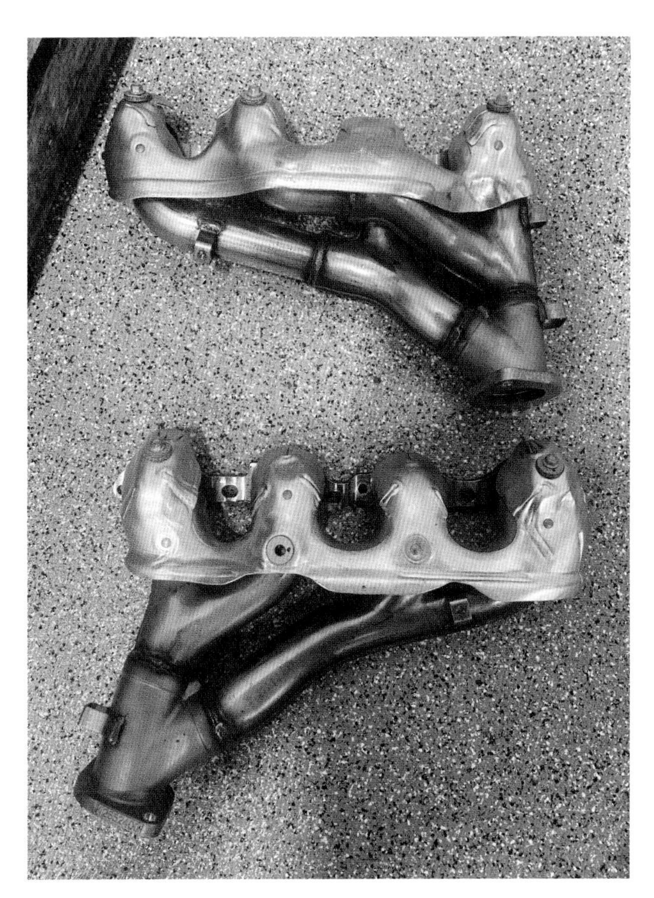

The use of headers allows the engine to expel spent combustion gases efficiently, allowing the engine to run more efficiently. Instead of all four cylinders in each bank cramming exhaust into one manifold, the headers allow a mini-exhaust system for each cylinder itself, which then converges the exhaust in the collector, where it then is expelled through the cats (if so equipped), the mufflers and the remainder of the exhaust system.

It should be no surprise that some header designs are better than others from a performance standpoint. Full-length, long-tube headers will always perform better than short or midlength headers, if they fit your vehicle. Many take the stance of function over performance, so they choose to use whatever fits easily to help make the engine swap easier versus looking for every last bit of horsepower and torque left in the setup.

The main differences in header costs is going to come from the material used; most retrofit headers are built from mild steel. They often are then ceramic coated for the durability, heat shielding, and aesthetics. The ceramic coating will not completely prevent rust or corrosion, but it will significantly delay its onset.

Higher end headers are built from stainless-steel tubing, which requires no further treatments for durability due to the anti-corrosion properties of stainless. Usually stainless-steel headers are built for the person who cares not so much about cost, but the level of the performance gained and longevity of the product. High-end stainless headers are usually more well thought out and make more power because of features such as equal-length primary tubing and high-velocity merge collectors (HVMC). The HVMCs smoothly transition exhaust pulses in the collector, which helps induce continuous exhaust scavenging from one exhaust pulse to the next and so on. Usually the benefits of these collectors are shown in the midrange torque curve. Stainless headers when available are usually about twice or more the cost of comparable mild steel headers, but you are also getting at least twice the features and upgrades. Stainless headers are available for the more popular retrofit applications, while inexpensive mild steel headers are more widely available for less popular applications.

Retrofit headers for the LS engines are built much like headers for newer cars, with the specific application in mind. The company uses an actual vehicle to build a mockup of the initial headers and develop the pattern from that initial test installation. Headers for a 1969 Camaro LS1 are not the same as the ones for a 1970 Chevelle LS1, but the same 1969 Camaro LS1 headers would fit a 1967 Camaro with a 5.3-liter truck engine or any other similar LS-based motor.

The only exception where standard off-the-shelf LS retrofit headers do not fit perfectly to the engine itself would be on something with LS7 heads, which have a large D-shaped exhaust port. This can be remedied by using the specific LS7 flange headers with larger primary tubing (1⅞ inches), or by using the O-shaped exhaust port gaskets with the LS1-style round-port headers.

As mentioned in the engine-mounting chapter, there are headers that fit specific chassis with specifically designed engine mounts. If the header kit comes with a motor mount option, it is usually better to use this setup as a kit than to purchase headers from Tom's Header Company and use mismatched engine mounts from Bill's House O' Mounts (fictitious names added for effect). Sometimes mismatching components will work in generously sized engine bays where header clearance is generous, but there are many instances where mismatched headers or mounts were used and compromises were made by modifying or "clearancing" the new header to fit adequately.

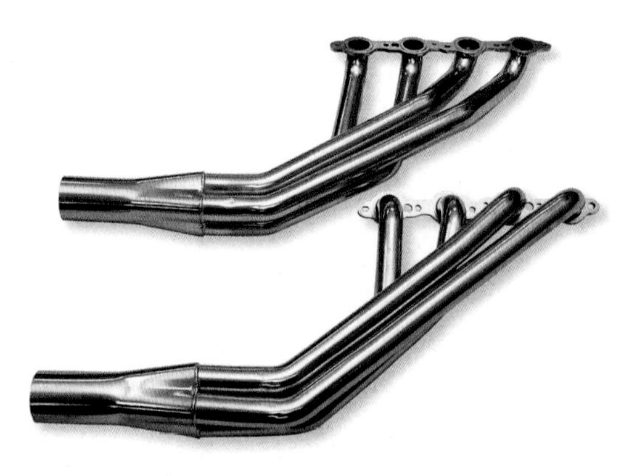

Swap headers come in many designs, and if you have any GM letter-coded chassis (that is, A-body, F-body, and so on) from the 1960s and newer, the chances are that you can easily find off-the-shelf swap headers that bolt in. Shown here is the popular Hooker LS conversion long-tube headers for 1967–1969 F-bodies. They also have A-body headers available. Hooker has its own motor mount plates designed for these headers, although the headers will work with other plate systems.

For street rod LS build-ups that would typically use block hugger "hedders" with a Gen I SBC swap, Hedman has similar block-hugger LS hedders. There are certain frame designs and motors would not allow these hedders to physically fit. Generally speaking, these are for street rods though. *Hedman Hedders*

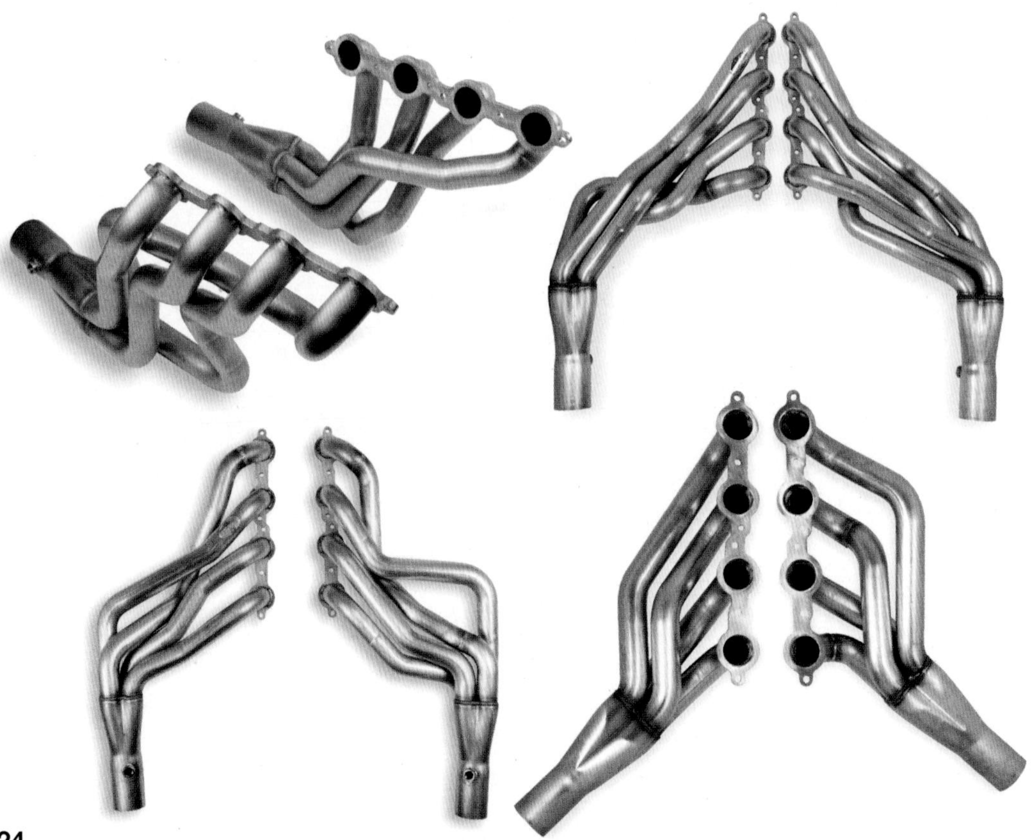

Hooker Headers is constantly adding LS swap headers to their catalog. Long tubes and mid-length headers are available for the majority of GM RWD LS builds. Mostly gone are the days of repurposing late-model headers into swaps, but these are swap-specific headers that fit your swap vehicle exactly. *Holley*

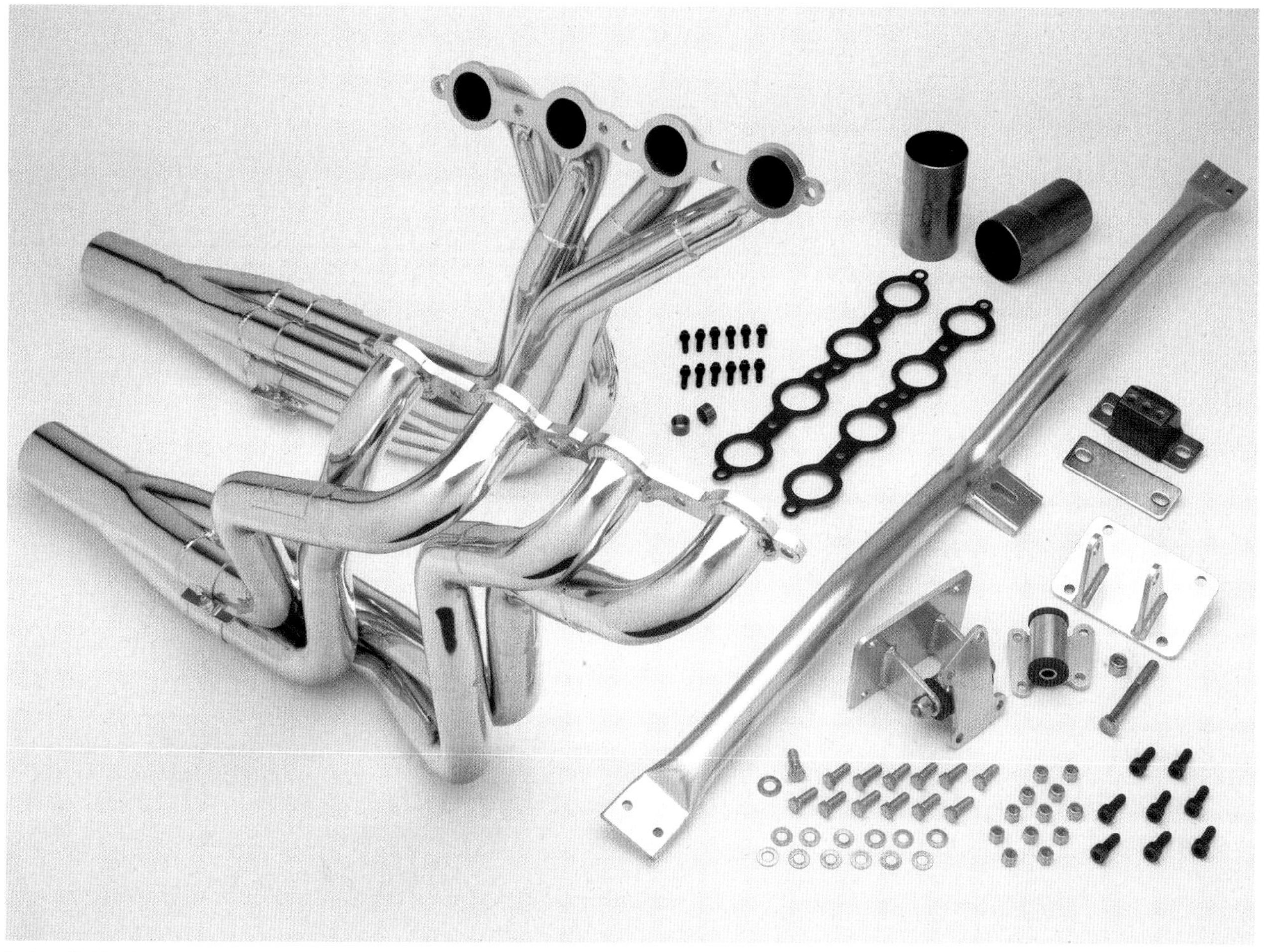

Hedman Hedders is teamed up with BRP Muscle Rods for header, motor mount, and transmission crossmember kits. The benefit to using a complete kit such as these is that all the homework has already been done. All you do is assemble the included parts and you are good to go on each subsystem. The G-body LS swap components are shown here. *Hedman Hedders*

Header companies are always adding new products, and the swap world has benefited from this. Companies such as Kooks, American Racing Headers, Hooker Headers, Speed Engineering, and Stainless Works have been adding to their LS swap catalog for years. Some headers are stainless, while the black-painted or ceramic-coated headers start off with a mild-steel design.

Long tubes rarely have components other than the headers themselves. It really depends on the vehicle. There is a good chance that you will have to build the Y-, H-, or X-pipe to finish or adapt your headers and manifolds to your exhaust system.

Common LS swaps obviously have the most parts available. Applications such as the 1967–1969 Camaro/Firebird, 1970–1981 Camaro/Firebird, and 1982–1992 Camaro/Firebird are some of the easiest for which to find exhaust kit components. For example, Hooker Blackheart complete stainless exhaust kits are available for all three generations of this platform. These allow the installer to completely finish their exhaust system at home in their driveway, without requiring a visit to an exhaust shop.

For more budget-minded folks, sourcing headers and then finding an exhaust shop to finish the exhaust system may make the most sense. In my 1991 Camaro 1LE, I have Speed Engineering long-tube headers, but there is no off-the-shelf Y-pipe available. In this case, I need to trailer the Camaro to the exhaust shop for Y-pipe fabrication. From that point, any cat-back exhaust (stock or aftermarket) for the 3rd-Gen and 4th-Gen Camaro/Firebird can be grafted in.

Speed Engineering has many of the GM LS swap vehicles covered in its catalog. Some applications have options such as mid-length headers, which have multiple vehicle fitments, or swap-specific headers that fit a specific application, which includes all common GM swap trucks, all generations of F-body, and even GM A-body, X-body, and G-body swaps. As mentioned, these cover the engine side of the swap, from the collector to the muffler tip, and are intended as an exhaust-shop or DIY-type exhaust build.

Speed Engineering also offers these 1¾-inch mid-length "universal" swap headers (P/N 25-1066). They're known to fit 2nd-Generation 1970–1981 Camaro/Firebirds, as well as 1970–1981 A-body and 1975–1979 X-body cars. I believe the best usage would be the C10-style GM trucks, as they seem to fit all truck year models from 1960 up.

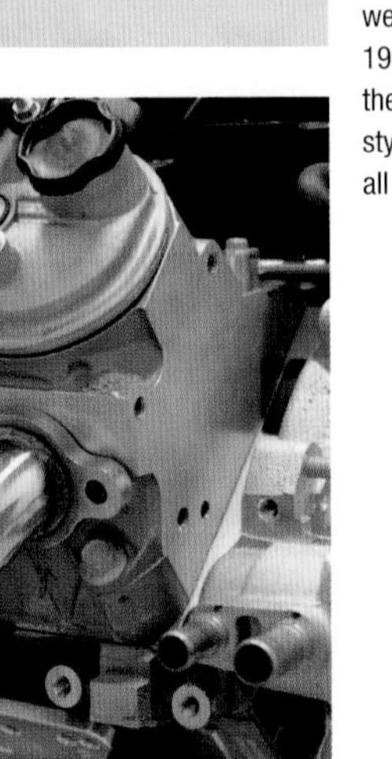

American Racing Headers offers several swap-type headers in stainless steel for GM muscle cars. The difference between ARH headers and the typical long tube headers are going to be mainly in material and header collector designs. ARH offers 309 stainless-steel 1½-inch or 1⅞-inch headers with a high-velocity merge collector. Additionally, ARH can match the exhaust port flanges to your design of engine, such as the D-port LS7 heads if used.

(TOP AND ABOVE) For 1982–1992 3rd-Generation Camaro/Firebird swaps, the Speed Engineering 304 stainless-steel long tube $1^{7}/_{8}$-inch headers (P/N 25-1035) work very well. These offer a great-fitting and budget-minded header kit. As with most LS swaps, you will need to fabricate or modify a Y-pipe or dual exhaust for these headers.

(ABOVE BOTH) Kook's Custom Header company manufacturers headers for some of the more popular LS swap vehicles, this application is a 1967 Chevelle (GM A-body). As with almost any long-tube header design, the header installs from the bottom. Take note that due to variations in motor mounts, steering columns, suspension choices, and/or transmission selection, you may have a header that requires extra work to properly fit the vehicle without interfering with other components. Dimpling a header tube or grinding a transmission case for clearance is the norm when dealing with LS swaps.

Non-GM LS swaps are not as straightforward as the GM-to-GM swaps. Due to the rather limited number of customers, they tend to be rare and costly. The headers that work for this configuration are often sourced from the company making the swap mount kit or subframe motor mount adapters. Examples would be Lexus, Toyota, Subaru, Mazda, Nissan, and BMW swap projects, among others.

Luckily for these car owners, they can find all the parts needed for their swap from one source instead of scrounging from multiple sources. While there are a couple different specialists for these cars, I suggest trying Sikky Performance. They have a well-designed system for headers to match their existing engine-mount and oil pan solutions. Headers from Sikky are USA made and engineered to fit the specific swap application as if the vehicle came with an LS already.

Sikky is also adding some GM vehicles to the catalog with their system of mounts/oil pans/headers that all match the same application, instead of a hodgepodge of parts from multiple sources. This method of building completely matching components alleviates issues caused by mismatched components from multiple sources that can lead to installation frustrations. Pre-engineered components from one source can prevent many headaches during your swap.

Installing the Hooker first Gen F-body LS1 retrofit headers is performed by coming through the bottom of the engine bay from under the vehicle. You will need to hold the header and gasket in place while starting the header bolts. Then tighten the header bolts from the center out to 22 lb-feet. Remember to retorque after the engine has been through a heat cycle or two.

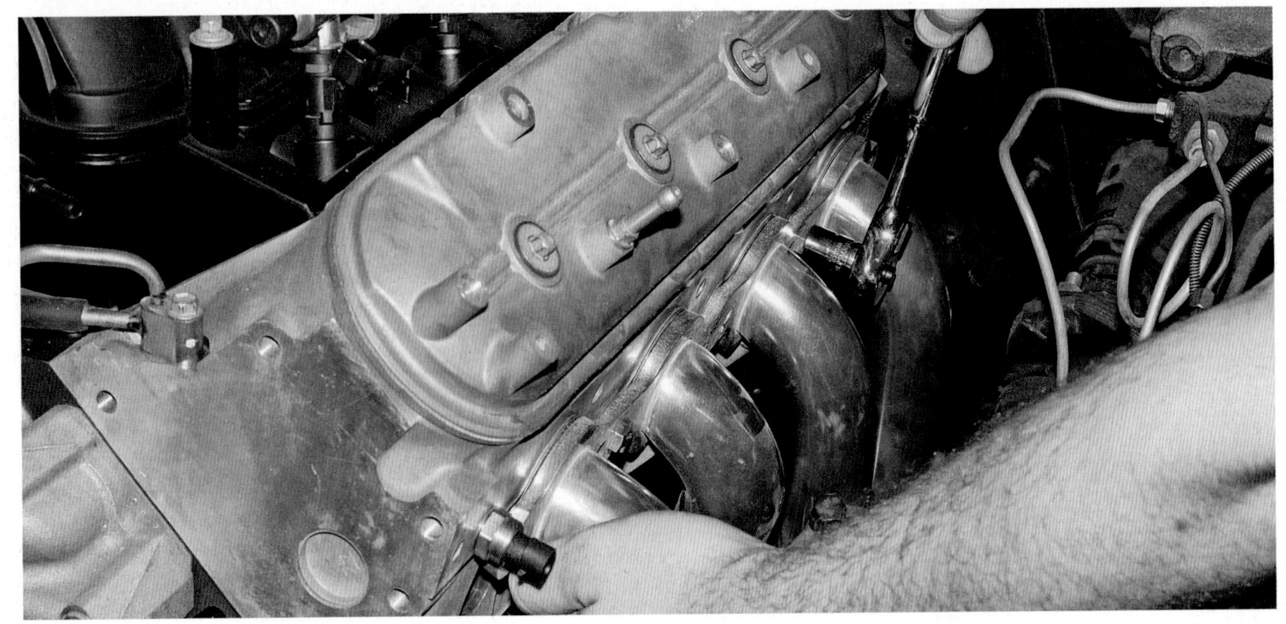

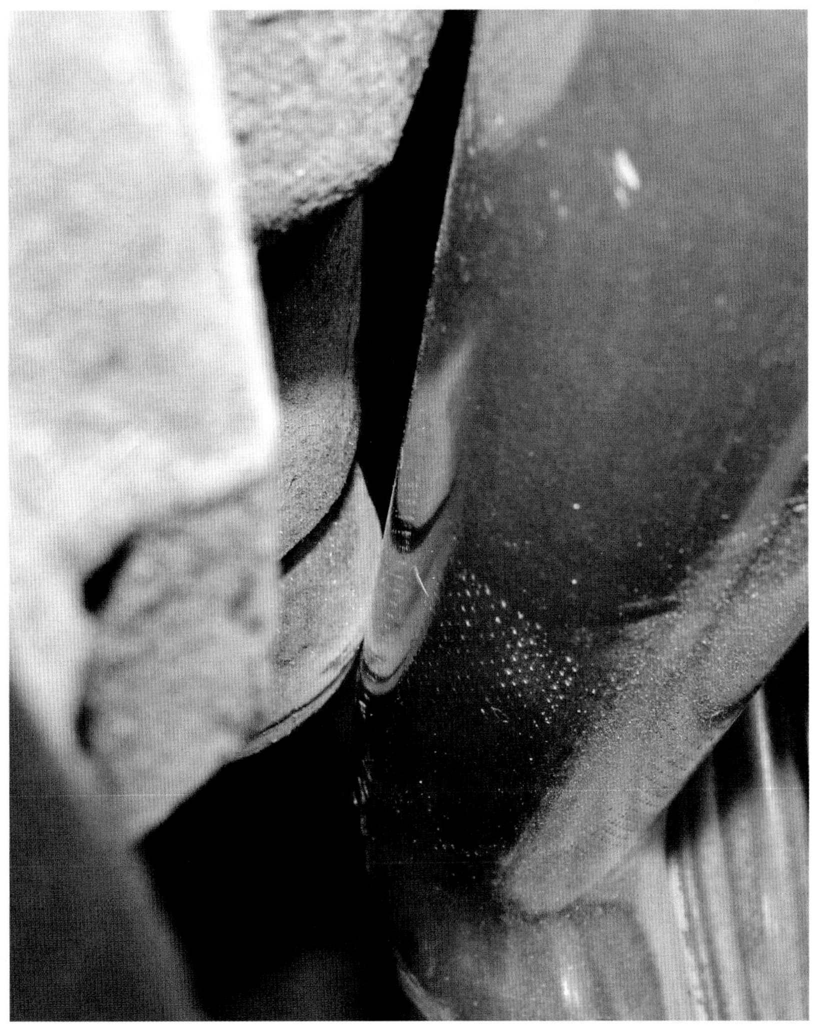

The Hooker LS conversion headers require the matching motor mount plates to alleviate interference issues from arising. In this first Gen, I used the standard Trans-Dapt mount plates instead of the Hooker mount plates and ended up with a small clearance concern on the stock steering pitman arm. The best answer for this is to use the correct motor mount plates which are a 1-inch setback design—alternatively you can dimple the header tube if desired for an easy fix, but only if the header is close to fitting and just needs a smidge more room. Ideally ½-inch of clearance is recommended, but as little as ⅜-inch is ok.

One manifold that seems to be flexible and fit a few different chassis setups is the 2010–2015 Camaro L99/LS3 manifold and catalytic converter package. On this first-generation Camaro, the driver's side is a little close to the factory power steering box, which will require a small clearance adjustment with a grinder.

As with most swap projects, the most versatile side of the engine for exhaust setups is the nonsteering side. The passenger side manifolds and even most non-swap header kits fit perfectly on the passenger side with no modifications required. Shown here is the passenger side 2010–2015 Camaro manifold on a first-generation Camaro.

Chapter 6
Engine Accessories and Cooling Systems

The engine accessories are the items bolted to the front of the engine and driven by the crankshaft pulley via a belt. These are the alternator, power steering, water pump, and air conditioning. In bare-bones configuration, the minimum required would be the water pump and alternator, if you wish to drive any significant distance. Additionally, you typically cannot reuse any accessories from your previous non-LS engine, which makes the front accessories a conversion-specific item.

In this section we will discuss the various stock donor options, basic aftermarket kits, and aftermarket show kit options for the engine-accessory drive system. We will discuss not only what is available for purchase but also how to install and connect these items to your swap project vehicle.

Further complementing the engine accessory drive system, we discuss cooling system designs, quirks, and requirements specific to LS-series engines and what route may best suit your needs.

The accessory drive and cooling system are key components of your swap. Both work together to provide proper cooling, battery charging, power steering, and air conditioning options. Shown here is my 1991 Camaro 1LE accessory drive from Dirty Dingo, which uses truck/5th-Gen Camaro accessory spacing and reused components from a donor engine.

Older GM cars with the three-bolt SBC-style mounting adapters typically cannot use a low-mount A/C compressor. As you can see, the compressor and bracket interfere directly with each other.

The CTS-V A/C compressor is a little smaller and tucks closer to the engine than the Corvette C6 A/C compressor. This was advantageous in the 2008 Saturn Sky LS3 swap, which has minimal real estate in this location. Still, the frame needed some minor modifications for the A/C lines.

FACTORY LS ENGINE ACCESSORIES

The various chassis configurations in which you find the LS engines means a different package for each stock vehicle chassis. Corvette C5 and C6 Y-body 5.7-liter engines receive a different setup than the 4th-Gen Camaro/Firebird F-body 5.7-liter engines; 4.8-, 5.3-, 6.0-, and 6.2-liter GMT800/GMT900 Suburban, Tahoe, Silverado, and Sierra trucks have different setups than other cars. GTOs, G8 GTs, and CTS-Vs have differing accessories than the 2010–2015 Camaros. You get the idea.

The reason for differing setups is space constraints; what fits in one setup may not fit in the next. Another difference is how the crankshaft pulley and damper are offset. Basically, the distance from the engine determines where the belt routing is located. Corvette dampers are the closest to the engine, LS1 F-bodies are middle spacing, and the truck/SUV and 2010 and newer Camaro have the farthest offset from the engine. The truck pulley spacing usage in the 2010–2015 Camaro SS allows the L99 6.2-liter engine to use variable valve timing (VVT), whereas the Camaro ZL1 with the LSA engine utilizes the "Corvette" crankshaft pulley spacing—with an extra belt drive for the supercharger itself.

Where the pulley offset will bite you is in its inability to interchange. You cannot use a truck pulley and damper with Corvette bracket accessories; likewise, you cannot use Corvette dampers with F-body or GTO accessories. The accessory-bracket offset needs to match the damper offset if you are planning on using a stock system. You can mix and match certain accessories themselves with bracket solutions from ICT Billet or Dirty Dingo Motorsports, depending on which accessories you have.

You will need the damper offset to match the water pump and pulley offset, or use ICT Billet water-pump spacers (such as when using the C6 LS3 water pump on 4th-Gen Camaro accessories) to bring the pulley into alignment. Earlier car water pumps (pre-2005) were not much different from each other; whereas late-model water pumps are much more specific to the vehicle. The lower-profile Corvette C6 water pump, for instance, is specific to that model and cannot be used with trucks or F-body systems out of the box. The 2010–2015 LS3/L99 Camaro water pump is closer to the truck water pump dimensions than the previous fourth-generation LS1 F-body water pump, but are a much better choice for the lower profile car intake than the truck water pump.

FACTORY ACCESSORY DRIVE SYSTEMS

The original LS1 F-body accessory drive is a compact simple setup that is quite common. If you have an F-body take-out engine with these accessories, you can use the alternator by itself with a short belt or with the power steering. The F-body accessories require a Gen III F-body or C5 water pump and the F-body spacing crankshaft damper. *Courtesy of General Motors*

C5/C6 accessories are similar with the water pump as the main difference. LS2+ C6s use a single-belt-width alternator to be more low-profile, while Gen III C5s share the same water pump as the Gen III F-bodies. C5/C6 accessories are closer to the engine than the F-body and truck setups and use the shallowest crankshaft damper. The accessories themselves fit a variety of applications and are identical to the GM front engine accessory drive (FEAD) system. *Courtesy of General Motors*

Trucks and SUVs receive a narrower but taller setup due to the generous underhood truck engine bay dimensions. The truck water pump is specific to trucks, and although there are a couple of versions, they all use the same truck pulley spacing. Truck crankshaft damper pulley location is the farthest from the engine. The truck setups other than midsize SUVs (Trailblazer size shown in image) use the older style power-steering pump with the metal housing, which coincidentally works well in 1988–1998 GM C/K Trucks/SUVs with stock power steering lines. *Courtesy of General Motors*

GTO Accessories are similar to the F-body in layout, but there are minor differences, such as alternator size, belt routing, and power steering reservoir design (remote tank). The GTO accessories also use Gen III F-body water pump and crankshaft damper spacing. The 2004–2006 CTS-V is also of a similar but slightly different spacing design. *Courtesy of General Motors*

One mixed combination that can be helpful is to use the LS3 Camaro (2010–2015) water pump with a truck damper. The belt paths are the same. The advantage here is that the Camaro water pump has the upper outlet exiting from the front of the pump on the driver's side as opposed to the truck outlet coming out the top of the pump on the passenger's side. This can simplify upper hose routing because many radiators have the upper outlet on the driver's side.

The reason earlier car water pumps are more flexible is the large pulley width; it is the width of two belts, so it can

The CTS-V and ZL1 LSA Supercharged 6.2-liter engine is one of the newer engine designations, but that doesn't keep it out of the hands of engine-swap enthusiasts. The accessory drive is available without A/C provisions in the kit for some cost savings. The main accessories (sans blower) will work on any engine as well and is similar to the 2004–2006 CTS-V accessory kit. This setup uses a C6 water pump and damper belt spacing. These kits are hard to find and may be best to source with a takeout engine donor. *Courtesy of General Motors*

As one of the most notorious engine designations around, the Supercharged LS9 does not disappoint diehard enthusiasts with a rated 638 horsepower. If you have the LS9 crate engine or the LS9 supercharger kit itself (fits all 6.2-liter heads), you will need this entire accessory setup to drive the 11-rib blower setup water pump and crankshaft damper, which is also required due to the wide blower belt setup. *Courtesy of General Motors*

After an eight-year hiatus, the fifth-generation F-body made a comeback with a retro look that borrows much from the first-generation F-body. The LS3 and L99 accessory drive is spaced forward the same distance as the truck and SUV setups, due to allowing clearance for the L99 VVT set and timing cover. This makes this water pump and crankshaft damper more similar to the truck designs than other previously available passenger car systems. *Courtesy of General Motors*

work with both car accessory setups. This is true as long as the belt tensioner matches the remaining accessory application. The truck water pump is the exception, as its pulley is the farthest from the engine and can be spotted by the threaded fan-clutch drive in the middle of the pulley.

One of the major clearance concerns you may run into with the low-mount LS accessories is frame clearance for the A/C compressor. In factory configuration the A/C compressor sits very low on the engine under the heater hoses on the passenger side. Issues arise from inadequate motor-mount adapter plate clearance and from the frame rail itself. Some are better than others in that trucks receiving the swap often have more room than older cars in this location. Notching and reinforcing the frame rail is an option if you decide to use the GM compressor location in vehicles that do not have clearance; otherwise, look into an aftermarket top-mount A/C setup if you require it.

Note that you will likely need to drill the additional alternator hole if you have an iron-block engine swap. Many of the low-mount style kits require that this extra bolt hole be added for additional bracket-support mounting. You can use your bracket as a template, but ICT Billet has an easy-to-use template to mark your block. Follow along with the pictures of drilling and tapping your block that are found elsewhere in this chapter.

If you have no A/C plans, then you do have it much easier, as the fewer options you are installing, the simpler the swap becomes. In fact, if you do not require power steering either, you can just use the LS1 F-body alternator bracket (or one of the many aftermarket alternator brackets) with a short belt to drive the water pump and alternator and call it a day. Lucky you.

The BRP Hot Rods LS motor-mount system (both engine-side and frame-side mounts) is designed around the stock LS accessory setups. With some cars you will have to cut and notch the frame with a weld-in kit to gain A/C compressor clearance, while the remainder of the accessories (PS/ALT) usually have little issues. If you use the larger GTO alternator and you have clearance problems, you can simply use the smaller LS1 F-body alternator to gain additional room.

When installing the low-mount F-body alternator onto an iron-block engine assembly, there is only one bolt out of the front three that is pre-drilled and tapped. The short middle bolt near the idler pulley is not retained, but the upper long bolt hole casting boss can be drilled and tapped so that two of the three frontal alternator bolts are used. Gen IV F-body alternators and the like also have a rear support bracket that adds further structure, so if you are worried that there is one less frontal bolt, keeping the extra rear support will help out. The alternator requires two bolts minimum to stay put and withstand the strain of the drive belt, and the rear brace adds extra support. Although the middle short front bolt will be missing, the alternator and bracket are quite sturdy without that bolt.

GM PERFORMANCE PARTS FRONT ENGINE ACCESSORY DRIVE

General Motors Performance Parts (GMPP) now called Chevrolet Performance, has released a few different accessory kits based on production vehicle accessories. There are C6 and LS6/LS2 CTS-V accessory drive kits currently available. More are on the way for the LSA and LS9 crate engines, specifically. These kits use the factory production replacement pieces (A/C compressor, PS pump, alternator) and come with all accessory pulleys, bolts, and hardware for installation to the engine itself. An instruction manual is also included, detailing the installation procedures, bolt locations, and torque specifications.

These kits are around $1,000 and can greatly simplify your engine installation if these are feasible for use in your project. You will need extras to finish the installation specific to your vehicle, such as power steering fittings, hoses, A/C hoses and components, and the alternator wiring harness and battery charge cable. The kits do not come with the water pump or crankshaft pulley or damper but do specify which GM part number damper is required. You can use a new GM pulley, used pulley, or one of the many aftermarket balancer kits. The CTS-V kit and the C6 kit use the shallowest crankshaft damper, which is the Corvette C5/C6–style dimension.

Note that, because these cars have been out of production for many years, the availability of OEM components has dwindled and many items may be discontinued. Alternatively, these items are available from online sources, or you can piece together a kit using a mix of OEM and aftermarket components as needed.

GMPP Corvette C6 Accessory Kit (LS2/LS3/LS7)

The C6 kit is similar to the C5 accessory drive in design, putting all of the driver's side accessories on the same bracket, which is then bolted to the cylinder head only; no block bolts are in place. The C6 kit puts the alternator in a top-mounting position far away from potential interference with the frame and the power steering in a lower mounting position at the bottom of the alternator/PS bracket. This would give extra clearance near the lower side of the block on the driver's side of the engine, as the C6 Corvette required clearance in this area due to the rack-and-pinion steering location. This is also the reason the accessories are shifted as close to the engine as possible on C5/C6 Corvettes: for steering rack clearance.

The A/C bracket bolts to the passenger side of the block, but like all production LS-based low-mount A/C locations, it may have frame or chassis interference, depending on the vehicle application. The A/C is on a completely separate system and belt, so its use is not necessary for the remaining accessories of the kit. If A/C is not required; just leave it off.

Since the Corvette kit shifts everything so close to the engine, the reuse of VVT sets and timing cover is not possible. There just is no physical room available with this low-profile system as equipped. If VVT is a requirement for a passenger car, look into bolting on a 2010 or newer Camaro setup. Another option is to buy or make accessory-drive spacers that shift everything farther out for additional clearance. Mast Motorsports and ICT Billet have built similar setups and spacer kits for VVT engines.

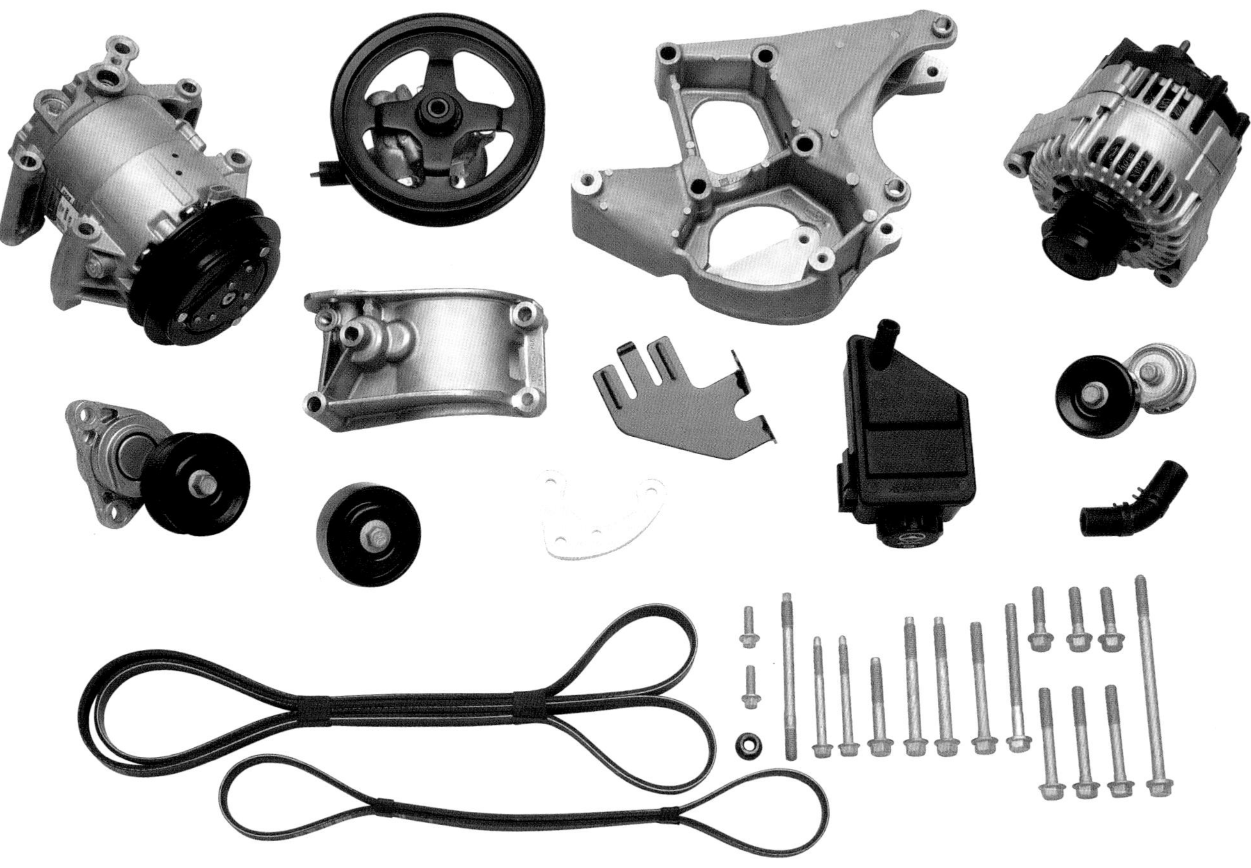

Another setup that GMPP has made available is the C5- and C6-style FEAD setup, part 19155066. This is identical to the production pieces and requires the C5 or C6 water pump and crankshaft damper to finish out. Unlike the previous GMPP FEAD kit, this kit does not bolt the alternator or the power steering to the block. Rather, the alt/PS bracket bolts to the driver's cylinder head. The illustration below provides basic component location for the belt drive. *Courtesy of General Motors*

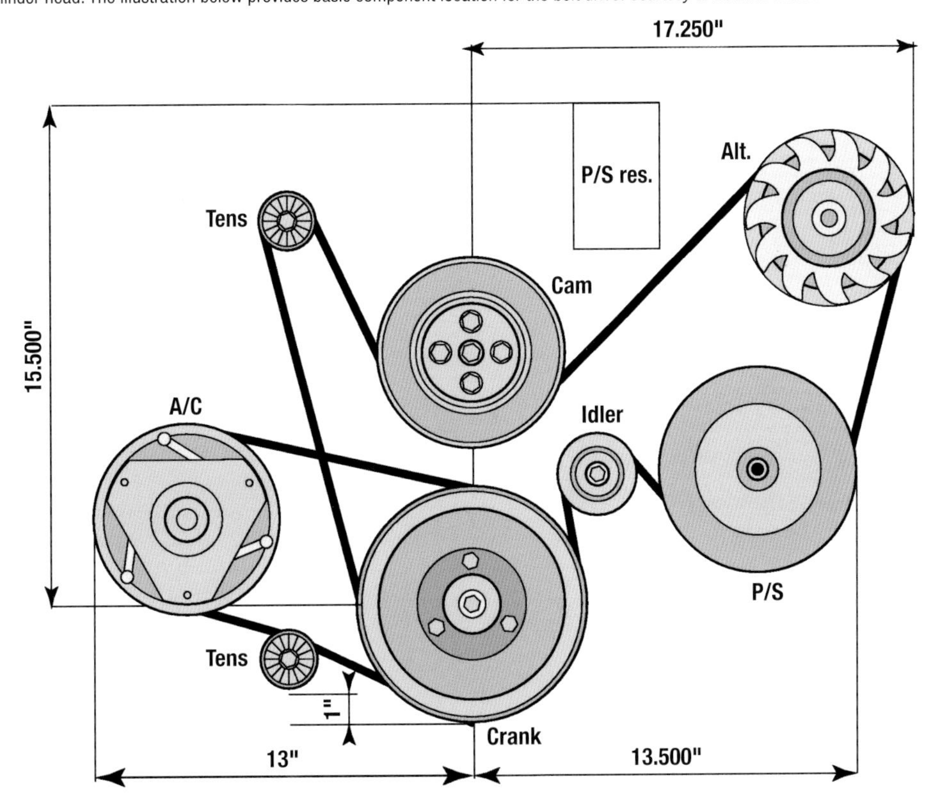

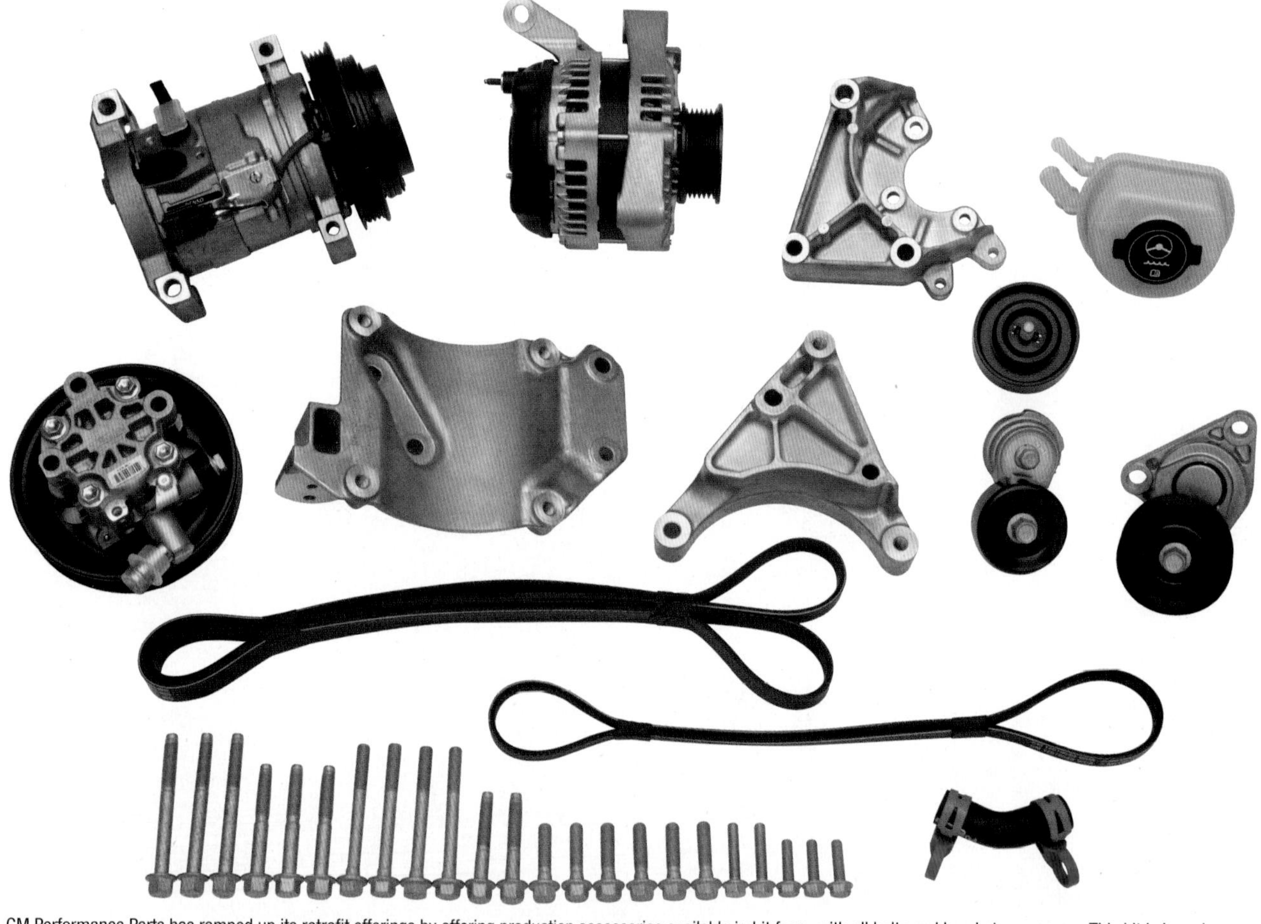

GM Performance Parts has ramped up its retrofit offerings by offering production accessories available in kit form, with all bolts and brackets necessary. This kit is based on the 2004–2006 CTS-V LS6 and LS2 accessories and is GM part 19155067, which is available from any dealership. Things you will need to finish the GM FEAD setup are water pump, crankshaft damper, hoses, and wiring. This kit will not fit iron blocks without first drilling and tapping the block. Specing info for this kit below. *Courtesy of General Motors*

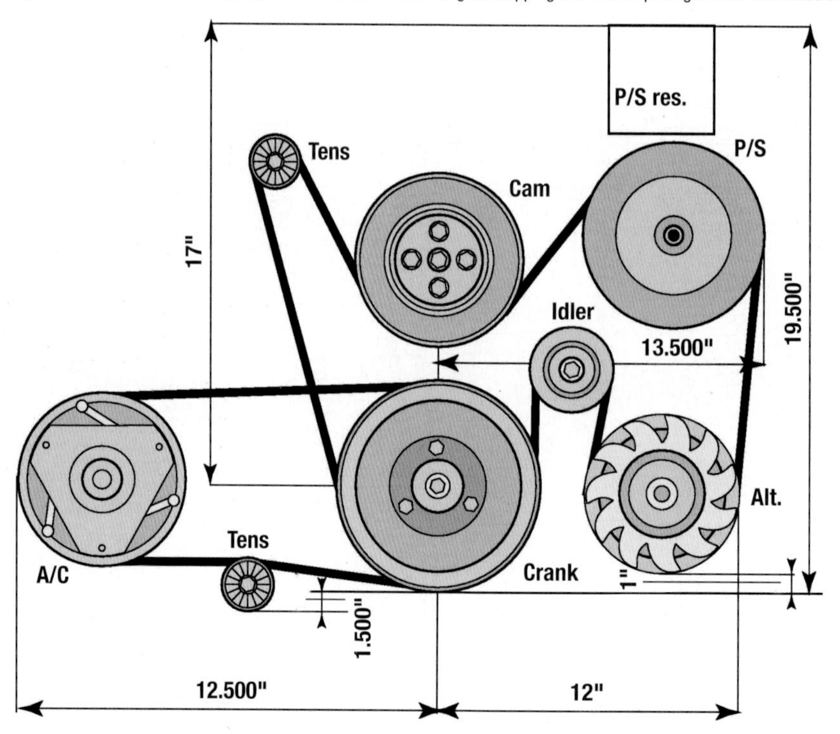

GMPP CTS-V Accessory Kit (LS6/LS2)

Using a setup design similar to the F-body, GTO, and Pontiac G8, the CTS-V kit puts the power-steering pump on top near the front of the driver's side cylinder head, and the alternator is mounted low. This kit is meant for aluminum blocks, but with the same modification described earlier performed to iron blocks (for the F-body alternator installation), it can be functionally adapted to any LS-based engine. This kit is based on the factory accessories for a 2006 Cadillac CTS-V with the LS2 engine and requires the same crank pulley.

A/C is also an issue with frame interference and motor mount clearance. If these aren't an issue dimension-wise for your swap vehicle, they can be adapted easily with fabricated hoses and lines adapted to your existing A/C system. Notching the frame rail near the passenger side of the engine is always an option to allow extra A/C compressor clearance.

The 6.2-liter LSA from the Gen II CTS-V and 5th-Gen Camaro ZL1 uses a similar front drive. The obvious addition to this is the LSA supercharger, which requires a few add-ons for the supercharger-specific drive belt. If you have one of these engines, lucky you, but also check out Holley (P/N 20-163BK) for any additional brackets needed that may not have come with your engine.

Note that several components of these OEM derived kits have since been discontinued by the manufacturer, so alternative items may be required if you are able to piece together this setup. In the Saturn Sky LS3 swap using the CTS-V accessory drive, I had a hard time finding a power steering reservoir, for instance. These items are usually available used from eBay or online communities.

Other GMPP Accessory Kits (LSA/LS9)

GM Performance Parts is releasing the accessory components required for both the supercharged 6.2-liter engines for Cadillac CTS-V LSA and Corvette ZR1 LS9 crate engines. The LSA kit can be had without the A/C system to save a few bucks, so this could be an option instead of the C6 or LS2 CTS-V kits, which come with everything.

If you are using the blower from the LSA on a 6.2-liter engine, you will likely require the LSA kit along with the CTS-V crankshaft pulley and water pump, as the blower is driven by a dedicated blower belt. The same external parts would be required for the ZR1 crate engine but for another reason: the ZR1 LS9 uses an 11-rib belt for all accessories, including the blower, which requires all accessories to match the 11-rib setup.

DIRTY DINGO MOTORSPORTS

A familiar name in the LS swap game, Dirty Dingo has been supporting LS swaps since 2005. Along with its innovative slider engine mounts mentioned earlier, the company offers a full lineup of accessory drive brackets and kits to support just about every swap you can think of. The engine mounts are made of ¼-inch high-carbon plate steel, and the accessory

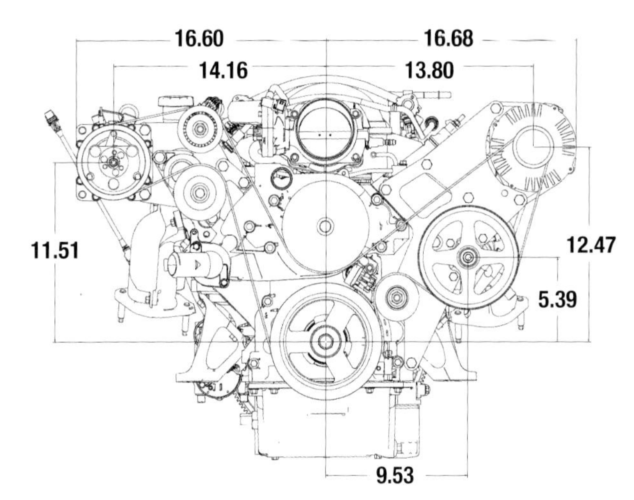

KWiK Performance builds various front-accessory aluminum brackets for painless mounting solutions by using water jet plates and aluminum spacers and F-body–style power steering and alternators. Additionally, the top-mount A/C solution saves a lot of headaches by conveniently mounting the compatible A/C compressor in a known working location with minimal interference. *KWiK Performance*

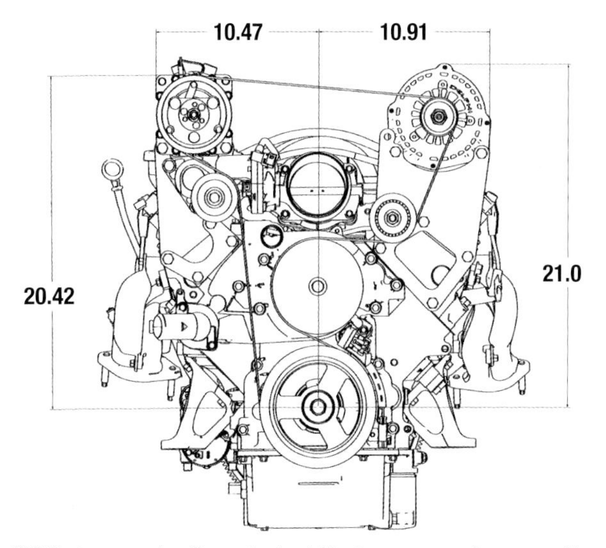

KWiK Performance also offers a street-rod–friendly narrow-mount accessory-drive setup. These mount the accessories high up in the engine bay so that narrow engine bays do not interfere. *KWiK Performance*

brackets are made of ⅜-inch-thick 6061-T6 aircraft billet aluminum and CNC lathe-cut spacers and come with grade-8 mounting hardware.

Dirty Dingo Motorsports has a variety of accessory drive solutions and options. If you need an alternator bracket, you can get just an alternator bracket. If you need A/C only, you can buy the A/C-only kit by itself. If you have absolutely nothing for your complete accessory drive, they can get every component you need to finish out your accessory drive plans. Dirty Dingo accessory brackets are user configurable to whatever crankshaft pulley kit you have, offering components for all three GM pulley-spacing designations.

(continued on page 142)

GM C6 ACCESSORY KIT INSTALLATION

Installation of the GM front engine accessory drive is straightforward and uses common metric hand tools such as 13-millimeter and 15-millimeter sockets and wrenches. The first part of the installation requires putting the required water pump and crankshaft damper in place.

The first official installation component is the main accessory-drive bracket. Line up the bolt holes with the matching cylinder head and start each bolt by hand a few threads before tightening. Don't forget the power-steering bracket, which is installed under the two uppermost bracket bolts.

Installing the bracket is easy if it is completely bare, as is this one. Torque the (four) 15-millimeter-head bolts to 44 foot-pounds, and that is all that is needed at this point.

The first accessory to install is the power steering. These (three) components—reservoir, supply hose, and PS pump assembly—are not assembled as the pictures show. I pre-assembled them beforehand, but they can be installed individually as well. There are (four) power-steering pump bolts and one U-shaped bracket that hold the pump in place. The bolts must all be started in place before torqueing.

Install the reservoir and supply hose if not already done. The reservoir aligns and clips into place downward over a tab. The supply hose must be installed at the same time; the spring-tensioned clamps can be relocated last.

Torque the (four) 13-millimeter-head bolts to 18 foot-pounds by turning the power steering pulley to gain access to the different bolt locations.

Some chassis setups may have some clearance issues no matter what parts you use. In this instance the stamped upper control arm was close to the accessory belt, so some trimming and grinding will be necessary to prevent potential belt wear. A better idea is to purchase tubular upper control arms so that no grinding is necessary.

Next, install the alternator using the (two) supplied alternator bolts. Torque the (two) bolts to 44 foot-pounds. If you have the wiring in place, you can connect the alternator charge wire and the alternator push-in connector.

Finally, install the main belt tensioner and idler pulley, then route the belt and install. The belt tensioner has two indicator marks on the housing, and one indicator that shows if belt-length or tension is correct. The moving indicator should be midpoint between the two range marks. If using a non-C6 water pump, you will need a shorter belt, as the included GM belt is intended for the C6 LS2–style water pump.

GM CTS-V ACCESSORY DRIVE INSTALL

The GM Performance CTS-V accessory drive installation is simple and straightforward and comes with step-by-step instructions. The installation can be used with Corvette LS3 or CTS-V LSA belt spacing and is best used with aluminum blocks for the supplemental low-mount alternator bracket bolts that mount into the engine block. The power steering bracket (with reservoir) mounts to the cylinder head, then the power steering pump is bolted on last.

HOLLEY MID-MOUNT ACCESSORY DRIVE INSTALL

The Holley Mid-Mount kit (P/N 20-185-BK) is an innovative design that works with all engine designations by mounting all accessory components to the water-pump housing itself. After removing the old accessories and factory crankshaft pulley, the new Holley-specific pulley can be installed using an installation tool and a new crankshaft bolt, followed by the new water-pump housing/manifold, to which everything attaches.

Once the water manifold is in place, the water pump insert from the C7 Corvette is installed, followed by the new alternator and power steering pump with the supplied banjo to -6AN steel tube assembly. The power steering pulley must be pressed on using a specific power steering installer (usually found for rent at local auto parts stores).

The last accessory component to install is the top-mount A/C compressor, which easily bolts into place. Finally, the belt tensioner and belt are installed along with any optional dress-up covers. Holley makes this front-drive kit in three finishes: an OEM style cast aluminum, a polished aluminum option, or a matte black, as shown. I chose black to match my black Holley valve covers and to offer an incognito appearance: black is engine bay camouflage.

(continued from page 137)

If you have stock accessories, you simply pick which accessory and belt spacing you have and then choose between a couple of different kit configurations. This is used in the 1991 Camaro 1LE swap example in this book. These brackets are great if you bought a takeout engine assembly and you want to reuse most of the components from your donor engine, but may not have room for that truck accessory-bracket orientation. These kits can reuse F-body, truck, Corvette, and other main components, which saves you a little bit of coin to spend elsewhere in your swap.

For my 1991 Camaro LS swap, I had a GM truck front-accessory bracket from a 5.3-liter truck alternator and power steering pump. The truck accessory bracket put the alternator into the hood, so I had to find another solution. I contacted Dirty Dingo, who told me they had a bracket kit that reused the truck accessories I already owned (P/N DD-LS-ALT-PS).

Mast Motorsports has built a few custom setups with a top-mount A/C compressor, which is also compatible with the Magnacharger blower setup. The Magnacharger kit is not made for a top-mount A/C system, but with some creative brackets and idlers Mast made it work. *Mast Motorsports*

This bracket kit uses truck pulley spacing (for the power steering pump metal reservoir clearance), so I also needed a water pump to match. Some applications can use the truck water pump but, because this is a car application, I procured a 2010–2015 Camaro water pump to have the correct "truck" belt alignment and still be a "car"-style front drive. This Dirty Dingo bracket kit is made for both small- and large-case alternators and it has dual bolt patterns for the different size alternator case bolt pattern. The large-case alternator uses a longer accessory belt as well. Note that, if you have complete 4th-Gen, Corvette, GTO/G8, or CTS-V accessories, these are a different part number and pulley spacing. If you have A/C, you can also pair the Dirty Dingo top-mount A/C bracket kit with either a GM R4 compressor or an aftermarket Sanden compressor.

It's not a huge win, but one of the perks of building later-model 1980s and 1990s GM swap projects (in lieu of 1960s and 1970s projects) when using the truck-style accessories is the ability to reuse certain items. For instance, because these vehicles come with the GM R4 A/C compressor and the Type-1 power steering pump with metal reservoir, you can use factory or OEM replacement A/C and power steering hoses with minor tweaks to your swap. This idea would work on vehicles such as 3rd-Gen Camaros, 1988–1999 GMC/Chevy "OBS" trucks, and other vehicles such as B-bodies (Caprice, Impala SS) that use similar components. There are even thread adapters available from fitting companies to adapt lines to other applications. If all else fails, you can always use -6AN to metric O-ring adapters as well.

ICT BILLET

ICT Billet has been adding a slew of new products, specifically for LS swaps, for many years. It has several options for front accessory bracket solutions and some good guidelines and complete parts lists for well-established swaps on its website. These components often fill a void in the market by having a solution to a common swap task, whether it's adapting fuel lines, power steering hoses, or, as in this chapter, front accessory drive brackets. I counted over one hundred different accessory drive configuration parts numbers! In this manual, we will cover some of the more unique and basic accessory drive options, referring to ICTbillet.com for specific accessory drives and belt sizing needs.

All of the ICT kits are built from aerospace-grade U.S. billet aluminum and, like many popular bracket kits, complement parts you may already have. If you have a donor engine from a Pontiac G8 GT, for instance, ICT has a matching bracket that reuses the G8 alternator and power steering pump. Likewise, if you have a Corvette or Camaro takeout engine, ICT has brackets for these as well. Reusing your existing major parts saves you quite a bit of money to spend on other things for your project. If you have a hodgepodge of parts, it is very likely that ICT Billet has a solution to make these match your crank-pulley spacing.

ICT Billet has three basic spacing options that follow the factory options for pulley and water-pump spacing. Their parts lists denote this by adding a -1, -2, or -3 to the end of a 6-digit accessory-bracket component part number. For example, the general spacing of a kit can have the same design dimensions, but the -X added to the end pairs it with the correct crank pulley and water-pump spacing that you have.

- 1 is the Corvette, CTS-V, G8, and LS3 crate crank-pulley spacing
- 2 is the 4th-Gen F-body Camaro/Firebird and Pontiac GTO crank-pulley spacing
- 3 is the 5th-Gen Camaro and GM Truck/SUV crank-pulley spacing

While you would have to physically change the crankshaft pulley to change the spacing, with the water pump choices you can space these further out to align with the next further out accessory drives by using ICT Billet water-pump spacers. This allows more universal fit for certain water pumps and applications.

For example, if you want to make a (-1) Corvette water pump fit the (-2) 4th-Gen F-body spacing, you add a ⅞-inch (P/N 551690-86) water-pump spacer kit. The .700-inch (P/N 551690-71) spacer kit would make a 4th-Gen water pump compatible with a (-3) 5th-Gen or truck crank pulley. You can also add a 1.5-inch (P/N 551697) spacer kit to the Corvette water pump and make it compatible with the truck/SUV spacing. Note that you cannot go the other way, making a (-3) 5th-Gen Camaro or truck/SUV water pump compatible with a (-1) Corvette or (-2) 4th-Gen F-body accessory spacing; that is impossible.

KWIK PERFORMANCE

Sometimes you may have one or two accessories gathered up but need a customized approach to mount what you have. KWiK Performance offers a few different and economical options for retrofit applications, including a top-mount A/C compressor kit.

KWiK Performance has options for mounting the alternator, power steering, and A/C separately, but also tandem in a complete bracket kit. You can option your kit for A/C and alternator, but no power steering, or you can have the alternator and power steering brackets first, then add A/C later.

The KWiK brackets are not cast aluminum like the OEM components. They are water jet cut from 6061-T6 aluminum for a clean appearance while offering a lightweight and strong setup using solid and proven spacer tubes for mounting. The KWiK setups can be obtained for most common applications of water pumps and crank pulley offsets such as F-body/GTO, C5/C6, and the truck applications, saving some costs by working around such items you may already own.

The KWiK setup can be as simple as adding an A/C compressor bracket and leaving the rest of the OEM brackets and accessories. KWiK also offers more complete system packages such as their "truck pull-out package." This system is for the builder who buys a salvage-yard truck long block, needs a new water pump and damper, and also wants the shortest LS configuration. This package includes a Corvette damper and water pump, high-output alternator, power steering pump and reservoir, A/C compressor, and, of course, the accessory brackets.

For the street-rod crowd, KWiK Performance has a high-mount accessory kit setup that offers a narrow package for use with early street rods such as '32 Fords with the side hoods in place. This accessory kit is narrow, and from the edge of the top-mount alternator to the edge of the top-mount A/C compressor is less than 22 inches. This kit can use any LS-based alternator, including the Corvette alternator.

Another narrow bracket is what KWiK calls its "high-and-tight" compressor bracket. This mounts the 7176 mini compressor in front of the cylinder head and requires the use of a truck or an LS3 Camaro damper. This has proven popular with builders of 1947–1954 Chevy trucks that have narrow inner fenders. This bracket makes a very compact system when teamed with the OEM alternator/power steering bracket.

KWiK Alternators

KWiK alternator brackets have a dual bolt pattern that accepts both the Corvette alternator with its unique bolt pattern and the truck and LS1 and LS3 alternators that share the more common mounting bolt spacing. Alternators from LSA supercharged engines (Cadillac CTS-V and Camaro ZL1) have a different pattern, with a third mounting boss; these do not fit the KWiK brackets.

KWiK Power Steering Pumps and Hoses

KWiK's most popular power steering pump configuration is a Type II pump combined with a factory Corvette remote reservoir. This provides additional clearance around steering boxes and upper control arms. The pump is similar to an LS1 Camaro pump in flow and pressure relief settings. It comes with a custom 90-degree hose barb for easy connection to the remote reservoir. It also has a 5.5-inch OD 6-rib pulley already installed.

The KWiK bracket will accept other Type II pumps such as the factory LS Corvette pump, but the integral hose barb points in an awkward position for connecting to the remote reservoir, and the 6.6-inch diameter pulley may cause clearance issues in tight engine bays.

An LS1 Camaro pump can be used by removing the attached plastic reservoir and pressing in a custom 90-degree hose barb offered by KWiK. The Camaro 6.0-inch diameter pulley will need to be pulled slightly forward for proper alignment or KWiK can supply their 5.5-inch pulley with the correct offset.

One issue builders of classic muscle cars face is how to connect a newer Type II pump with a metric O-ring pressure

The KWiK brackets make use of the commonplace Sanden-style universal compressor with multiple mounting tabs. This compressor is not a variable displacement compressor like the newer compressors, so it easily works with conventional air conditioning systems. *John Loftin*

port to a steering gearbox with SAE inverted flare fittings. The KWiK answer is to provide conversion fittings that convert everything to -6AN threads and then assemble hoses with matching AN fittings. KWiK hoses are made from steel-reinforced black rubber for the pressure hose, nylon-reinforced black rubber for the return hose, and a ⅝-inch oil-resistant feed hose from the reservoir back to the pump.

These hose kits are preassembled and ready to install in most front- and rear-steer GM cars and trucks. They come with a combination of conversion fittings to take care of either pre-1980 gearboxes with SAE inverted flare fittings or the 1980+ with metric O-ring ports.

KWiK A/C Compressors

Since the first edition of this book, KWiK has expanded its A/C bracket offerings. The most popular is what they call a "wide-mount" bracket, which mounts the larger Sanden-style compressor, often referred to as the 508 style. Part numbers are available to fit all three LS belt paths.

KWiK also offer a "high-and-tight" bracket that mounts the smaller Sanden SD7B10 (7176) compressor directly in front of the cylinder head. This works well in narrow engine bays, but the bracket only works with truck or LS3 Camaro dampers with the most forward belt path.

AFTERMARKET A/C COMPRESSORS

The most popular aftermarket AC compressors are made by the Sanden Corporation or one of many copycat suppliers. Early Sanden nomenclature for the most common compressor was "508," which originally stood for five cylinders and eight cubic inches of displacement. Later, Sanden changed the 508 model to SD5H14. The "5" still refers to the number of cylinders and the "14" is displacement in cubic centimeters divided by 10. There are also four-digit numbers that give clues about which pulley is included and the hose fitting configuration; the SD5H14 with a seven-groove pulley and hose connections pointing up and out of the back plate is a 4514.

Vintage Air discovered that, in rare cases, the five-cylinder compressor could develop a slight harmonic vibration at certain rpm, but the seven-cylinder version eliminated this. Consequently, Vintage Air's recommendation is to use the seven-cylinder version with all LS engines. This compressor was originally called the "709" (currently the SD7H15), with a four-digit designation of 4665. Because this compressor only costs about $10 more than the five-cylinder version, KWiK has made it their standard offering.

The "mini" compressor also has seven cylinders and is labeled the SD7B10, with a four-digit designation of 7176. It's common to refer to the mini compressor as a "7176" to help avoid confusion with the larger seven-cylinder SD7H15.

The Sanden-style compressor features a seven-rib pulley; where, the LS belt drive uses six ribs. Be sure to align the belt to the proper grooves in the A/C compressor pulley so that the belt does not wear prematurely. *Tom Munsch*

For some applications, such as supercharged engines that have brackets, tensioners, and pulleys using the real estate on the passenger side of the engine, KWiK offers a "low-mount" bracket that places the smaller Sanden compressor back in the OEM location low on the passenger side of the engine, where it's driven by a dedicated four-rib belt.

Since this "mini" compressor is about 1.5 inches shorter than an OEM compressor, it may clear engine mounts or crossmembers that would otherwise be a problem in this low location.

HOLLEY PERFORMANCE ACCESSORY DRIVE

Holley Performance has been around the performance game for a long time, so they know how to meet the needs of their customer base. When I wrote the first edition of this book, there was little support from the major aftermarket companies for LS swaps, and that left a hole for many smaller, niche companies that wanted to expand into this area. Fast-forward to today and all the major companies support LS swap enthusiasts. This includes the Holley Group, which has moved ahead with offerings like a few variations of core accessory-drive configurations that cover just about any swap project you could imagine.

There are three main categories offered by Holley. One is the Low-Mount, whose design resembles an aftermarket version of a 4th-Gen F-body accessory mount for location and spacing. The High-Mount is the next, mounting the alternator and A/C higher for frame or suspension clearance. The third option is a new idea: the Holley Mid-Mount accessory drive. This kit uses no brackets, instead employing a unique water-pump housing to mount the alternator, power steering, and A/C compressor (it can also be used without power steering and/or A/C). The main version we cover here is the Mid-Mount kit after introducing the Low-Mount and High-Mount kits and features.

Each kit is available in three different cosmetic treatments. Most of the brackets are cast aluminum and machined, much like the OEM parts. Accessories are available finished in natural, black, or polished to best match your underhood powerplant theme. Additionally, Holley can supply the alternator, power steering, or A/C compressor units as well as any needed brackets, hardware, or adapters separately.

As with most companies, current information, configuration options, and the latest technical knowledge for this evolving market are available in Holley's online catalog.

If you are using a GM C5/C6 Corvette accessory drive and a late-model power steering box with metric fittings, you may be surprised to find out a factory replacement 1997–2004 Corvette power steering line fits with little effort. The opposite side return line is not under pressure so the use of ⅜-inch oil return line (such as rubber transmission cooler line) is as easy as cutting to length and clamping into place.

Holley Low-Mount

The Holley Low-Mount (P/N 20-162) is made for the 4th-Gen Camaro/Firebird crankshaft-pulley and water-pump spacing as delivered. To use it with the truck and 5th-Gen Camaro crank pulley/water-pump spacing, you add a spacer and a longer bolt kit. To use the Low-Mount with Corvette/G8 spacing, you must replace your crankshaft pulley with one of the compatible choices, then replace or space your water pump to match.

Alternately, there is a Holley Low-Mount LSA 6.2-liter (P/N 20-163) complete kit made for the supercharged LSA crate engine that may fit your application. The LSA kit uses the Corvette/G8 crank spacing and includes the supercharger tensioner, all S/C brackets, and idler pulleys; these add a bit more to the basic price of the regular Holley Low-Mount. (Note that you need an aluminum block to mount the three alternator bolts with the LSA kit.) While it may not be cost effective to buy the entire kit for a naturally aspirated build, this is definitely a nice option. You can also source the brackets individually to use the Holley Low-Mount kit with Corvette/G8 spacing, which is a common spacing for 6.2-liter GM crate engines, supercharger or not.

Ideal swaps for the Low-Mount are street rods and hot rods with limited chassis room or frontal height. This is a tucked-in, narrow design allowing extra clearance, especially around the lower areas where the A/C compressor and power steering pump/reservoir are located. The alternator is in the same location as a 4th-Gen setup. Obviously, there are many swaps where the Low-Mount A/C compressor would not physically fit, but for these you can add a top-mount A/C bracket. Check the drawings and measurements from Holley.com against your car's dimensions and review before you buy the Low-Mount A/C-equipped front-drive kit.

The Low-Mount components can be purchased as brackets only—intended for you to supply your own accessory components—or as a complete kit with all the accessories, bolts, and pulleys required for installation. You'll need the latter if you have no replacement parts and are starting from scratch. This option is nice, as many 4th-Gen donor accessory drives may be harder to find, and having a similar option from the aftermarket such as Holley is a great solution.

Holley High-Mount

The High-Mount is most similar to the C5/C6 Corvette design. The alternator and A/C compressor are mounted higher up in the engine bay, but not above the intake manifold. This nets a wider accessory drive, so check the dimensions and diagrams and make sure the fitment works for your application. The Corvette-style power steering pump is mounted below the alternator, so it's basically a flipped orientation compared to the Low-Mount configuration.

Any block choice would be compatible with the High-Mount kit, as the main accessory bracket bolts directly to the left cylinder head. The top-mount A/C bracket bolts to the right cylinder head, and nothing bolts to the block itself.

The GM truck accessory drive in full-size trucks is the older metal housing design that is a little easier to fit into earlier-model trucks. The hoses are known to directly fit in 1988-1998 C/K truck and SUV applications. *John Loftin*

Shown here are the three different LS crankshaft damper/pulley configurations so that you can see the difference in belt spacing. On the left is the C5/C6 damper/pulley, shown in the center is the F-body/GTO damper/pulley, and lastly the truck/SUV/2010 Camaro-style damper/pulley is shown to the right.

Unlike the Low-Mount, this High-Mount kit can fit all three crankshaft and water-pump spacing options via separately available add-on install kits. The main kit comes with the accessories and brackets, but you need to buy a separate installation hardware/spacer kit to match the crankshaft pulley/water-pump spacing you have. For instance, if you have the main kit (P/N 20-137) and an LS3/L99 from a 5th-Gen Camaro with truck pulley spacing, you need to add the long spacing kit (P/N 21-3). Luckily, even though these add-ons are necessary, they are not unreasonably expensive.

Holley bolt hardware and spacer kits required, in addition to base accessory kit:

- Corvette/G8/Standard Alignment add-on: P/N 21-1
- 4th-Gen/GTO/Middle Spacing add-on: P/N 21-2
- 5th-Gen/Truck/Long Spacing add-on: P/N 21-3

Like the other Holley accessory drives, you can purchase individual components separately to build your own kit. The most likely deletion from the kit would be the A/C portion, though one of its nice features is that Holley offers a GM R4 compressor. This means all those GM 1980s and early-1990s LS swaps that had a GM compressor can now reuse the R4 compressor—and likely the factory-equipped GM A/C manifold hose—for an easy A/C installation (not the norm for LS swaps). This is a huge advantage for keeping A/C intact for your project.

Holley Mid-Mount

One of the more innovative front-drive kits to hit the swap world is the Holley Mid-Mount kit. The kit is complete and does not require a specific crankshaft pulley or water pump because it includes these as the focal point of a complete, standardized system. This setup has no extra brackets; all the mounting points are on the Holley Mid-Mount specific cooling manifold/water-pump housing, which is unique and a huge asset as it fits (almost) everything. There are also the options of an LS wet sump, which a crate LS3 6.2-liter would have, or the LS dry sump like a crate LS7 7.0-liter would require.

There are a few different versions that you can use in your swap, from simply mounting an alternator as a race setup (with A/C delete), to alternator and power steering, to the full kit, which comes with everything including an A/C compressor. Much like the other Holley front-drive parts, you can buy individual components, but due to the specific pulley-spacing requirements, it is probably best to get the kit. If you need to replace something later, view the individual parts as a service replacement.

To tuck in the accessories closer to the engine's centerline, the Mid-Mount uses a specific cast-aluminum water-pump housing (with a C7-derived water pump insert) to mount everything. This brings the Sanden SD7 A/C compressor and Holley 150a alternator directly in front of the cylinder heads instead of off to the side, or on the side of the block. This helps with underhood clearance in most engine bays. The Series II power steering pump is mounted to the left side of the engine under the alternator location and conveniently comes with a -6AN metal adapter to add ease of access and assembly to your power steering gearbox.

An interesting and useful bracket for A/C compressors in 1988–1995 OBS GM trucks is the stock aluminum SBC/V6 air conditioning compressor bracket. There are bolt holes that line up to the stock LS cylinder head from this bracket, and with some external bracing, this makes the stock 1988–1995 A/C bracket and original A/C compressor line up perfectly to the remaining truck accessories. Not even the A/C hose needs to be changed.

The power steering reservoir is the component furthest from the engine centerline, but lying only 12 inches out; by comparison, the exhaust manifolds are wider. If there were an item to interfere with the chassis, this would likely be the area to check, and you can always convert to a remote reservoir to save underhood real estate if needed. Holley offers the complete dimensions and measurements on its website. Note that, because the A/C and alternator components are now in front of the cylinder head, the length of the entire engine assembly may be an issue if you have a very compact engine bay. This isn't a typical problem for most GM swaps, but some of the import car swap projects may need to measure carefully.

The Mid-Mount has a specific crankshaft pulley spacing, which only fits the Mid-Mount kit, and there are two crank pulley options available: a cast-iron pulley like the OEM version, or an upgraded SFI-approved ATI damper. The only other main option to choose is cosmetic: natural aluminum, blacked out, or polished. Follow along with the install sidebar (page 141) as I install the black Holley Mid-Mount (P/N 20-185BK) onto a 1967 Camaro build.

VINTAGE AIR FRONT RUNNER SYSTEM

For those with a lust for functionality and aesthetics, the Vintage Air Front Runner accessory kit delivers. Vintage Air, in addition to producing A/C kits for older vehicles, makes several billet front accessory drive kits in an easy-to-install kit for your LS-series engine swap. They make a kit for every LS engine that fits a plethora of engine bays. The Front Runner kit offers a compact package helping to not crowd the engine bay space. All versions of the kits tuck close to the engine, and at about 21 inches wide by 17 inches tall and only 7 inches from the face of the block to the front of the pulleys, you definitely know that this is a compact system. They use ultra small components such as the 140-amp alternator that tucks in front of the cylinder head fitting the dimensions allowed perfectly.

The Front Runner kit comes complete with the ATI damper, brackets, pulleys, accessory belt, alternator, power steering, water pump, and it wouldn't be from Vintage Air without including the air conditioning. All of these components can be obtained with either a bling high-polished finish down to the last bolt, or a black anodized OEM-style late-model finish for a high-tech look. With several pages of detailed instructions, the kit can be installed with normal hand tools. Correspond this with their well-known under-the-dash heating and A/C unit and you can

(continued on page 152)

VINTAGE AIR INSTALL

Vintage Air is well known for its under-the-dash A/C and heating kits for many older vehicle designs. The company is known as "the Inventors of Performance Air Conditioning." Since many of their HVAC customers build LS engine powerplants, Vintage Air built its front runner accessory drive system to cater to the needs of the swap crowd. The kit comes with everything, down to the last bolt required to bolt the kit to the engine. You will need to use some anti-seize on the bolt threads to ensure ease of future removal.

To use the Vintage Air Front Runner setup, you will need an ATI damper for the three-bolt crankshaft pulley mounting provisions. The ATI hub is a press fit to the LS crankshaft and requires a long M16 × 2.0 threaded rod, nut, and large spacer for installation. The ATI damper bolts to the hub last and must be clocked correctly to the matching dimple marks, then is retained using (six) Torx bolts.

After the damper, you can install the (six) 8-millimeter–1.25 × 130-millimeter water pump studs in lieu of stock LS water pump bolts. The front runner system is also mounted to the water pump studs. Install the water pump gaskets, then finally set the early (bolt-on pulley style) water pump into place. No extra bolts hold the water pump, so for now it is just sitting in place waiting for the brackets.

The first bracket installed is the lower support bracket (also for power steering), which bolts to the F-body alternator location bolts. Using (two) 10-millimeter–1.50 × 70-millimeter bolts and ⅜-inch flat washers, only hand tighten the bolts for now. If using an iron block, the top support bracket bolt is not used, so only install the lower bolt.

The main accessory bracket is now installed over the water pump and studs. Use the 5⁄16-inch flat washers, then the 8-millimeter–1.25 12-point nuts, but do not fully tighten.

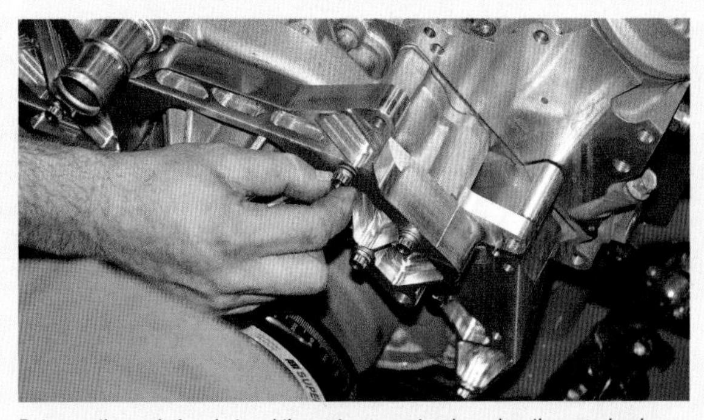

Between the main bracket and the water pump tensioner location, an aluminum spacer is installed with a 10-millimeter–1.50 with a 70millimeter bolt length. Then another 10-millimeter–1.50 × 70-millimeter-length bolt is installed through the main bracket into the top of the lower support bracket.

(continued on page 150)

VINTAGE AIR INSTALL *(continued from page 149)*

The alternator is installed next, using a 8-millimeter–1.25 × 40-millimeter bolt and $\frac{5}{16}$-inch washer for the top bolt, and a $\frac{3}{8}$-inch–16 × 3.250-inch bolt and $\frac{3}{8}$-inch washer for the bottom location. Once the alternator bolts are hand-tight, you may final torque all of the already installed bracket nuts and bolts to specifications. The water pump M8 nuts and alternator M8 bolt are torqued to 22 foot-pounds, and all the larger $\frac{3}{8}$-inch and M10 bolts are torqued to 37 foot-pounds.

Install the included A/C compressor using (two) $\frac{5}{16}$-inch–18 × 4.250-inch bolts, spacers, washers, and locking nuts. Torque the two bolts to 28 foot-pounds once both are installed.

If installing power steering with the front runner accessory kit, that is our next component needed. The VA kit uses the Type II plastic or remote reservoir PS pump. If using your own pump, note that some pumps have threads in the pump-mounting bolts. If so, these must be drilled out. If using a VA-supplied pump, there are no issues. Bolt the pump to the lower support bracket using (two) $\frac{5}{16}$-inch–18 × 2.750-inch and $\frac{5}{16}$-inch flat washers and torque to 28 foot-pounds. The pulley is installed last.

Along with the water pump pulley, the crankshaft pulley can be installed. The crankshaft pulley is held into place using (three) $\frac{3}{8}$-inch–16 × 1.250-inch bolts and $\frac{3}{8}$-inch washers. It's also a good idea to use some thread locker on these bolts. Torque these (three) bolts to 25 foot-pounds.

If using power steering and your pulley was not pre-installed, now is the time to do so. You will need a power steering pulley installer and remover kit. The centerline of the pulley where the belt rides to the back of the PS pump measures at 3.80 inches when fully installed. The rear of the pulley to the back of the PS pump measures 3.29 inches when installed. Usually these bottom out around the correct location when using a correct pulley installer.

Install the water pump pulley next, using the pulley and (four) 5/16-inch–24 × .750-inch bolts and 5/16-inch washers. I recommend a bit of thread-locking compound on these bolts. They are torqued to 22 foot-pounds, but it may be easier to torque once the belt is in place.

The tensioner setup is specific to the VA system and must be installed correctly for proper belt tension. There are three main parts to the tensioner: a spacer, alignment dowel, and the tensioner itself. The alignment dowel must be installed on a main bracket, the spacer aligns to the dowel and tensioner bolt hole, and then the tensioner aligns to the spacer holes, facing the water pump. Torque the bolt to 37 foot-pounds, and then install the accessory belt.

Since we are using the polished kit, VA included a matching A/C clutch cover and belt-tensioner cover for aesthetic purposes. These covers round out the kit and make the tensioner and A/C clutch match the remaining accessories.

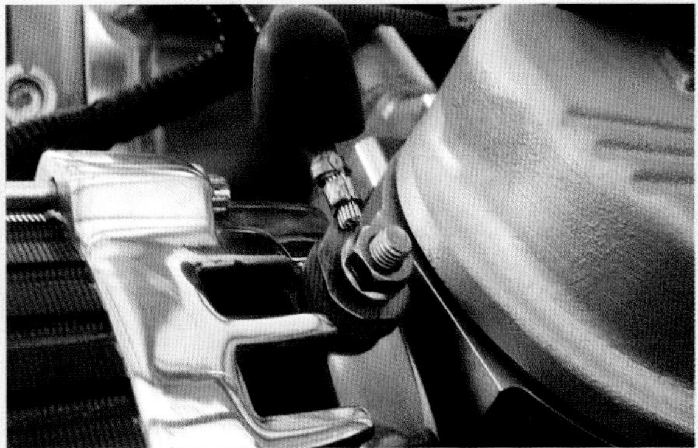

When the Vintage Air front runner installation is complete, you are left with a high-quality and complete accessory drive. From this point, you will need to finish the remainder of the installation by mating the individual accessories to components on the vehicle. The alternator, for instance, uses a single wire hookup for simplicity and only requires that one charge wire to be integrated into a 12-volt battery tie in. The A/C compressor needs hoses to connect both to the condenser and evaporator, while the power steering would require a pressure line and return line fabricated.

(continued on page 152)

DIRTY DINGO ACCESSORY BRACKET INSTALL

The Dirty Dingo bracket kit (P/N DD-LS-ALT-PS) for the 1991 Camaro uses the truck spacing 5th-Gen Camaro crankshaft pulley and water pump. This allows reuse of the common donor truck power steering and alternator, which is helpful because the OEM truck bracket itself sets the alternator too high for it to clear the hood. First the truck power steering pump is bolted to the front bracket using supplied bolts, then the brackets and required spacers are installed on the engine together. The idler pulley and alternator are installed next and all bolts are torqued.

The truck power steering pulley that had been removed is now reinstalled using a PS pulley install tool, and the belt tensioner and belt are installed. A note on alternators: At first I used the truck alternator, but decided against this because I needed it for a different swap project (2010 Jeep JK). Luckily the Dirty Dingo brackets are also designed for use with a 4th-Gen F-body alternator, and I had an extra one on hand that I could swap with the truck version. The four-pin F-body alternator needs an ignition resistor wire to start charging; ICT Billet P/N WPALT31 is what I used.

(continued from page 148)

turn your frostbite-in-winter, heatstroke-in-summer ride into modern-model user-controlled creature comforts.

Vintage Air's A/C compressor sits above the water pump, much like some of the other retrofit company accessory kits. Locating the A/C compressor away from the motor mount and frame areas helps alleviate extra work in and rewelding the frame for clearance. Not that cutting and welding is bad, but if you can avoid it and have to buy swap parts anyway, sometimes buying the right part can keep the gray hair away.

There are a few different options, and the kit can be ordered with or without power steering in addition to exterior finish selection. As far as engine choice, you basically have the LS7 kit to allow for the LS7 dry-sump oiling dimensions, or you can order the regular LS-kit, which is for everything else, or to make it easy, non-LS7. The kits all come with A/C and alternators, but you have the choice of PS delete ($225 less), or if you have your own PS pump, you can further get the kit with PS provisions and no PS pump.

Another well-thought-out option for the builders who have VVT-equipped engines is a kit from Vintage Air that is an add-on to the standard LS kit (spacer kit add-on P/N 176005). It includes spacers for the water pump, crank pulley, and power-steering bracket, an additional set of water pump gaskets, and includes longer studs and bolts to accommodate the additional 1.66-inch length. The reason for the extra spacing is due to the VVT timing cover's intrusive presence behind the water pump.

WEGNER LS DISTRIBUTOR CONVERSION ACCESSORIES

This is more of an information topic than how-to, but since many people are coming from carbureted setups into mostly EFI engines with this manual, I thought a few installers would like to know their accessory options should they choose to install the Wegner distributor conversion kit on their LS engine. This could be the easiest way to sneak into the LS engine phenomena without needing the LS EFI engine knowledge and electronics. The conversion kit is made to use a distributor and carburetor.

Because the Wegner distributor conversion comes with the necessary front cover and camshaft extension pieces, you may already know that the OEM accessories for the LS engine will not bolt on. The timing cover is extended from the block for the extra room of the fuel pump lobe and distributor gear on the camshaft snub extension—in addition to providing the fuel pump and distributor mounting locations into the timing cover itself.

Wegner does offer some solutions to the distributor conversion. The easiest solution is to use their extended Stewert water pump and alternator kit. This water pump is extended to allow timing cover clearance. It also provides a conventional SBC cooling system arrangement for radiator hose locations, SBC thermostat provisions, and an easy alternator bracket mounting location.

The alternator (SBC-style) mounts to the passenger side of the water pump using a billet bracket. The alternator still uses a serpentine belt drive and auto adjusting tensioner to make life a little easier. This conversion also requires an ATI crankshaft balancer, mainly for its flexibility in bolting on a conventional three-bolt crankshaft pulley. It is much easier and flexible to make a pulley the correct depth than to redesign a new crankshaft damper.

Wegner also builds full kits allowing power steering, alternator, and A/C with their distributor conversion kits. This allows another option for complete street-bred LS conversions. Wegner has adapted the LS engine for circle track racing by using mechanical fueling and ignition, and in doing so has opened the door to allow the retrofit crowd to use the LS-based engines for non-EFI and non-electronic LS retrofits. Who would have thought!

ACCESSORY INSTALLATION TIPS

Installing your accessories should be done after the engine is in place. If you are using GM-based accessories, you can install the water pump and crankshaft damper at any time during the accessory installation.

When using aftermarket brackets, you typically install the damper and water pump first. One of the reasons to install the water pump last on the GM setups is the ease of access to the power-steering bracket bolts when the water pump is out of the way. If the power-steering pump is assembled to the mounting bracket, it's far easier to use a wrench to install the various pumps while not worrying about water pump interference. You can easily add the WP afterwards.

If you are using the C5/C6 accessories, make sure to install the driver's side coil pack rail before mounting the alternator; there is a hidden coil pack mounting bolt under the alternator that is hard to tighten when the alternator is in place. All remaining bolts are no problem. It also is beneficial to have the portion of the engine wiring harness that extends to the coolant temperature sensor in place; it is even a tight fit for wiring.

Here's a trick if you're using the stock C5/C6 A/C compressor and bracket. Since the bracket bolts to the block with only one bolt and the other mounting bolts are through-compressor bolts, start a few threads using the compressor mounting bolts so that you can properly align the A/C bracket. Skipping this means you will be pulling it back apart to align it anyway. Since there is only one bracket bolt, any minute misalignment will not allow the remaining through bolts to thread into place.

GTO alternators are a bit larger than their F-body counterparts. If you are using a GTO alternator and have limited or no clearance, a swap to the smaller F-body alternator might do the trick. If there is no room for the low-mount F-body alternator, you may have to go with a top-mount alternator bracket kit that bolts to the driver cylinder head, such as the C5/C6 accessories or aftermarket brackets.

Sometimes you will have a power-steering pump without the power-steering pulley in place. After making sure the pump shaft dimensions match your new pulley (C5/C6 are smaller than all others), you will need to press the new pulley into place. This is recommended to do after the PS pump is mounted to the PS bracket and engine to help create a sturdy working area. To press the PS pulley on, you will need an installation tool. A long threaded shaft, nut, and washer would make the minimal required setup, although there are tools specifically for this purpose that you can rent or borrow from your local auto parts store.

Some pumps are threaded standard ⅜-inch diameter threads, and others are a metric thread. To install, first set the pulley into place, followed by the installation tool. Turning the nut on the installation tool draws the pulley into place. The stock pulleys will fit flush with the end of the pump driveshaft, while aftermarket-intended depth varies, depending on manufacturer. Some will bottom out and be aligned, while some you have to align with the existing pulleys to be perfect. This can become quite tedious to get just right. It could involve pressing, removal, and repressing several times until proper alignment is achieved.

If you have pulley alignment issues, you can check by using a Gates laser belt alignment tool. These can be found online or at parts stores and attach via a magnetic base in the pulley grooves. Using these laser alignment tools can help you find issues that your eyes cannot see. If you have

IDENTIFYING CRANK DAMPER AND WATER PUMP DESIGNS

There are three different crankshaft damper dimensions, not counting the different dampers required for the supercharged 6.2-liter found in the Cadillac CTS-V LSA and Corvette ZR1 LS9 engines. The main OEM damper and pulleys are the 1998–2002 F-body (LS1), 1997–2010+ Corvette C5/C6 (LS1, LS2, LS3, and LS6), and the 1999 and newer GM Truck/SUV (4.8-, 5.3-, 6.0-, and 6.2-liter), from these most retrofit accessory kits are meant to be paired with (some accessory kits require the aftermarket ATI Damper for the bolt-on pulley versatility).

To identify which OEM damper you have is rather easy. If it is still on the engine, you can take a measurement from the front-most face of the damper to the timing cover and compare to the measurements in the illustration for identification. If you have the damper off, compare it to the off-engine picture identifications.

As far as water pump designs, all early car LS1 engines (pre-2005) use the same design, a wide pulley with the upper water pump outlet on the passenger side of the engine facing forward. Pre-Gen IV trucks are similar, except they have a mechanical fan clutch threaded stub and a passenger-side water pump outlet that is mounted upward. This style would interfere with the low-profile passenger car intake manifold, so it is best kept matched to the truck intake manifolds.

The 2005–2012+ C6s, 2005–2007 CTS-Vs, and 2008–2009 Pontiac G8s use the low-profile LS2- and LS3-style water pump. The difference with this water pump is that it is designed to fit with a low-clearance engine accessory belt. The LS1 F-body and truck dampers are incompatible with the low-profile water pumps.

Gen IV trucks also went this same route with the narrow pulley design, made to match the corresponding truck crankshaft damper. Also, in electric fan mode, some of these do not have the mechanical fan threaded stub on the pulley. It is hit or miss whether one would have one or not, but it would only matter if you wanted to use the mechanical clutch fan from the trucks. Most swappers go with electric fans. You can also shorten this threaded stub via machining for electric-fan clearance.

The 2010–2015 Camaros have a different water pump design, which is similar to the truck design but with a driver's-side mounted water pump outlet. These only match the truck or 2010–2015 Camaro crankshaft damper.

This is the 2009 and newer Caddy 6.2-liter LSA water pump, which has a driver's side upper hose outlet. This water pump would be ideal for the LSA crate engine, as it provides the necessary bolt holes for the supercharger belt idler and tensioner brackets. *Courtesy of General Motors*

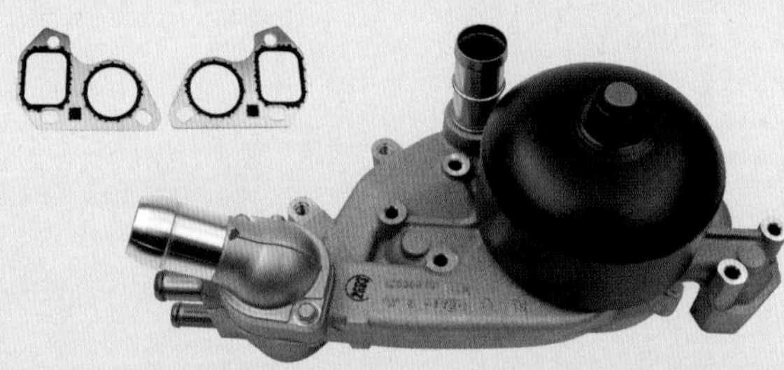

The original LS1- and LS6-style water pump fits both F-body and C5 and C6 accessories and features a conventional passenger-side upper hose outlet. Later models of the dual application–style water pump used an LS2+ removable thermostat. The pre-2003 OEM water pumps have the thermostat integral to the thermostat housing. *Courtesy of General Motors*

alignment problems, you may need to shim or machine the item to align perfectly.

Once the accessories are in place, you must integrate the new accessories with your swap project vehicle. For the alternator, this involves connecting and routing the wires. The power steering requires a pressure feed and return hose setup, and the A/C requires high-pressure low and high-side refrigerant hose fabrication.

ALTERNATOR WIRING

The OEM alternator is quite easy, only requiring a voltage-sensing wire and a large charge wire. The stock charge wire is 6-gauge and can connect either directly to the positive terminal of the battery cable, the starter positive feed, or depending on application can be tied into a battery junction box. Do this only if larger-diameter wiring such as 2-gauge is present to that point. The best location is directly on the battery of course, but as stated this can change depending on vehicle and the convenience factor. If there is a battery junction box 2 feet from the alternator, this is likely a good location to tie into while avoiding excessive wiring being crisscrossed through the engine bay.

The other wire needed for the alternator is the charge signal or sensor wire. Many engine wiring harnesses will have this feature pre-installed if intended for an LS swap, if you are using a reconfigured donor harness, you will have to build your own signal wire. This wire simply energizes the alternator into charge mode when running and voltage is present.

The simplest form of this circuit is through the dash-located alternator light on pre-1998 vehicles, a time before dash gauge signals went through the ECM first. In the older setup, voltage is run through the bulb and into the alternator. If the alternator is not charging, the wire is grounded into the alternator and the dash charge light will illuminate. If the alternator is charging, it submits 12 volts through this signal wire and cancels the 12 volts coming the other direction. If not using a charge bulb, an 82-ohm 5-watt resistor can be mounted inline to a key-on ignition wire to signal the alternator that the key is on. This simulates

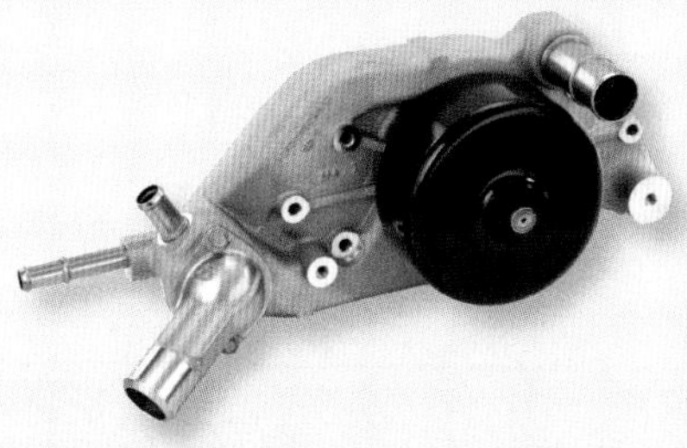

The ZR1 LS9 water pump has wider, more durable bearings to support the extra belt load induced by the supercharger. This is ideal for LS9 crate engines. *Courtesy of General Motors*

The 2010 and newer Camaros' dimensions are similar to the trucks' pulley offset from the engine. The fifth-generation Camaro LS3 and L99 water pumps will only work for 2010 and newer Camaro accessory-drive setups.

If you have one of the short C6 LS3 water pumps and need to use it with a different belt spacing, such as 4th-Gen or 5th-Gen crankshaft pulleys, water-pump spacers from ICT Billet to get the correct new spacing can be employed instead of buying a new water pump. These are available in ¼-, ½-, ¾-, ⅞-, and 1.5-inch thicknesses. Additionally, the ¾-inch or larger spacers have a convenient ⅛-inch NPT threaded port, which can be used for a coolant gauge temperature sender or as a coolant air bleed recirculating port.

In the chapter text, we discussed the difference between the standard C5/F-body water pump (left) and the lower profile LS2 water pump (right). This picture demonstrates that difference. The LS1/LS6 water pump will fit a multitude of applications using the C5 and C6, GTO, and F-body belt spacing. The LS2+ design will only work on C5 and C6 belt spacing due to the lower profile single-belt pulley.

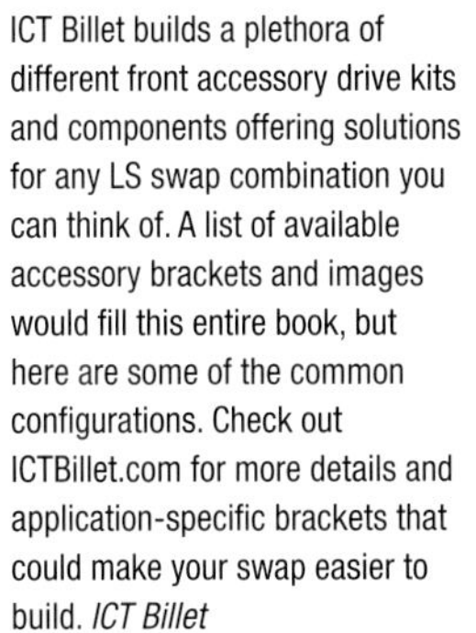

ICT Billet builds a plethora of different front accessory drive kits and components offering solutions for any LS swap combination you can think of. A list of available accessory brackets and images would fill this entire book, but here are some of the common configurations. Check out ICTBillet.com for more details and application-specific brackets that could make your swap easier to build. *ICT Billet*

REMOVING AND INSTALLING CRANKSHAFT DAMPER

Often the crankshaft pulley that comes with your donor engine may not be correct for your new application. To remove the stock crankshaft damper you will need a three-jaw puller or a dedicated crankshaft damper remover. The stock crankshaft bolt is a torque to yield design and not intended to be reused with the new crankshaft pulley. If your accessory kit doesn't have one, make sure to grab a new one from your GM dealer.

The stock crankshaft pulley bolt is very tight, so grab a buddy and have him lock the flywheel down with a long screwdriver or pry bar. ICT Billet makes a flywheel lockdown tool that mounts in place of the starter, but if you have no access to one of these, you must improvise using a pry bar. Using a six-point 24-millimeter (or 15⁄16-inch socket) and a long breaker bar, the crankshaft pulley can be removed using some muscle. It is a normal right-hand threaded bolt, so loosening it counterclockwise removes it. If you have access to a beefy impact gun, you can also loosen it in this manner.

Once the crankshaft bolt is removed, use the three-jaw puller to remove the crankshaft pulley. You may need to thread the old crankshaft bolt in halfway to have something for the puller to press against. If using a dedicated damper puller these will have a pushrod that pushes against the center of the crankshaft with the bolt removed. The pulley will be tight at first, but once budged, it will start coming loose the more you tighten the threaded puller bolt. Once you get to where the pulley is visibly loose and just holding onto the crankshaft by a hair, be ready for it to break free and make sure to not let it drop.

If you bought a used engine, this is a good time to replace the timing cover gasket and front crankshaft seal with a new replacement seal. Your local GM dealer or aftermarket parts suppliers will have gaskets and seals available. The GM seal is meant to be a dry install, meaning no oil or lubricants are to be put onto the sealing surface itself; it forms a reliable seal without lubrication. Once the seal is replaced, you may move onto installing the new pulley or damper that matches your accessories.

The easiest way to install the crankshaft pulley is by using an M16 × 2.0–millimeter threaded rod and washers to press the damper into place. Some people have had good luck using a longer threaded bolt, but this does put extra frictional strain on the internal crankshaft threads, whereas a threaded rod and nut setup does not. Several companies make an installation tool, and you may even be able to find one at your local parts store to borrow or rent.

Using the installation tool, press the damper on as far as it will go, then remove. Take the old crankshaft bolt and tighten it to 250 foot-pounds of torque to seat the new pulley into place, then remove it. Install your new crankshaft bolt with a dab of thread Loctite, torque it to 37 foot-pounds, then tighten it an additional 120 degrees unless your accessory kit specifies their own torque specification.

the resistance the bulb itself creates in the same wire. Many older GM vehicles will have this wire pre-existing in the front engine lamp harness, as it operates with the original alternator as well.

Modern factory alternators are ECM regulated voltage control (RVC), specifically the 2005 and later two-wire-connection Delco GM alternators that control via programmed duty cycle for various voltage outputs. These default to 13.7 volts charging when unplugged, so if this is enough and you do not have a lot of electronics to operate, you don't need to do anything extra. To retain factory operation, these can be used with the correct ECM and calibration, but they can also be used with aftermarket ECUs by outputting a 5-volt 128hz output at 70 percent duty cycle, which nets about 14.5 volts from the alternator.

When using the Vintage Air kit, the alternator included is a self-energizing one-wire alternator. The only wire required to be hooked up to the vehicle is the charge wire itself. Using 6-gauge wire and connectors, connect from the charging post of the alternator to a 12-volt connection source such as the battery, fusebox, or starter battery post. Make sure to use a fusible link smaller than the wire size you use to protect the charge wire and alternator in case a short to ground should occur. *(continued on page 160)*

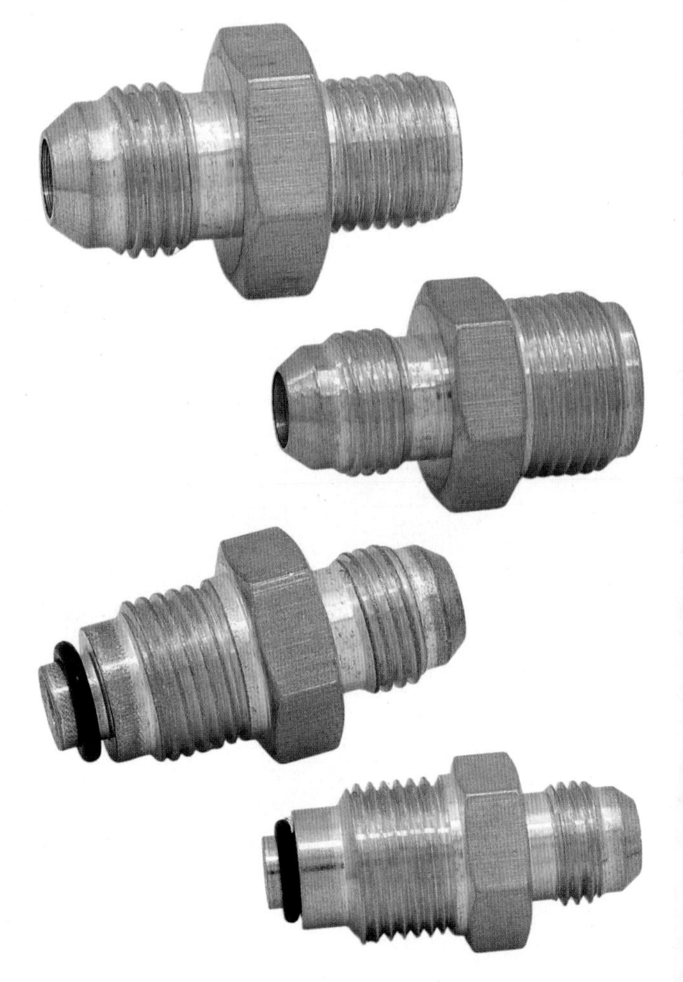

For swaps, you will often need to make custom power steering lines to connect your old steering box to your new power steering pump. ICT Billet has various fittings to convert these metric or flare fittings to -6AN, giving you options to make a high-pressure steering line that works for you. Crimped-on steel hose ends are preferred for power steering pressure lines, but some AN fitting companies make DIY steel hose ends as well. *ICT Billet*

MEASURING BELT LENGTH

Unless you buy your accessory drive setup in 100 percent kit form or use an OEM GM setup, you will have to measure the belt length to determine what belt length is needed. The alternative is buying a few belts at a time until your luck strikes and you find the right one by chance. Why do that when measuring the belt length is an easy task that can be done with minimal outside help? To measure the belt length, you must have all accessories in place and a length of wire, yarn, clothes line, or the like.

Route the "hose measuring tool" around the pulleys (you can use duct tape to hold the line in place if you run out of hands) in the same belt routing as recommended by the manufacturer with the ends meeting close to the belt tensioner. If you have no belt tensioner and have an adjustable alternator, put the alternator about ½ inch from the end of travel on the tighter side. The trick here is to compress the tensioner while keeping the string tight for accuracy. Once you feel good about it, either cut or mark the string and remove. Then simply measure the length of the string and take this to your parts store. Most accessory belts are going to be six-rib belts. The stock A/C belt is four-rib, if used.

Often you cannot just order a belt length for the car you pulled an engine out of and have it work unless you have every accessory from that vehicle. If you have PS deleted, or a different bracket system, you will need to find the belt length. In this image, since I used the older LS1/LS6 water pump with the GM FEAD accessory kit, the included belt length was slightly too long and needs to be shorter.

Using rope, string, wire, or clothes line, wrap the "belt-measuring tool" around the pulleys just like you would a belt following the same routing. Put the end of the belt near the tensioner as that is where you need to be marking the rope.

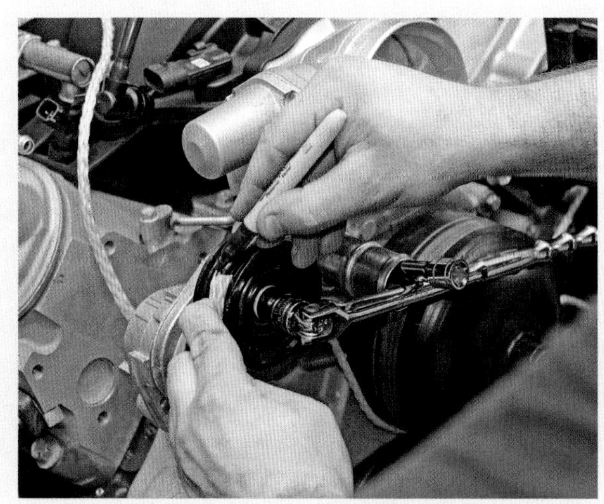

Holding tension on the belt tensioner and on the rope to keep slack out of the system is sometimes tedious. You might need a third or fourth hand or figure out a way to lock the tensioner while you mark the rope. With no slack in the device, mark the rope where the end of one side overlaps the other side. It doesn't matter if one side is cut or not, or if you mark both ends. The point where they both line up is your two "A–B" measurement points.

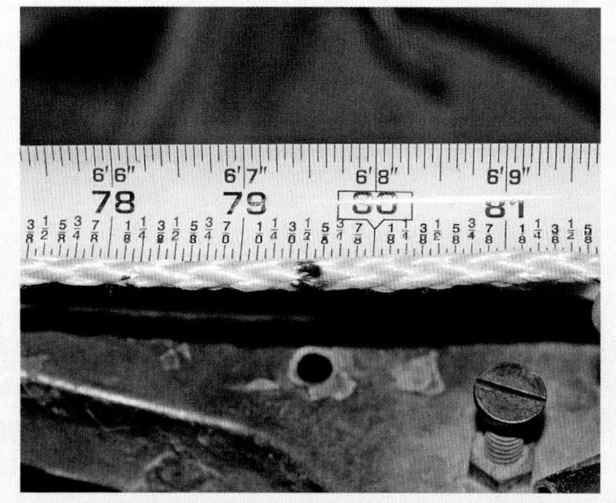

After marking the device, measure between the two points. This one came out to a 79.5-inch belt length. Take that measurement to your favorite parts store. It is no use looking up the application, as you just determined the exact belt length requirement to start with. If it is too tight or too loose at this point, you can usually get belts in ½-inch increments either direction.

DRILLING IRON-BLOCK ENGINES FOR F-BODY STYLE ACCESSORIES

All iron-based truck engines are missing two holes near where the F-body-style alternator would be mounted. One is missing completely (the one for the bracket itself), while the other has the boss casting, but it is not drilled. Drilling and tapping the boss allows the ability to use the F-body and similar alternators.

The first thing to do is mock install an alternator bracket to mark the location. I find it easiest to remove the steel dowel that clamps the upper alternator in place and then mark the block using a black permanent marker. It is an almost perfect fit, so the hole can be centered exactly where the alternator bolt is by using the marker.

Once marked, use a center punch to tap a divot in the block as a starting point for a ⅛-inch pilot hole drill bit. Keeping the bit aligned perfectly with the block is a chore, but you can use the opposite long alternator bolt as an angle reference.

After successfully drilling a pilot hole, step up to the correct $^{21}/_{64}$-inch drill bit by following the pilot hole through the block. Drill at a medium speed and only to the depth needed to facilitate the alternator or accessory bolt length by comparing the drill bit depth to the amount the bolt sticks out of the back of the alternator and bracket. Be sure to use a cutting lubricant to assist the drill bit and keep it from getting too hot.

POWER STEERING HOSE ADAPTATION

Power steering requires two hoses, one of which is quite straightforward. The other is a little more complicated. The return hose is the easier of the two. You simply need a ⅜-inch-diameter hose that is oil resistant, such as a transmission oil cooler line. You connect the return side of the steering rack or gearbox to the ⅜-inch hose fitting on the power steering reservoir and add clamps.

The pressure line will have to be custom made on most cars, though trucks sometimes can use the existing SBC power-steering lines. The pressure line can be made like the OEM line with high-pressure PTFE hydraulic hose with pressed-on ends that match your vehicle's steering. It's entirely possible that the pump side is the 16-millimeter metric O-ring fitting, while the steering unit uses the older-design ⅜-inch flare fitting.

hen you're building flexible braided AN hoses for ur power steering, the last thing you want to do is stall a generic rubber hose and a hose clamp for your w-pressure power steering return hose. ICT Billet has e solution with this 3/8-inch hose barb fitting to -6AN apter (P/N F06AN375CP), so you can have a nice atching -6AN power steering feed and return line.

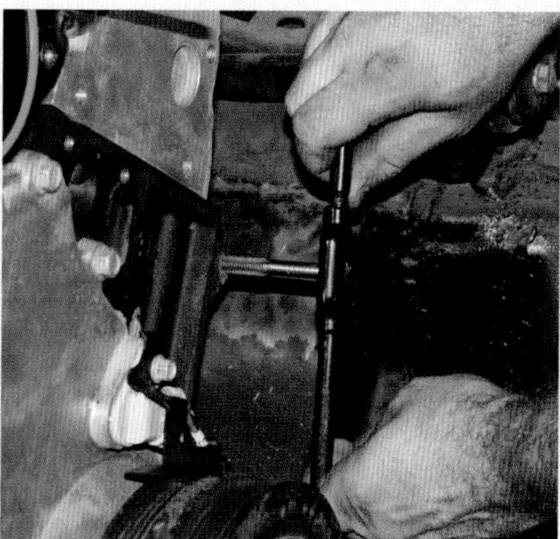

Next, use the M10 × 1.25–millimeter thread tap and start cutting threads. The first couple of threads prepare the bolt hole and allow you to get the tap situated in a straight cutting pattern. Turn the tap about one-half turn at a time once threads start cutting, then as resistance is felt, back off a bit. Then repeat. Use cutting oil, and if you feel too much resistance, remove the tap and use compressed air to clean out the debris. If the hole is straight, you will easily be able to tap the hole through to the bottom.

Before installing the alternator and bracket, blow the newly threaded hole out one last time. If you did everything right, the alternator will bolt right to the block and bolt on as if it were a stock aluminum block. Torque the bolts to 37 foot-pounds.

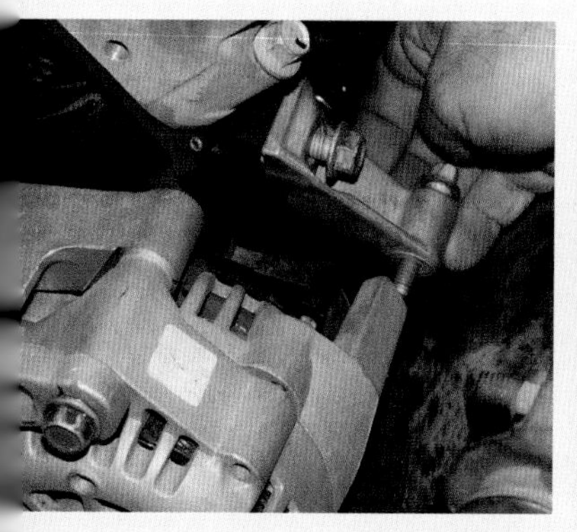

:e the alternator is in place, you can add the stock k-brace alternator bracket as extra support. It just s to the back of the alternator and to the side of the k easily.

The alternator is securely mounted with the (two) front through bolts and the back-brace bracket. The middle bolt won't be used with the iron-block applications, but it is not necessarily required when using the other (three) remaining bolts, including the support bracket.

If you need to drill your truck iron block for the low-mount alternators as found on 4th- and 5th-Gen Camaros, you can use the alternator bracket itself to mark the location for drilling, as shown. But ICT Billet has a better solution with this template (P/N 551445). It bolts between the existing lower block hole and a cylinder head accessory hole to offer a perfect starting location for a pilot drill hole. *ICT Billet*

You can use adapters for these, or check with a local hydraulic shop for custom, OEM-appearing lines. Make sure to add a few extra inches to your measurement for extra routing convenience. You do not want to come up short on a custom steering hose, and a little slack to allow engine movement won't hurt either.

The other way to make power steering hoses is by using AN hoses and fittings. Russell, Aeroquip, and other hose manufacturers offer high-pressure PTFE hoses for pressures exerted by the power-steering system. Additionally, you can find just about any combination of AN adapters to adapt your new pump to AN sizes or the power-steering gearbox to AN sizing. At this point, it's just a matter of building a long enough AN hose, with the right angled ends and connecting the two. AN simplifies this install by not relying on subletting the hose-making task.

Many times during a coolant fill procedure you cannot get all the air in the engine block to vent. There are several ways to help get the air out, including pulling a vacuum on the cooling system and letting the vacuum draw in coolant. This works well, except not everyone has that capability. An easier DIY helping hand can come from drilling the thermostat or thermostat housing with a small drill bit. This allows the cooling system to burp easier and fills the engine block with coolant, both when off and when running.

If your swap project doesn't use a heater or you are adding it later, don't just cap and plug those heater lines. Use a trimmed-to-fit Dayco (P/N 87629), which is a 180-degree bend hose that allows coolant in the block to circulate past the thermostat while warming up. Later, when winter arrives and you need that heater again, remove the bypass hose and just add heater hoses like normal.

(continued from page 156)

ENGINE COOLING SYSTEM

Every water-cooled engine requires a cooling system of some sort to carry combustion and friction heat away from the engine. The cooling system consists of the water pump, radiator, cooling fan(s) thermostat, radiator hoses, and coolant working together as a team to keep engine temperature stabilized and controlled.

The cooling system functional description would be that the water pump circulates coolant through the block first, then the cylinder heads where it absorbs heat. As the coolant heats up to the thermostat opening temperature, it is diverted from the engine through the upper radiator hose back into the radiator to be cooled. The coolant is returned to the water pump through the lower hose, which is where the thermostat is located. Note the thermostat is not located on the outlet like the SBC engines; it is in the inlet side of the water pump.

STOCK LS RADIATORS

In the accessory kit section, we determined which water pump you have or need to obtain. These all work the same, so it's just a matter of which one is on your engine and driven by the belt drive.

You can reuse the stock radiator if it's in good shape or a donor radiator from a newer vehicle or custom-built radiator fabricated from aluminum. A custom radiator will bolt right into place, assuming you have a conventional setup for radiator mounting (like how the stock radiator previously bolted in). A donor radiator would need some thoughtful fabrication or mounting solutions.

Along with the radiator installation, you may wish to install electric fans. Most of the time it's because your water pump was not designed for a mechanical fan mount (unless you have the truck version). Even so, the electric fans are more than capable of providing additional airflow at low vehicle speeds and are highly reliable and fairly easy to wire into the vehicle.

If using a donor LS-style radiator, you must figure out a way to mount it to the core support of the swap vehicle. In the case of the 1967–1969 Camaro, where the radiator bolts to the core support, to use a 1993–2002 F-body radiator, you either have to make a lower and upper mounting bracket or modify the plastic end tanks.

Modifying the tanks for bolt-on usage is easy. Cut off the A/C condenser tabs flush, then drill an oblong ¼-inch hole in the flat left behind. Then, using the modified radiator, find where it needs to be mounted on the core support. If you have other items in the way on either side (such as a battery), you can offset the mounting location by a few inches to compensate. Once a location is determined, mark and drill the core support for the new donor radiator mounting holes. I recommend starting with one bolted into place, then level the radiator before drilling the remaining bolt-hole locations and install the other three bolts that are left.

LS1 F-BODY RADIATOR AND FAN INSTALLATION

Some builders have cut their radiator core support to use these condenser tabs, but that requires some extra bracing to hold the radiator upward. In these next photos we show how to install a donor (cheap!) F-body radiator without much trouble.

The first thing to do is to cut the (four) stock condenser tabs off, not the entire tab, just the slotted portion. This will leave a flat mounting surface to modify.

Next, drill the (four) tabs to about 5⁄16 inch. You can either tap these holes for a bolt or leave them unthreaded and use a locknut for mounting purposes.

Line up the radiator to your core support and mark the first bolt. It helps to have a small carpenter's level handy to make sure the radiator is installed level in the chassis. If you have the battery in the stock location, you may have to relocate it or simply shift the radiator over to the driver's side a bit for clearance. Drill the core support to about the same or one size larger bolt size as the radiator. I drilled the core support 3⁄8 inch and for use with 5⁄16-inch bolts and nuts.

I find it helpful to line up the first bolt hole so that some weight is supported while locating the remaining bolt holes. Align the radiator evenly, drill the (three) remaining bolt holes and tighten the bolts equally. You don't want them tight to the point there is strain, rather just enough to not flop around. You can also rubber insulate the bolts if you wish to isolate the radiator from vibrations.

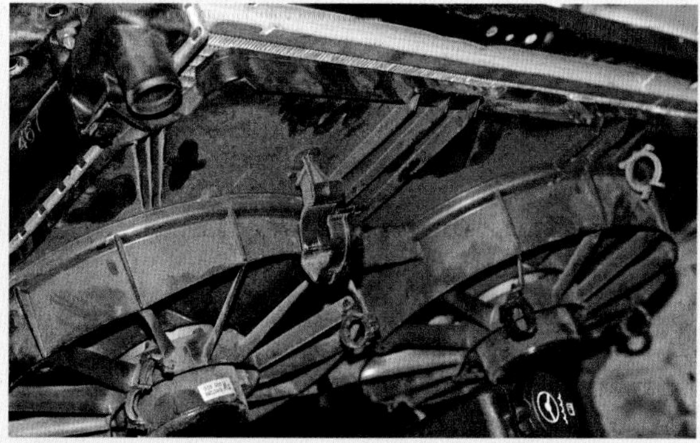

After mounting the radiator, the F-body fans simply fall into place on the engine side of the radiator. From here, install the fan wiring and start locating radiator hose dimensions. Also, if you do not like the way the top of the radiator looks, you can cut an aluminum or stainless-steel plate to make a radiator trim cover to hide it.

COOLANT AIR-BLEED SYSTEM

Since the LS-based engines use standard coolant flow routing (block first, then heads), and no coolant goes through the intake manifold such as in the SBC, there is a high spot in the cooling system where steam and air can collect. GM's solution to this is a coolant bleed pipe. Simply, this is a self-bleeding cooling system function where any air in the engine is purged and bypasses the water pump. In stock form it is connected to the highest point in the radiator. As long as coolant gets to circulate through the coolant bleed pipe and out into another radiator hose, or water pump, or anywhere where coolant circulates, this system will work adequately.

Some builders just block off this air-bleed system and have no ill-results. I recommend using it, as it is not an inconvenience, especially considering the amount of other labor-intensive work involved with a swap. This is an easy task.

If you use an LS1 F-body radiator, you can simply use the stock provisions for the coolant air-bleed built in the radiator itself. The coolant-bleed pipe outlet is ¼ inch, while the radiator connection is ⅜ inch, so you likely will need to adapt the hoses together. BMR Fabrication offers a ¼-inch to ⅜-inch hose adapter for an easy fix, but if you are creative, you can make one yourself. Some builders have even used 5/16-inch coolant hose stretched over the ⅜-inch fitting and clamped securely on the ¼-inch side. This is not the way I would normally do it, but I see a ton of them assembled this way with no issues.

Some builders will have fabricated aluminum race-style radiators built for their swap vehicle. Certain LS swaps are so common (e.g., first-generation F-bodies) that many companies offer pre-built LS-swap radiators for older cars. When going this route and spec'ing your radiator requirements, you can have the radiator specialist add the coolant-bleed pipe fitting. Many do it for no extra charge or if charged, it is a minimal out-of-pocket expense. Often it is included if the radiator company builds LS-swap radiators often.

Another option is to place a T-fitting into your upper radiator hose or heater hose to connect the air-bleed hose into. Trailblazer SSs have such a fitting in a factory configuration of the bleed hose routed into a heater hose.

One company that offers radiator hose–sized fittings to adapt the coolant-bleed hose into is www.jagsthatrun.com, which offers an aluminum inline adapter for the bleed pipe, with a manual bleed valve, or even with a sight glass to see coolant fill and flow. If you have one of these hose fittings, all you do is connect your coolant bleed hose into the fitting and you're done. JTR also has a manual coolant bleed fitting that helps during the initial coolant bleed process.

ICT Billet has radiator hose adapters that have a ⅛-inch NPT to which you can add a temperature sensor or coolant bleed port. Additionally, if you have ICT Billet water-pump spacers, these also come equipped with a ⅛-inch port for a coolant bleed hose. I like using the radiator hose adapter style, as it purges the cooling system at a higher point in the engine bay.

The last method of tying the coolant bleed pipe into the cooling system to maintain functionality is by drilling and tapping the top or side of the water pump itself and installing a ¼ × 27–inch pipe fitting with a 90-degree ¼-inch hose barb. Before drilling and mounting any fittings, make sure nothing is going to interfere; you may have throttle linkage or accessory interference to take into account. Also, a brass fitting may not be as aesthetically pleasing as a stainless-steel fitting, so if you are going for looks, make sure you get a fitting that doesn't stick out like a sore thumb.

After drilling and tapping the water pump, apply some sealant to the threads of the new fitting and wind it into place. Pipe threads are tapered; the more you tighten them,

When building your radiator hoses, you will likely need to cut and shorten donor hose lengths or Frankenstein a hose together. ICT Billet has what you need for this. While they offer many different hose splice connections and reducers, the one I needed for the 1991 Camaro required shortening and splicing together a 1¼-inch upper radiator hose. I used P/N AN627-20X125, which also comes with a convenient ⅛-inch NPT port that I used for my coolant/steam bleed hose connection. Make sure to use coolant spec or AN hose for the air bleed, because vacuum line will not work.

Another version of the coolant-bleed fitting is shown here. There are a few sources of these inline aluminum adapters, such as jagsthatrun.com. If you do not use an in-line fitting, you can use an LS retrofit radiator, or drilling and tapping the water pump near the upper hose outlet also works really well. *Tom Munsch*

the tighter the fit. Don't go too tight as you risk damage to the water pump or fitting. Once the fitting is in place, connect the coolant-bleed pipe to the new bleed fitting using ¼-inch coolant spec hose, and clamp it into place. Take into account the clearance of moving parts or sharp edges; if your routing is questionable, you can also wrap the hose with some protective tubing.

Even though the stock system bleeds air and steam when the vehicle's cooling system is full, sometimes during a fresh coolant fill or when refilling a system after draining, the cooling system becomes air-locked. What this means is the cooling system needs to be burped but cannot do so by itself because the coolant itself is trapping air inside the engine block. Two things that help during a refill are removing the coolant temperature sensor first. Fill up the cooling system until coolant drains out of the open coolant temp sensor hole or remove the front bolts on the coolant-bleed pipe and fill the system up until coolant seeps out of the coolant-bleed holes in the cylinder heads. These two things will help prevent air lock if they work as intended.

If an air-locked cooling system gives persistent troubles during the first startup, sometimes more drastic measures are necessary. This could be due to the radiator's location in relation to the engine location or for any other reason, but keep in mind that stock newer car cooling systems give repeated trouble during coolant fill-up procedures as well. Many dealerships will use a coolant fill tool that creates 20 to 25 inches vacuum in the cooling system, removing the air. Once the vacuum is pulled, a valve is opened and pre-mixed coolant fills up most of the engine. This is a reliable method of ensuring a good coolant fill, but since not everyone has access, there is one other thing you can do if all other methods did not help.

This method involves taking a popular SBC thermostat modification and carrying it over into the LS thermostat. Sure the LS thermostat has a small bleed-valve orifice that is supposed to help air escape, but it's often not as functional as the next procedure. If having problems with an air-locked cooling system, some builders have found out that adding a few small drilled holes in the thermostat helps immensely with good coolant flow and preventing initial startup air lock. Drilling a $\frac{3}{16}$-inch hole near the perimeter of the thermostat will solve this issue. I tend not to do this until it is required and then only as a last resort. The additional holes do not hurt the functionality of the cooling system but help a bit by always circulating a bit of coolant. These holes help even more when you do not have a heater or heater hoses. You can loop the hose fittings using a Gates P/N 18777 (or comparable).

The heater hoses function as a thermostat bypass, allowing hot water to circulate through the block until the thermostat opens. Without coolant flow, the thermostat would operate lazily, as hot water would take longer to heat soak the cooling system if coolant is not flowing. This is especially so when you factor in that the thermostat is on the inlet to the water pump rather than the outlet as in older engines.

USING CORVETTE OR F-BODY DUAL FANS

Use of the fourth-generation F-body radiator is a huge benefit because it gives you the ability to easily use the readily available F-body dual-fan assembly. These are a proven design that can be controlled easily either with a wiring kit (such as the one Painless Performance offers) or by some creative wiring using relays and the LS ECM to control the fans operation. In the used marketplace, these dual-fan kits can be priced anywhere from $100 to $150, which is quite a bargain considering their effectiveness and proven durability.

The factory dual-fan setup with GM's ECM control uses a low-speed and a high-speed system via three relays controlled by the ECM. The ECM grounds the relays according to fan temperature calibration.

The factory five-pin relays are a very reliable setup. You can find relay banks that are pre-wired to add to your harness to have low-/high-speed control of your fans. You can also wire your own relay bank, but be sure to use high-quality relays to avoid premature failure. I use waterproof 60/80amp 12-volt relays with 12-gauge wiring, which cost about $15 each, so it cost $45 total in parts to build my own reliable three-relay high-low fan control kit.

You may wonder why three relays are needed when there are two fan speeds. Basically, low speed of both fans is controlled by activating only relay number 1 while keeping relays number 2 and 3 off. This puts the fans in series mode (low speed) by sending 12 volts into the left-hand-side fan first, through relay number 3 (which is still off) and then into the second right-hand-side fan, which has the sole grounding point. This is due to the five-pole relay's ability to transmit voltage through its naturally closed circuit when not energized. This is the fifth-pin on most relays that is rarely used. Low-speed fan operation shares a single voltage line in-series through both fans, hence the slower fan speed.

High-speed fan operation is both fans operating in parallel, where each has its own 12-volt circuit. To operate a high-speed fan, all three relays must be grounded and

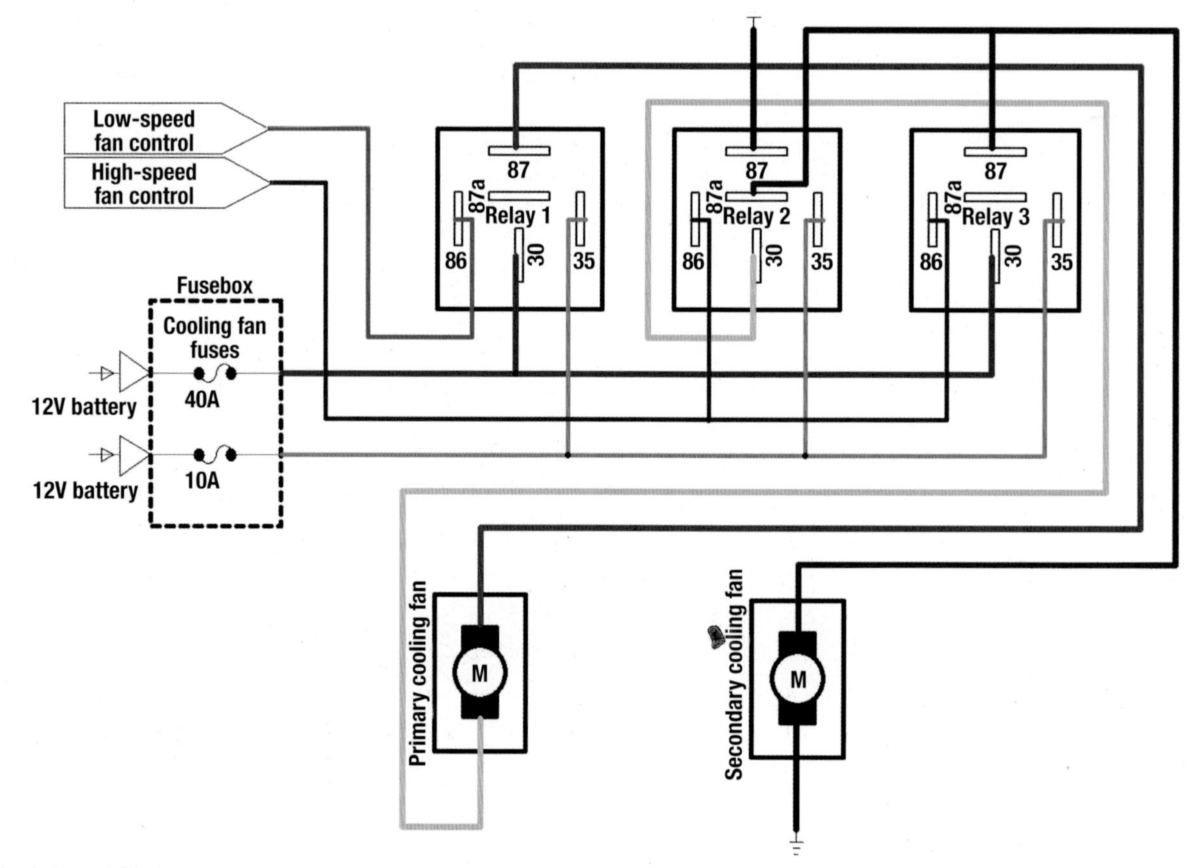

Here is the stock F-body–style two-speed fan wiring schematic. This three-relay system has a low speed and high speed that operate by changing the ground path direction of relay 2. In low-speed relay 1 is energized, providing power to the primary cooling fan first, then the remaining voltage is passed through relay 2 (which is off) and onto powering the secondary cooling fan. The two fans share a single voltage source through one relay and operate slower. In high-speed fan operation all three relays are energized, but the difference is that each cooling fan has its own individual voltage source instead of being shared such as in low speed.

activated through the ECM. Relay number 1 is on exactly the same as it is in low-speed mode, but since number 3 switches the ground to a direct chassis ground instead of "grounding" through the second fan, fan number 1 becomes its own circuit. Look at relay number 3 as the left fan ground control relay. Its primary purpose is to switch the way fan number 1 grounds (through the second fan in low-speed or directly grounded for high speed). Relays number 1 and 2 control 12-volt power to the fans. Fan number 2 has a permanent ground path and does not have a switching ground. This fan wiring schematic can be adapted to other dual-fan setups as well as when using the Gen III Warren-style ECM.

Alternatively, you could just wire-up the fans through a much easier and simpler two-relay system for a true full-speed on/off operation to have only one fan come on at the low-speed temperature, with the second tag-teaming at the high-speed fan calibration. The ECM fan calibration can determine at which temperatures one or both fans are turned on or off. As long as both fans are on full blast at the desired temperatures, it really doesn't matter if you use a two- or three-relay system. In this configuration, it is likely that the low-speed single-fan operation can easily do most of the cooling work, and when conditions are necessary, the second fan jumps in to help out.

The Gen IV–style ECMs control the speed of the fans via pulse-width modulation (PWM), similar to some later-model electric fuel pump systems. The fan speed initiates at a small percentage of maximum when less cooling is required, then ramps up to full 100 percent when more airflow is required.

Since most builders will use a simple on-off electric fan, we won't be covering the requirements to have a PWM fan assembly. (You can buy aftermarket PWM fan controllers for swap-specific radiator fans if needed.) Rather we'll cover how to use that signal to control a single or dual fan. You would have to let your tuner know that you want to use a non-PWM fan, in which case the PWM output becomes a relay ground. If you use a dual-fan setup with a Gen IV ECM, you will have to Y-split this ground signal wire into two separate relays on the ground side of the activation circuit. If using the GMPP Gen IV harness kit, this uses a pre-wired single-power wire output that you can hook directly to the power side of your single fan or Y-split it as power for both fans. The GMPP harness comes with a pre-wired fan relay for simplicity.

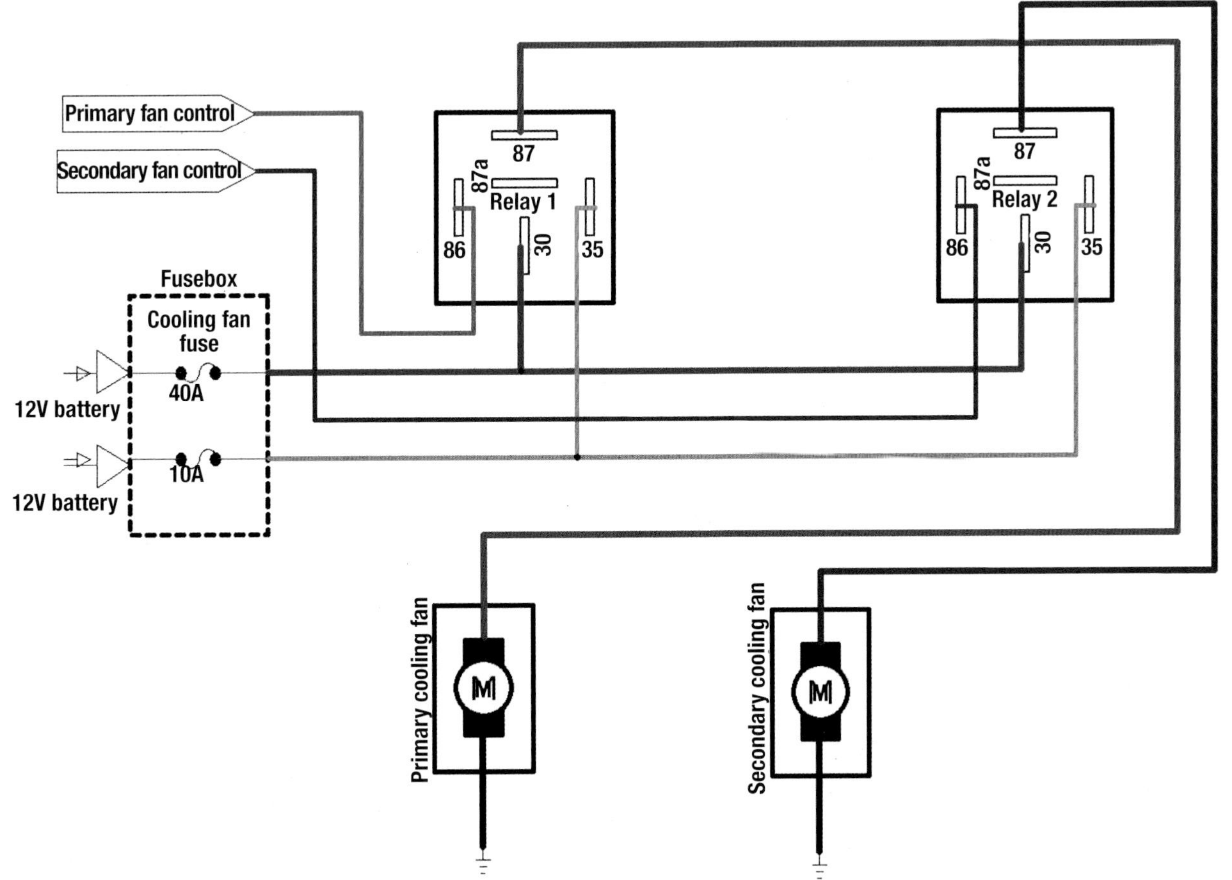

In this schematic the F-body cooling fan controls are functional as a one-fan or two-fan setup, each running at full speed. The difference is that the primary fan operates alone, then when cooling demands require more airflow, the second fan jumps in as well. This is a simpler and more understandable setup for the novice to wire in if the three-relay system proves to be too complicated.

Dual-Fan Speed Operation by Relay and Signal

Fan Status:	Relay 1	Relay 2	Relay 3	ECM Command:
Both Fans Off =	off	off	off	None
Both Low Speed in Series =	on	off*	off	Grounding Primary Fan Circuit Only
Both High Speed in Parallel =	on	on	on	Grounding Primary and Secondary Fan Circuits

*Relay 2 Provides ground path direction and changes both fans from low speed to high speed by using series or parallel operation.

When Relay 2 is "off," Fan 1 grounds through Fan 2, causing both to share a single voltage and grounding point; this is series operation and inturn causes slower fan speed.

When Relay 2 is "on" it changes the ground path to a direct ground and both fans have their own voltage source and grounding path, offering full fan motor speed.

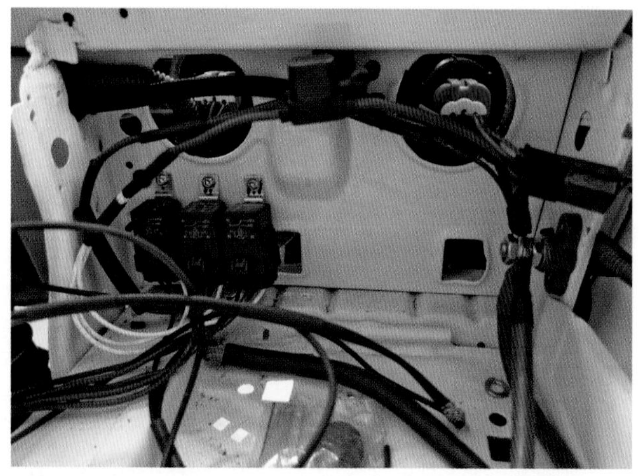

I used 4th-Gen F-body cooling fans on the 1991 Camaro, because the original cooling fan was an anemic single fan that would hardly cool the original SBC engine. Using the wiring harness schematic in this chapter, I built my own fan harness and three-relay bank to give me OEM-style low- and high-speed fan functions. The Holley Terminator X controls both the fan speeds using pins B3 (H) and B10 (G) on the eight-pin Input/Output Auxiliary Harness (P/N 558-400) included with Terminator X kits. If you need a complete, stand-alone fan wiring kit for your dual fans that includes everything you would need, check out Painless Performance P/N 30118.

USING AFTERMARKET FAN KITS

If you don't have access to the stock fans, or your new radiator does not have provisions for bolt-on fans, there are many companies that have single- and dual-fan kits. Be-Cool has dual-fan kits that match their radiator kits, Spal has single and dual universal fans with (preferred) or without shrouds, and Vintage Air has fan kits with fan shrouds, fan shrouds by themselves, and a couple of different options for large single fans that move some big air. Any of these kits can be used with the stock ECMs and wiring schematic or can be controlled independently from the ECM.

GEN III VS. GEN IV ELECTRIC COOLING FAN CONTROL

Gen III engines in F-bodies and C5s come equipped with dual fans, operating in a low- or high-speed mode. Gen IV engines in applications such as C6s and CTS-Vs come with a large single fan with an ECM controlled fan control module. The difference is that there are several fan speeds instead of the two low and high speeds that are typically found. Gen IV engines in other vehicles such as the 2010 Camaro have twin fans that operate much like their older counterparts with just two speeds; low and high. The ECM can be programmed for either setup, but will require some wiring if you are swapping from the single-fan design to the dual-fan design or vise versa.

Where some extra confusion may enter is when using Gen III–style dual fans with a Gen IV ECM that is not wired for dual fans (such as the GMPP wiring kit) since it only has a single-fan control. To solve this problem, the dual fans need to be wired through two relays with the single Gen IV wire controlling the ground or positive side of both relays. There is no simple way to control the fans on a low-high speed without an external cooling fan switch with the Gen IV ECMs.

Now, approaching it from the other direction and using a single electric fan with the Gen III controller, you can use a single relay and only use one of the two settings or outputs for fan control. You would need a beefy fan like the 18-inch Vintage Air setup to really move some air if only a single aftermarket fan is used.

TUNING FAN TEMPERATURE SETTINGS

Setting the ECM-commanded cooling fan activation temperatures can easily be accomplished by anyone with the HP Tuners software and its associated tuning cable. Even when using the Gen III truck ECM, you can add the electric cooling fan feature with tuning as early trucks came with a mechanical fan. You will need to add the wiring, and then basically "turn" the fans on in the ECM calibration. This calibration is simple to change so that the novice tuner can even calibrate the cooling fan temperature ranges with no dire effects. Anytime you replace the stock engine thermostat with a lower temperature thermostat (say a 160 degrees F instead of a 195 degrees F), you must recalibrate the fan settings to match, otherwise the engine will operate at the same temperature as before during slower stop-and-go speeds below 45 miles per hour and only run cooler at higher speeds.

The new thermostat will begin to open at 160 degrees F, but this is not where you want to begin operating the fans. You want them to come on and off above this temperature, otherwise they would stay running nonstop. You need a range above the thermostat to control the temperature. Since most if not all thermostats are from Hypertech, you can follow their fan setup instructions, which are as follows:

- Stock 195 degrees thermostat
  Fan 1 on/off at: 209 degrees F/200 degrees F
  Fan 2 on/off at: 219 degrees F/210 degrees F

- 180-degree Hypertech PowerStat
  Fan 1 on/off at: 194 degrees F/185 degrees F
  Fan 2 on/off at: 204 degrees F/195 degrees F

- 160-degree Hypertech PowerStat
  Fan 1 on/off at: 184 degrees F/175 degrees F
  Fan 2 on/off at: 194 degrees F/185 degrees F

When tuning the fan temperatures on a dual-fan setup, there are two settings: low speed and high speed. The low speed comes on after the engine gets up to temp, and if that provides enough cooling airflow for the conditions, then the high fans won't be necessary. In the tuning table you will see that the fan may come on at 212 degrees F and off at 202 degrees F. This is because you are using the fan to bring down the temperature, so you start turning the fan on at the higher temp and off at the lower temp setting, not the other way around.

With LS swaps, the stock car's radiator often can't keep up. For the Saturn Sky LS3 swap, the tiny four-cylinder stock radiator couldn't cool down a 600-horsepower LS swap, so I ordered a Ron Davis performance radiator (P/N 1-38SOL09). This radiator has two rows of 1-inch-wide tubes, netting a huge 2¼-inch core—which sure trumps the stock unit. It also comes with a performance Spal single fan and custom shroud. Note that new upper and lower brackets are required to provide engine clearance, which calls for some minor mounting fabrication.

AFTERMARKET RADIATORS AND FAN KITS

As is the case with many performance components, there are certain parts that are flexible enough for use in a multitude of engines in a certain chassis. Radiators and cooling fans are one of those flexible items. Properly designed cooling components such as these can support an SBC, BBC, or in this case, an LS-based engine platform. Some aftermarket radiators are specifically made for swaps, but *any* radiator can work provided it offers enough cooling. Any aftermarket race-style radiator is meant for high-cooling demands, so you can be assured that you are building a more than adequate cooling system.

There are a few differences in these radiators. The main items are the tank and radiator materials, along with core thickness. The stock radiator in your older car may have been a copper core and possibly only a one-row or two-row setup. This was likely more than sufficient since the engine may have been a run-of-the-mill 305-cubic-inch, 350-cubic-inch, or the like. Aftermarket radiators are built with overkill in mind to handle anything you throw at them.

Modern radiators use an aluminum core for high efficiency, low weight, and low cost. The race-bred aftermarket radiators expand on the aluminum strength and also use fabricated aluminum end tanks. Plastic-end tank radiators such as the stock LS1 are just clamped into place with tabs that are bent over the plastic, using a gasket for the seal. This is problematic with high-horsepower setups that are prone to expelling combustion into the cooling system and further pressurizing the cooling system. Often race cars that use the plastic tanks will blow the end off a radiator due to this pressure. They then end up with a real aluminum tank–equipped race radiator afterward.

The benefit of an aftermarket radiator comes into play when you want a no-hassle bolt-in cooling system. Many companies such as Ron Davis, Dewitts, Be-Cool, Howe, Champion, and others offer true bolt-in radiators for your many applications with the option to have single or dual fans included. There is no cutting, modifying, or cussing at them. Just take it out of the box and bolt it into place, like the stock radiator you removed. Of course, this is only the case with common RWD GM vehicles and the common LS swap vehicles that have already been sorted, but with some core dimensions on hand and your application made known, any radiator company will custom build and fabricate one to your specifications.

The dual fans that come with radiator kits are normally intended to be operated separately from the ECM using a temperature switch that grounds a fan relay at a certain temperature. With some wiring skills, you can wire these into the ECM just as a stock LS-equipped vehicle operates. Either option will work fine, but with the ECM method, you have more control in on-off temperature settings if you can flash your ECM.

Some aftermarket low-mount A/C compressor options exist that use a smaller compressor driven off the main accessory belt. These shift the compressor forward and away from engine mount interference often experienced with low-mount OEM compressors. Where these applications fit will vary, as you still need to make sure frame clearance doesn't pose an issue. *ICT Billet*

Chapter 7
Transmission, Torque Converters, and Clutches

Earlier we touched on properly mounting and mating LS-series engines to the transmission. In this section, we will discuss the actual transmission descriptions and what it takes for each setup to function with the LS engine and harness you may already own. From two-speed Powerglide transmissions to six-speed automatic and manual transmissions, this section should help determine which transmissions are compatible, which transmissions are not, and how to get your engine and transmission to function smoothly together.

There are many factory and aftermarket transmissions that will bolt to the LS engines. The good news is that the engine package is flexible enough with the SBC bell housing pattern to easily allow bolting up any GM transmission. The big plus with the LS engine is that the rear-wheel-drive versions retain the standard SBC and BBC bell housing bolt pattern introduced more than 50 years ago. The only difference is that the late-model transmissions bolt to the oil pan structure and require a longer torque converter hub or spacer due to the shorter LS crankshafts. Using an older SBC-style manual transmission, such as a Muncie, requires the use of a different pilot bearing to take up the offset from the shorter rear crankshaft flange.

While this manual is mostly an LS engine swap manual, the LS transmission choices are popular additions too. The huge benefits of a modern four- or six-speed overdrive transmission are sought after even with non-LS–based engines.

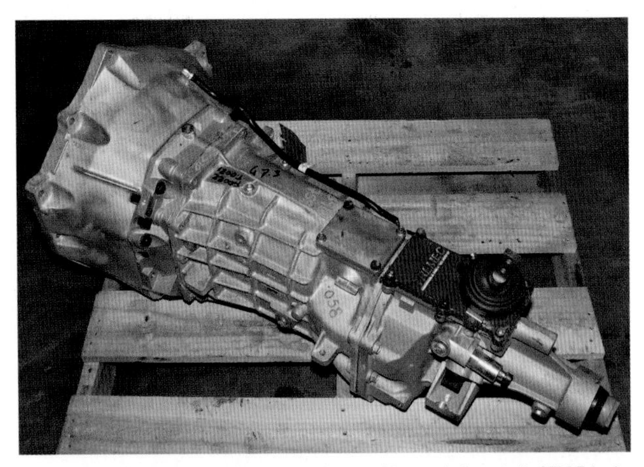

There is also a huge demand for six-speed manual transmissions, both OEM F-body donor and aftermarket T56 offerings. The main external difference between the OEM T56 and aftermarket T56-style transmissions is mainly the shifter location, although there are many internal strength improvements and upgrades to the aftermarket designs that the OEM T56 does not have.

Manual transmissions are the easiest to use due to the minimal amount of electronics required. The automatics on the other hand are not as simple due to complications like the ECM and the fact that all LS OEM transmissions are electronically controlled. This can complicate the swap project. The four-speed automatics are the most compatible with any setup, but are also weaker and, to some, not the latest and greatest.

IDENTIFICATION AND COMPATIBILITY

There are many sources available to help identify older two-, three-, and four-speed non-LS transmissions, so it is unnecessary for this manual to help identify the difference between a Powerglide, TH350, TH400, or any of the many others. I would rather use that real estate to help get the installations detailed with some extra information. Likewise, for older four-speed manual transmissions. Can these older V-8 transmissions be used with LS-series engines? The short answer is yes, and we will cover the equipment necessary to use them.

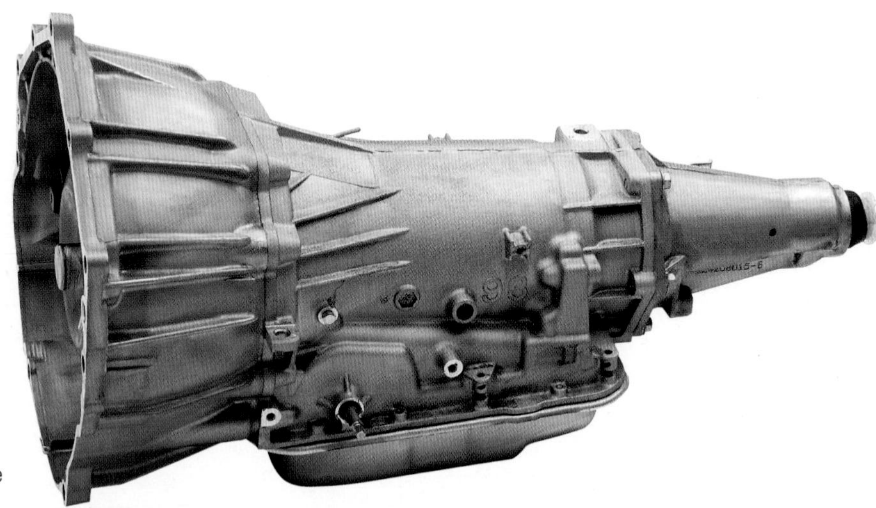

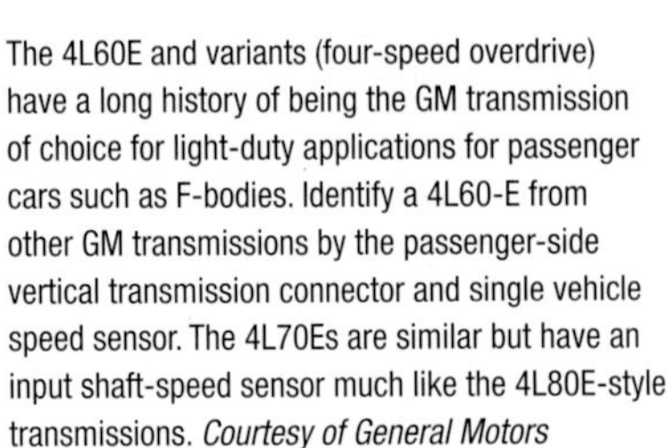

The 4L60E and variants (four-speed overdrive) have a long history of being the GM transmission of choice for light-duty applications for passenger cars such as F-bodies. Identify a 4L60-E from other GM transmissions by the passenger-side vertical transmission connector and single vehicle speed sensor. The 4L70Es are similar but have an input shaft-speed sensor much like the 4L80E-style transmissions. *Courtesy of General Motors*

The 4L60Es behind LS engines have a different torque converter and input shaft than the same transmission behind an SBC setup. The truck versions have a longer shift lever for a transmission-mounted PRNDL (Park Reverse Neutral Drive Low) switch, which may be easier to install and wire into your vehicle than a shifter-mounted PRNDL switch. *Courtesy of General Motors*

The 4L80E transmission and its variants represent the workhorse version of a GM four-speed overdrive transmission, comparable in strength to an older three-speed TH400. LS-style 4L80E transmissions share torque converter designs with SBC 4L80E designations, although when behind an LS-series engine such as the 6.0-liter, the flexplate is equipped with a hub spacer to center and support the torque converter hub. The 4L80E bulk connector is on the left side of the transmission, and it has both an input and output speed sensor. *Courtesy of General Motors*

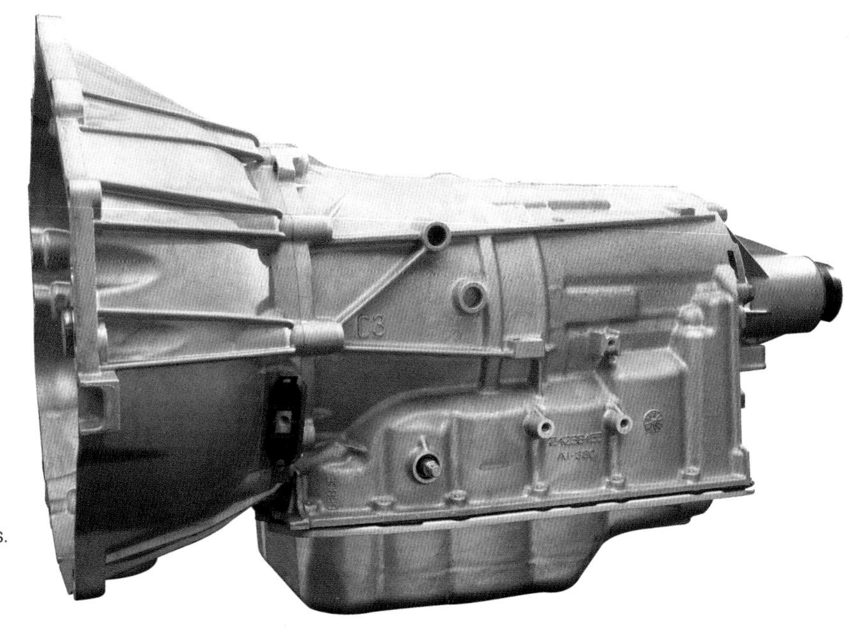

GM Powertrain developed a six-speed automatic design that is found in many 2007 and newer performance vehicles. There are two transmission designs, the 6L80 and 6L90. The 6L80 is found in light-duty trucks, Pontiac G8s and 2010–2015 automatic Camaros, while the 6L90 version is found behind heavy-duty trucks and the more powerful engines such as the supercharged 2009–2012 Caddy LSA. These have an internal TCM but will only communicate with automatic versions of the Gen IV ECMs. *Courtesy of General Motors*

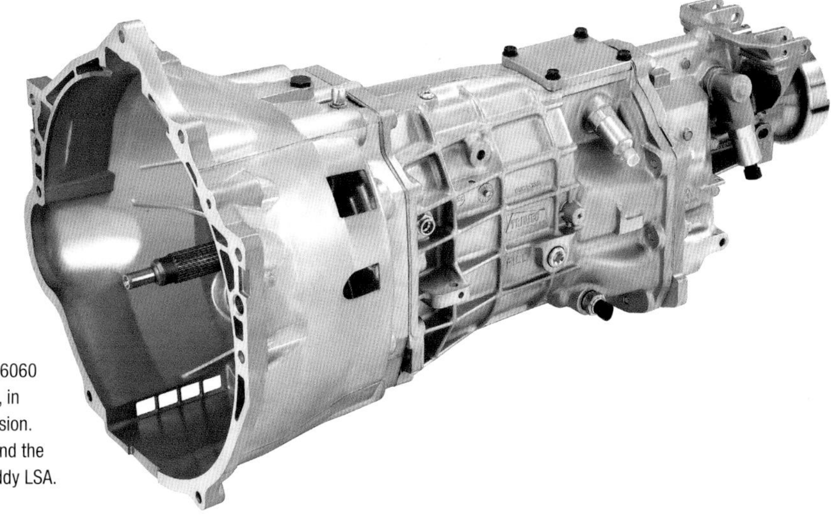

Another manual transmission to consider is the Tremec TR6060 six-speed transmission. It has some leftover T56 traits but, in general, should be considered a different style of transmission. This is the transmission behind the LS3 in new Camaros and the limited-availability manual transmission supercharged Caddy LSA. *Courtesy of General Motors*

The matching LS OEM transmissions we will identify with images are those of the electronically controlled variety that would come mated to the LS-series engines from the factory. These are the engine-mounted GM 4L60-E four-speed auto (and updated versions), GM 4L80-E four-speed auto (and updated versions), Tremec/Borg-Warner T-56 six-speed manual, Tremec TR6060 six-speed manual, and GM 6L80-E six-speed autos. The C5 and C6 Corvette rear transaxle drivetrain is not ideal for engine or transmission retrofits without extensive chassis and frame modification. The entire floor of the car would require extensive modification and, hence, the C5/C6 drivetrain will not receive a lot of attention in this book.

If you have a 4L60E and a Gen IV ECM/harness, you will need an external transmission controller such as the T42 GM OEM unit or wire in an aftermarket controller such as a TCI transmission control module. The GM TCM is factory equipped in 2005 and newer Gen IV engines with 4L60E transmissions.

The 6L80/6L90 transmission control module is located internally to the transmission, and it communicates with Gen IV ECMs via a network communication line known as CAN. While the Gen IV ECM does not control the transmission, it does need to be compatible and "know" which transmission it is mated to. You will need a similar ECM as the engine and transmission combo you are installing.

AUTOMATIC TRANSMISSION CONTROLS

The method you employ to control your GM electronic transmission will vary, depending on which engine harness and ECM you are using. GM Gen III ECMs can control GM four-speed electronic transmissions, both the 4L60-E and 4L80-E. If you are using a harness that came from a six-speed manual vehicle, you can morph the automatic add-on harness into the six-speed manual harness. Gen IV ECMs can control (or rather communicate with) compatible model year four- or six-speed automatic transmissions via the transmission control module (TCM) or transmission control unit (TCU).

Speartech Fuel Injection Systems offers a pre-terminated and marked add-on harness for a conversion from six-speed to automatic. The best thing about this is that it is extremely easy with the correct pin locations identified. The only splicing that will need to be done is the shifter park and neutral position switch and the power and ground connections for the transmission harness itself. These can be wired into existing splicer packs located in the engine harness or tied separately into dedicated 12-volt ignition power.

When using a 4L80-E transmission with a 4L60-E harness, Speartech also has you covered. They offer a transmission adapter plug that converts the 4L60-E connector to a 4L80-E plug in. You must also add the input shaft speed sensor into the ECM harness, though this, too, is an easy addition with only two wires needing to be added. ECM calibration is also required. The easiest way to accomplish this is to have your

ECM base tuning flashed to a truck transmission calibration that was equipped with a 4L80-E transmission. Other than the physical work of fitting the larger 4L80-E into your chassis (custom transmission crossmember, transmission tunnel interference, etc.), the electronics are fairly painless when compared to the rest of the LS-swap project.

GEN IV TRANSMISSION CONTROLS

Gen IV transmission controls have been removed from the engine ECM as mentioned in the ECM harness chapter. The transmission-control functions are carried out by the transmission-control module. Gen IV engines with 4L60-E TCMs are mounted near the ECM, while six-speed automatic transmissions have the TCM mounted inside the transmission itself. Communication between the ECM and either design TCM is carried over a network called a CAN cable. The Gen IV stock TCM is not compatible with the Gen III controllers, as the Gen III controllers do not communicate data similarly.

When using the Gen IV–style harnesses, the TCI/FAST TCM can communicate these sensor values through the CAN bi-directional Communications Link. The transmission controller is controlling the shifts and communicating back and forth with the ECM. The ECM sends the required inputs to the TCM over the CAN line.

When using a non-matching OEM engine and transmission package such as a Corvette C6 engine with an auto from a conventional RWD chassis, it makes sense to have the ECM match a like-year package. For example, a 58× LS2 with a four-speed automatic (4L60E style) using the E67 ECM with a T42 transmission control module can be base programmed to a 2007–2009 Trailblazer SS. On the other hand, the earlier 24× LS2 with a four-speed automatic will need an earlier base tune and ECM to match, such as that from a 2006 Trailblazer SS, or 2005–2006 Pontiac GTO with the same automatic transmission and T42 TCU.

Another option for use with any LS engine controller is a stand-alone transmission controller such as the TCI/FAST TCM for four-speed auto transmissions. When using a Gen III–style engine harness, the stand-alone controller does require certain inputs such as engine rpm, throttle position, engine load via MAP sensor, and vehicle speed from the transmission output shaft speed sensor. All of these can be either T-ed off existing sensors or accessed by tying into certain stand-alone engine harness outputs. GMPP harnesses and others and some other harnesses use these outputs in an optional bulkhead fitting made for this purpose.

HOLLEY ECU TRANSMISSION CONTROLS

Holley's EFI systems have helped move the LS swap world forward by leaps and bounds. Holley EFI has also added four-speed GM electronic transmission controls to certain Holley ECUs. To use these transmissions, you will need the Holley Dominator ECU or the more budget-friendly Holley Terminator X Max. Both have extra inputs and outputs to control the GM transmissions using the mostly plug-and-play Holley P/N 558-405 in connector J4 on the ECU itself.

If you have the Terminator X Max, you likely ordered it for the 4L60/80 transmission control; if so, it will have come with the required harness. Using your handheld or Terminator X software, you simply add the transmission controls to the ECU software and it seamlessly integrates into your Holley EFI controller. You may need to tweak shift points and firmness for a custom transmission tune for your application.

GM Performance Parts offers an external transmission controller for 4L60E transmissions compatible with any ECM style, or even a carbureted engine with similar sensor outputs. While the Gen III ECM has the internal capacity to control GM 4L60E transmissions also, you would need to wire in an add-on 4L60E harness to a manual-equipped engine harness if you want a 4L60E-controlled by the factory ECM. *Courtesy of General Motors*

Another option for an external transmission controller is from TCI. Their newest transmission control unit setup has the ability to control every aspect of a 4L60E or 4L80E when wired into the stock engine harness for engine sensor outputs. *TCI*

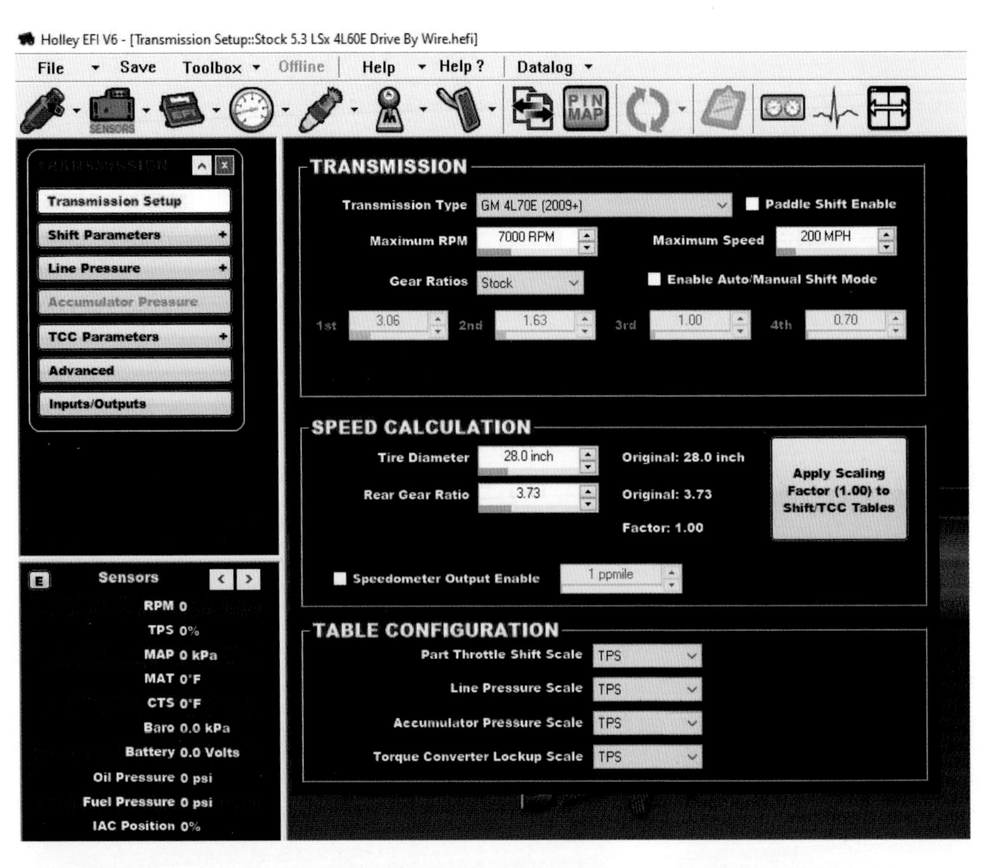

The Holley Terminator X Max not only has EFI engine management, but it can also control your four-speed electronic transmission, such as the 4L60E and 4L80E units. Terminator X software is easy to use. With an LED screen or a laptop connected, you can easily tweak things like shift timing, firmness, speeds, and torque converter clutch strategy.

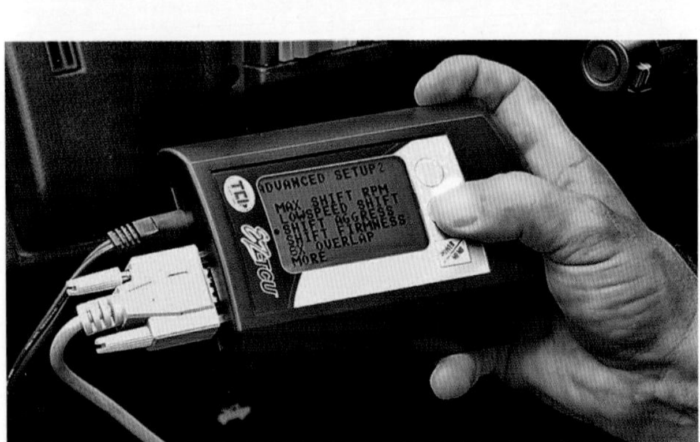

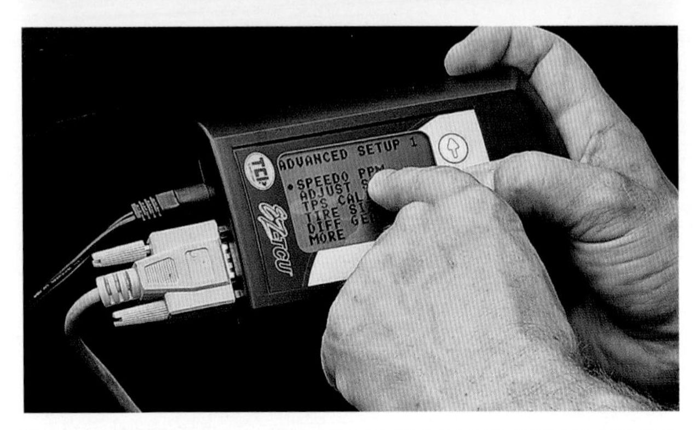

The newest TCI transmission control unit varies from the older designs that you had to program through a laptop. The latest generation TCU is able to be programmed through a touch screen. You can program shift points, shift firmness, and downshift points all though the handheld programmer. *TCI*

The Holley transmission harness (P/N 558-405) plugs into the J4 Holley connector at the ECU and the transmission connector plugs into the bulk connector on the side of the transmission along with the vehicle speed sensor for speed input. This harness fits with the Holley Terminator X Max and Holley Dominator ECUs only. *Holley*

Note that, while the Holley HP EFI and base Terminator X do not have electronic transmission control, this can still be added via aftermarket transmission controllers. It will just function as a separate entity and not in collaboration with the engine controller. It is best to start with the right ECU if you have transmission plans that align with the Holley compatibility. If you have a manual transmission or a "non-E" transmission, this is much less of a concern with EFI choice.

For SBC automatic transmission reuse such as Powerglide and TH200-/350-/400-style transmissions, you will need a converter hub extension if your torque converter isn't made for an LS engine swap. ICT Billet's offering (P/N 551165) extends the converter pilot into the back of the crankshaft for necessary centering and support. *ICT Billet*

Rather than taking the easy way out and using leak-prone rubber transmission line and hose clamps, the smarter and safer method for running custom transmission cooler lines is to use AN fittings and hoses.

Often when you couple an aftermarket torque converter with transmissions originally intended for SBC-derived buildups, you will run into a difference in torque-converter bolt patterns. The misalignment is shown here. You can redrill the flexplate and hope it works.

Or you can purchase the 4L80E-style flexplate and spacer kit for a bolt-in setup. Since the 4L80E shares the same torque converter as the older SBC trucks, the flexplate matches most aftermarket TH-style torque converters.

LS OEM TRANSMISSION INSTALLATION

The physical fitment of all the non-Corvette OEM LS-style automatic transmissions to the engine is painless. You will need an automatic flexplate flywheel that matches your transmission's torque converter and crankshaft flange bolt pattern. The majority of flexplates are similar and interchangeable, but there are a few small quirks to be aware of when dealing with the LSA and LS9 engines as these have extra flexplate bolts compared to the six-bolt flange commonplace to *all* other LS engines.

The conventional six-bolt (crankshaft flange) single-plate flexplate uses three torque converter bolts, which matches the stock setup on 4L60s and the 6L80s. The 4L80 and 6L90 use six torque converter bolts, so make sure you get the right flexplate if you need to make use of those three extra bolt holes. The easy way to do this is to order the automatic flexplate similar to the year and model of your engine.

The early OEM 4L80-E vehicles came equipped with a longer crankshaft due to the SBC-style torque converter that is retained when being used in the early Gen III

If your car already had a TH350 or TH400 transmission that you wish to reuse, you're in luck. The LS engines share the same SBC/BBC GM bell housing bolt pattern, although you will only be able to use five of the six bell housing bolt locations. Due to the shorter cylinder case of the LS engines, the bolt hole at the 2:30 position is not retained.

Check transmission to tunnel clearance with the engine in place and the transmission located as close as possible to its final resting place. The OEM LS transmissions are all dimensionally a little larger than original three- and four-speed transmissions. I had this 4L65E transmission in and out a half-dozen times in order to make more room without cutting the transmission tunnel when the 700R4 that previously occupied this space had inches of available room. When sufficient clearance is obtained, you can do a final install of the transmission and crossmember.

In the not-so-distant past, 6L80 and 6L90 transmissions were unpopular for LS swaps. There are always some nuisances to deal with, such as transmission tunnel clearance and crossmember fabrication, not to mention the CAN bus electronic requirements. You don't need to reuse donor cooler lines anymore, as ICT Billet has these 6L80/90 transmission cooler fittings (P/N 551121-6L) that adapt to -6AN, which can provide you with the ability to have custom transmission cooler lines made and routed as needed. *ICT Billet*

6.0-liter powertrains. Since GM still offered the Vortec L31 350-cubic-inch/4L80E combo in heavy-duty cargo vans and old-body HD trucks and that the 6.0-liter/4L80 combination was a year later as compared to the 4.8-liter/5.3-liter new-body trucks, GM made it easier to use the 4L80E with the LS engine by introducing a longer crankshaft into the 2000–2001 6.0-liter LQ4 engines instead of building a new torque converter. The longer crankshaft rear flange matches the SBC torque converter dimensions for the 4L80-E. Starting in certain 2001 trucks, GM used a different flexplate and a spacer instead of using a longer crankshaft. GM offers the same spacer kit with or without a flywheel for use with 4L80-Es or for any other SBC-style transmission and converter combination. The GM 4L80-E flat flexplate is 12551367, and the spacer is part number 12563532.

Note that very few early 1999 4.8-liter engines with the manual transmission also came with the longer crankshaft. These should be treated the same as those with the 6.0-liter long-crank issue. The early 4.8-liter engines have limited compatibility with other transmission choices and should only be used with the manual transmission they come with or the early design 4L80E transmissions (it seems pretty crazy to pair the 4.8-liter engine with such a beefy transmission). Luckily, 4.8-liter swaps aren't as popular as the 5.3- and 6.0-liter truck engine donors, but you may run into these in the wild as mislabeled 5.3-liter engines. Just check the rear crank flange length and compare it to the measurements provided.

All OEM LS transmissions bolt to the block using eight M10 × 1.5–millimeter transmission bell housing bolts. ICT Billet has a full bell housing bolt set (P/N 551652) if you do not have one. This set can function as the T56 transmission-to-bell housing bolts as well. Two of these are fastened to the stock oil pan, so if you are using an aftermarket oil pan, only six will be used. Keep in mind that you will want to fit your transmission dipstick tube into place if using a four-speed automatic before installing the transmission or engine assembly, as it is nearly impossible after the fact. If you are installing the transmission separately from the engine (engine first, then trans), you will need either a transmission jack or a common floor jack (and perform a balancing act).

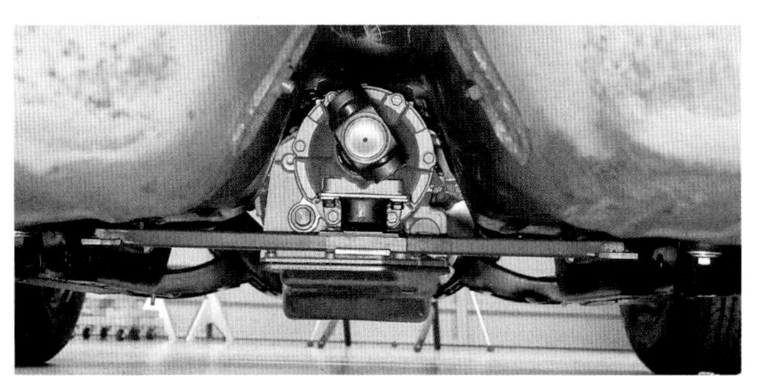

Transmission crossmember fabrication may be necessary when using a non-OEM (to chassis) transmission. Companies such as Competition Engineering have universal transmission crossmembers that give you all the pieces to assemble a crossmember in your custom location. Here, the 6L80 transmission crossmember was fabricated using pine as a template. Then a permanent crossmember was put together using plate steel and steel tubing. *Ian McDonald*

The old Turbo-Hydramatic–style transmission dipsticks will require an additional bracket to bolt to the transmission. That original location does not have threads in the block (save for the GM LSX block), but you can drill and tap the unused bolt hole in the transmission case with a larger and very short ½-inch bolt, add Heli-Coils, and use a short ⅜-inch bolt. That said, the easier solution is to use the ICT Billet dipstick adapter bracket (P/N 551323).

The factory LS transmission dipstick tube options bolt to the cylinder head or the next lower transmission bell housing bolt via a stud and M8 nut. The adapter is only required for the SBC/BBC-style dipstick tubes. Aftermarket flexible dipstick tubes are typically universal enough to be available for an alternative bolt hole.

If using a new crate transmission, make sure to pre-fill the torque converter with as much transmission fluid as it will take; usually a quart is enough. The four-speed autos can use Dexron III or better; the six-speed autos require Dexron VI, which is synthetic based and a little more costly than normal ATF.

The six-speed automatics are meant to be maintenance-free, so there is no dipstick provision to worry about on the car versions (the trucks still have a dipstick tube). You pre-fill the 6L80 and similar transmissions with Dexron VI via a small plastic plug on the passenger side of the transmission. Refer to the fluid capacity of the donor vehicle for exact fluid amounts, as these vary depending on vehicle and transmission pan design. Getting it close is all that matters before starting it, as the level needs to be set with the transmission fluid hot. There is what appears to be a small drain plug in the pan, but appearances are deceiving. This plug is for setting the transmission oil level once the engine is running and warm. The fluid level is set to the top of the threads of this plug and will drip fluid out when checked at a specific temperature.

When the transmission is in place and lined up, install and torque the bell housing bolts to 44 foot-pounds, but do not use the bolts to draw in the transmission to the engine. Get it as close as possible by hand pressure and make sure the torque converter freely spins without dragging the flexplate when fully installed. Then align the torque converter bolt holes with the flexplate bolts through the starter access hole. There is one oval bolt hole in the flexplate that is intended to be installed first. If you do otherwise, the other two bolts will not align. If installing an aftermarket converter with replacement 7⁄16–inch-diameter bolts, the oval bolt hole needs to be enlarged slightly. Use a Unibit to drill the hole if needed.

Once all three bolts are started, finger-tighten until all are fully installed but not tight, then go back and torque the bolts in rotation. It is recommended to check converter bolt fit and alignment against your flexplate prior to installing the transmission, as you will not want to remove the transmission after fully bolting it to the engine and remaining drivetrain to enlarge the bolt holes.

NON-LS AUTOMATIC INSTALLATION

As mentioned earlier, the LS engine family uses the conventional SBC-style bell housing bolt pattern. This opens the door for non-LS transmissions such as older mechanically operated and hydraulically shifted two-, three-, and four-speed automatics.

Why would you use such a transmission with a high-tech engine package? For one, this may be something you already own, so to save costs during the engine installation, it may help you finish the project by reusing some existing components. Additionally, your existing driveshaft length, transmission crossmember, and shifter provisions may be unaffected by the LS engine installation. Further, these transmissions are often inexpensive to purchase and rebuild while being really stout even when mildly built.

Like the 4L80-E transmission "factory adaptation," the older SBC torque converters will require a flexplate spacer kit to support the center hub of the torque converter. This is available from GM in kit form or as a spacer itself. Many aftermarket torque converter companies will offer spacers or extend the torque converter hub for the LS engines at your request at little additional expense. Check first with your converter source before buying the GM spacer kit, as it may not be necessary.

The 6L80/6L90 transmission pans provide the transmission level indicator by this "drain plug," which is actually the fill plug. You fill up the transmission fluid level using Dextron IV through this plug or the small access hole on the passenger side until fluid drips out of the fill plug when the transmission is up to operating temperature. All other transmissions have a conventional dipstick and tube.

Both the 4L80E flexplate and 1998 LS1 flexplate have matching SBC converter bolt-mounting holes. Without the matching holes, you will either have to redrill your existing flexplate on a drill press or elongate the stock LS converter bolt holes quite a bit. From my point of view, it's easier to buy the right part than modify, especially if you have no access to the correct modification tools or just want a stress-free bolt-in deal. The torque converter must be centered exactly and spaced evenly with the crankshaft centerline, otherwise the torque converter internal stator will be off-center and will eventually lead to possible vibrations and broken transmission parts. Usually the torque converter and stator support will suffer.

Another similar option to the GM kit is the TCI flexplate setup (PN: 339754), which comes with a slip-on converter hub extension and matching converter bolt pattern for either LS or SBC converters. It does cost a little more, but the TCI flexplate is SFI-approved for race use and very durable. I recommend the use of the stronger ARP mounting bolts with the TCI flexplate. For the GM spacer setup, you can use only the longer GM bolts. Hughes Performance also has an LS-compatible flexplate and torque-converter spacers for use with SBC torque converters. The Hughes flexplate is part number HP4004, and their torque converter hub spacer is part number HP3795. If using an LS-specific torque converter, the spacer is simply not used.

Either flexplate should work with any LS-style starter. The starter parts numbers vary between years and the wiring connects slightly differently if you compare older Gen III (bolt-on starter wire) to newer Gen IV designs (clip-in starter wire), but so does everything else in an LS swap. The flexplate/flywheel ring gear is located in exactly the same spot and is

Speedometer connectors may be different, depending on the year of the transmission. If you are using a TH400 and an electric speedometer, you can use the early 1990s 454SS pickup truck speedometer hardware to have an electric output.

In older cars that have cable-operated speedometers, if you wish to keep the speedometer functional but are using a late-model electronic transmission, you will need to either have the tail housing of your transmission fabricated for the vehicle speed sensor or you can add this handy Abbott Cable-X speedometer cable drive to operate the cable speedometer. *Abbott Enterprises*

the same diameter for all LS engines, so using the matching bolts and wiring should be all that is required. You can make it easier on yourself by simply ordering the same starter as year and make of your engine.

At this point you can finish the installation by fitting or mocking the transmission crossmember, then torque converter dust shields, and finally the starter. Once the correct driveline angle is set and the transmission crossmember is in place with no further changes necessary, you can measure for driveshaft length. Use this dimension to have a new driveshaft made if your existing driveshaft is incompatible or needs to be lengthened or shortened. More details on this later in the chapter.

CHECKING OR ADJUSTING POWERTRAIN ANGLE

To check powertrain angle, you need an angle finder. These are available from a number of sources. Central Tools offers one with a magnetic base that can be used for other purposes such as driveshaft or pinion angle later on if suspension adjustments are in your future. To measure powertrain angle, set the base of the angle finder on a flat object like the base of the transmission pan and take your measurement. You ideally want the powertrain angle to match the rear-axle angle and point slightly downward in a RWD rear-axle application, which is normally 3 degrees to keep the U-joints happy. If the powertrain needs adjustment, you will have to change the angle via the transmission mount location. Shimming the transmission up higher will result in a lower angle, while relocating the transmission downward will increase the angle.

TRANSMISSION COOLER LINES

Why bother with a transmission cooler? As a general rule, each 20-degree F drop in temperature can double the transmission's life. Convinced?

Transmission cooler lines are required for any automatic transmission. I prefer steel or AN lines as the rubber lines (even when made for transmission oil) tend to loosen over time. Steel lines require a flare nut and 45-degree double flare to seal properly to the transmission. AN lines can be made from -6AN hoses and fittings but would require a transmission cooler with -6AN fittings or flare to AN adapters to utilize with the stock transmission cooler in the radiator.

Depending on the transmission selection or retention, you may already have a transmission cooler in place with lines and fittings. If changing to an LS-style transmission, you will at least have to modify your existing lines. Worst case is you have to build new lines from scratch.

To do this, I recommend either steel replacement lines with flared fittings or building new AN lines using braided AN hoses. If you choose steel lines, they should be at least ⅜-inch diameter throughout and can run to the factory transmission cooler in the radiator. An even better option is to supplement the factory cooler with an external transmission cooler mounted in front of the radiator's fresh air-stream. If using AN lines, you can adapt into the stock radiator cooler, but it is usually cleaner and easier to use a Permacool or B+M transmission oil cooler specific to AN lines.

When using the radiator equipped with transmission cooler and building completely new lines, remember that the inlet feed is the lower fitting on the radiator, and the upper line is the return. When tying in a transmission cooler in series to the internal radiator cooler, you must tie it into the return line path, as the integral radiator cooler takes the bulk of the fluid temperature out (they are quite efficient), while the external cooler cools the fluid further using outside air. If reversed, it will still work. It just won't be as efficient.

LS 4L60E and 4L80E transmissions use quick-connect transmission cooler line fittings. In the right application you can use the entire system from the donor vehicle. If you use the same transmission and radiator, you can use the cooler lines, often with only with slight modifications. If anything, you may have to extend the rubber lines a few inches to mate up properly.

Try to avoid using full-rubber lines from the transmission to the cooler. Hose-clamped transmission cooler hoses have a tendency to leak or pop off, inevitably at the most inopportune of times. It's ok to use rubber transmission cooler lines where rubber line already existed, like on the flexible joints near the radiator or when adding an external cooler. These OEM end fittings have a better grasp on hoses than most fittings or cut steel tubing hose ends.

If the stock quick-connect fittings will not work, you can remove the stock quick-connect fittings and use either AN or flare fittings. The GM flare fitting for a 4L60E transmission is part number is 8637742; the part number for the AN adapter fitting is Russell 660441, which is a ¼-inch NPT fitting adapter to -6AN. Use a light addition of Teflon paste to the fitting threads.

Since we have mentioned the 6L80/6L90 installation, the transmission cooler lines on these bolts on to the transmission case much like a stock A/C hose, using a single central bolt. The lines bolt on the driver's side of the transmission, which makes having an OEM transmission cooler line just about mandatory to have an easier starting point to build the remainder of the lines. Billet adapters also can convert these to AN hoses if you do not have the OEM donor lines. Once the lines are purchased, you can adapt them the remaining distance to the radiator or external transmission cooler. In OEM form, they are routed under the bell housing and follow the right side of the oil pan until ending near the front of the stock A/C compressor location, making easy work for the bulk of the transmission cooler line distance. At this point you just connect the gaps between the cooler lines and the transmission cooler with transmission oil hose. In this configuration, the line closer to the engine is the return line back to the transmission, while the outer line is the supply from the transmission.

NISSAN 6-SPEED TRANSMISSION ADAPTATION

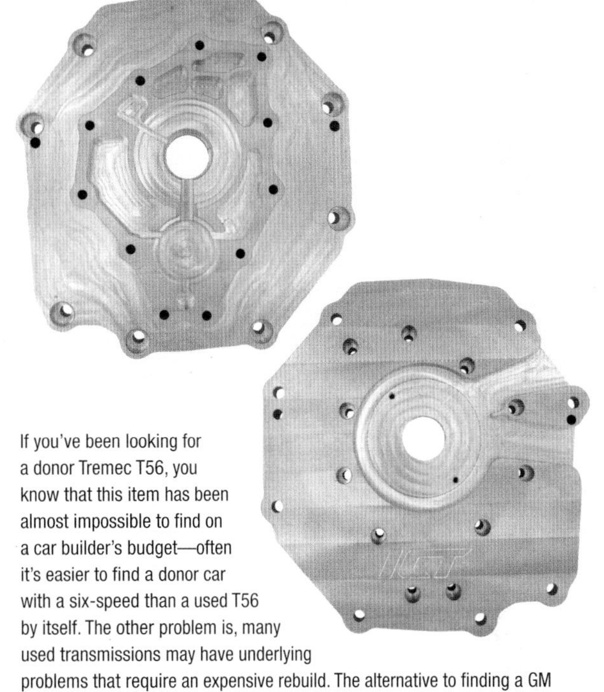

If you've been looking for a donor Tremec T56, you know that this item has been almost impossible to find on a car builder's budget—often it's easier to find a donor car with a six-speed than a used T56 by itself. The other problem is, many used transmissions may have underlying problems that require an expensive rebuild. The alternative to finding a GM six-speed is to go with a new aftermarket transmission such as the Tremec T56 Magnum, but this too can be a pricey path to follow. Here is an alternative solution from ICT Billet, employing a less expensive Nissan CD009 six-speed. *ICT Billet*

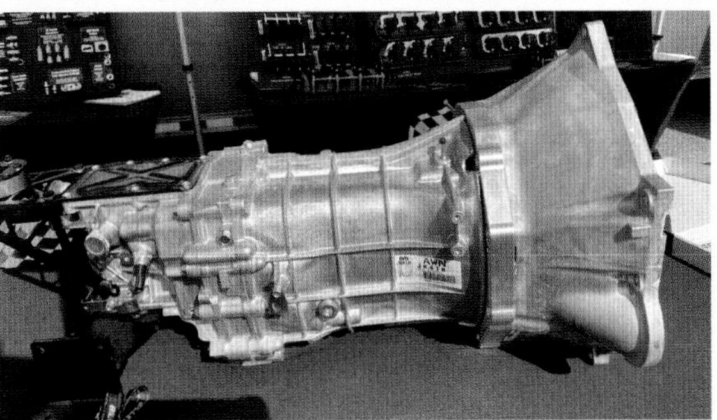

ICT Billet makes two versions of their billet adapter that configure the CD009 transmission to the LS engines, depending on the transmission variation you source. First, cut the old bell housing off at the second rib of the casting, then bolt on the new replacement front plate, which adapts the Nissan transmission to a replacement GM LS1 T56 bell housing. This provides a location to bolt on the factory GM clutch release bearing as well. The kit also comes with a new clutch disc and a pilot bearing to adapt the GM crankshaft to the Nissan input shaft.

AUTO SHIFTER CONSIDERATIONS

Shifters are available in a wide array of styles. You can use a modified stock shifter for your car, a donor shifter from an LS1 F-body for a 4L60E or 4L80E, or you can use an aftermarket automatic shifter from companies like B+M or TCI.

If reusing your OEM shifter and it has the wrong gear markings on it, there are companies that will modify the shifter indicator plate to match your transmission. Often you will have a two- or three-speed shifter that is swapped for a 4L60-E or 700R4 transmission, and in instances like this, the matching shifter indication plate adds that final touch if you are looking for that OEM "restoration" look many strive to have.

The main concern when using an aftermarket shifter with an electronic transmission is sending the correct signal to the ECM to indicate that the transmission is in gear to let the ECM know which idle speed engine target rpm is necessary. The stock neutral safety switch built into F-body shifters indicate park and neutral for starting purposes and activate the reverse indicator lights.

Aftermarket shifters have *only* park or neutral switches and the reverse light switch. There is nothing built-in to determine an "in-gear" signal to the ECM. There are a few ways to get around this. One is simply adding a pair of relays into the park, neutral, and reverse signal wires. When the shifter is not in any of these positions, it only has forward gearing choices left. The proper signal is obtained by reversing the operation of the relays to provide a grounded signal path for the ECM.

The other way, and possibly a plug-n-play option, depending on the harness used, is when using the 4L60E or 4L80E from a truck. All 4L80s have a park/neutral switch on the transmission case, and the truck version of the 4L60E family also has the park/neutral switch on the shifter side of the transmission. One could even convert the car version of the 4L60E to the truck version by installing a longer shifter input shaft and then bolting on the truck park/neutral switch. If this route is taken, it frees you to use any shifter you wish, because all electronics will be on the transmission itself.

The four-speed automatic transmissions are the only ones that require gear shift indicators to communicate shifter position to the ECM. The six-speed automatic transmissions have all required electronics internally. Now, even with the newfangled six-speed automatic, you still need to wire in your park/neutral switch for starting out of gear and your reverse light switch for obvious reasons.

In factory applications this is done over the CAN bus network. If you have a gateway module that decodes this, you can reuse the signal. If you do not have that capability, you will either need to use the aftermarket shifter or the Dakota Digital gear indicator (P/N GSS-3000), which has worked great in some swaps with 6L80 transmissions. The Dakota Digital provides everything needed and additional gear indicator perks if you happen to have Dakota Digital gauges.

Pressing the pilot bearing into place requires hammering it with either an installation tool or a large socket of the same diameter. An old transmission shaft works extremely well also. Gen IV bearings are a bit easier as they are not recessed as much as the Gen III–style pilot bearing.

CLUTCH AND MANUAL LS TRANSMISSION INSTALLATIONS

All OEM manual transmissions that have been paired with LS engines are six-speed models. They are all based on the Borg-Warner T56, which is now built by Tremec. F-body transmissions are the most common core transmissions to use, but also some of the more expensive versions, even as core purchases that need to be rebuilt. These transmission variations are also found in Pontiac GTOs and Cadillac CTS-Vs. The 2010–2015 Camaro, G8 GXP, and 2009–2012 supercharged CTS-V have a slightly different version (TR-6060), but the same issues arise when installing into older cars, including extra deterrents such as the need for an aftermarket shifter to correct the weird location of the remote shifter and a custom slip-yoke splined driveshaft.

There seems to be a lot of confusion on which pilot bearing is necessary for which manual transmission. The larger "LS2" version on the left is ideal for older SBC-style transmissions with a .400-inch shorter input shaft and for Gen IV–style transmissions (sans GTO) such as the TR6060. Every manual Gen III LS–derived T56 transmission uses the smaller Gen III–style bearing on the right. This is coincidentally the same dimensions as the SBC roller pilot bearing as well.

Note that there are two different T56 bell housings, which look identical: an F-body version and a Pontiac GTO version. These are interchangeable, save for the starter. If you use the GTO version, you need the slightly different GTO starter. I bought the GTO bell housing and later discovered my LS1 starter would not fit the GTO bell housing. Fortunately, you can grind the bell housing and starter, but I just ended up buying the correct starter to make it work.

First you have the mechanical portion of the installation. You will need to have a clutch, either stock or aftermarket replacement, and a pilot bearing on hand before proceeding with the transmission installation.

With your new clutch you should have received a clutch alignment tool. If not, go borrow or purchase a GM 27-spline clutch-alignment tool for the installation. To install the pilot bearing, you will need a hammer and pilot bearing installer. You can use a large socket if needed.

If using the small OEM LS1-style pilot bearing, drive it in until the rear face is aligned with the larger hub flat, it can go a tad farther but as close to this location is perfect. The larger Gen IV–style bearing taps into place in the converter hub recess of the crankshaft. Keep this in mind if you are moving from a manual to an automatic as it will need to be pulled out. The larger pilot bearing (usually required with older non-LS manual transmissions) just gets pressed fully into place until it stops.

After the pilot bearing is installed, the flywheel can be mounted. Line up the witness hole in the flywheel with the matching hole or dowel in the crankshaft, then use six flywheel bolts (or eight for LSA engines) and a dab of thread-locking compound if new bolts are not used (red or blue Loctite). If new bolts are bought, you don't need an additional thread locker. Torque in a star pattern to 74 foot-pounds. After the flywheel comes the clutch disc and the pressure plate assemblies. These use six bolts that are torqued to 45 foot-pounds. You must use a clutch spline alignment tool to center the clutch disc to the pilot bearing and crankshaft centerline. Once the clutch is installed, you can install the transmission bell housing; the stock T56 bell housing is bolted to the engine using eight M10 bolts, six in the block and two into the oil pan.

CLUTCH CHOICES

OEM "Donor" Clutches

There is a trend to stick with what GM provides for clutch choice, mainly stock LS1 and LS6 clutches. The few upgraded choices available that are better than the original LS1 type are the LS6, LS2, and LS7 clutches. The LS6 is marginally better than the stock LS1 clutch. The clutch disc itself is no different but does require a slight increase in pedal pressure. With the LS6 horsepower rating of 405 and LS1s with bolt-ons not far behind that, this really is not a good clutch choice unless your engine is stock and staying that way. The LS2 clutch handles about the same horsepower level as the LS6, but requires the LS2 dual-step flywheel rather than the flat LS1–LS6 style flywheel.

Last up for the LS engine is the replacement LS7 clutch setup. Factory rated at 505 horsepower, this clutch setup can handle significantly more power than the LS1/LS6/LS2–style clutch; it would be my go-to clutch for naturally aspirated street setups with mild power. Remember the LS7 Z06 Corvette is several hundred pounds lighter than the F-body and makes about 450 rear-wheel horsepower to the ground. So, it does not require the best of the best clutch to hold that amount of power in a chassis of that weight. This is an acceptable choice for a mildly modified LS vehicle that is set up for mostly street driving with limited race use, but there are much better and capable aftermarket clutches for just few dollars more if you are putting some big power to the ground.

The LS7 clutch bolts in as an OEM replacement. Even though it is a different design, with a larger clutch disc than what you may have removed, it has the same dimensions where it matters for the clutch actuator. For these transmissions, you reuse the same clutch actuator that the car originally came with, which fits all 1998–2002 F-body, 2004–2006 GTO, Pontiac G8, and 2010–2015 LS3 Camaro six-speeds.

Almost anything that applies to the Tremec T56 transmission clutch install also applies to the aftermarket variations such as the Magnum and the TR-6060, other than shifter and driveline variations. You may need a different shifter design or a different driveshaft yoke when you install these versions of six-speed manual transmissions as compared to conventional T56 six-speed installs.

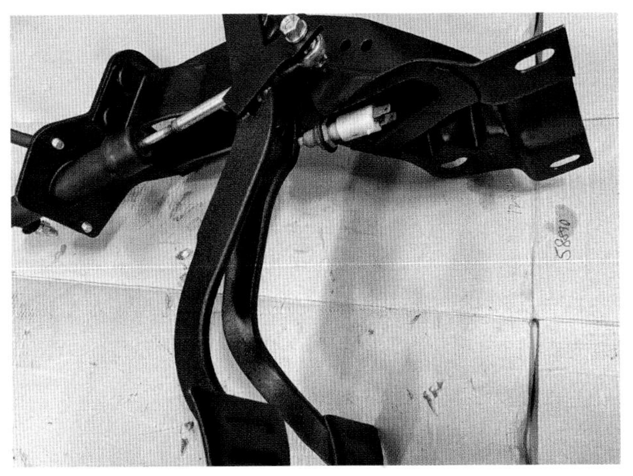

With LS manual transmissions, you can take advantage of modern clutch hydraulics and retrofit newer systems to the classic cars. Some kits are more complete than others, so a little fabrication or welding is sometimes required. On my 1967 Camaro, using the BMR Fabrication clutch master firewall plate (included in P/N TCK005), the adapter plate easily bolts into place on the lengthened brake master cylinder or brake booster bolts. Using the stock clutch pedal required welding on a mounting tab to bolt the clutch master cylinder rod onto the pedal.

McLeod Clutches offers upgraded performance clutches if you need something much better than stock. The twin-disc RST clutch bolts onto the stock or aftermarket flywheel and provides a like-stock pedal feel and more comfortable driving experience than an aggressive single-disc clutch. The RST is rated for 800 horsepower, which handles most naturally aspirated LS swaps and some mild boosted builds. There are several configurations, but the one I'm using behind my 427ci engine in the 1967 Camaro is P/N 6405507, which comes with a billet steel six-bolt flywheel.

AFTERMARKET CLUTCH ASSEMBLIES

Upgraded clutches are nothing new. It is common sense that if you increase the engine power and torque, you need to upgrade everything in between your flywheel and the ground to match. Aftermarket clutches function off two main principles, clutch friction and clutch pressure. Increasing one or both allow higher torque loads than your typical stock clutch. The clutch system is not much different from braking functions, although instead of slowing down, you are accelerating the vehicle.

Upgrading the friction material alone increases the torque capacity due to the higher coefficients of friction. Friction material determines the torque capacity and drive-ability of the disc. You can liken friction material to different sole materials in shoes. You may have noticed some shoes are more sticky or slippery than the next on varying surfaces. Note the difference between athletic running shoes and hard-bottom formal dress shoes on a smooth surface. The difference in traction is described as the coefficient of friction.

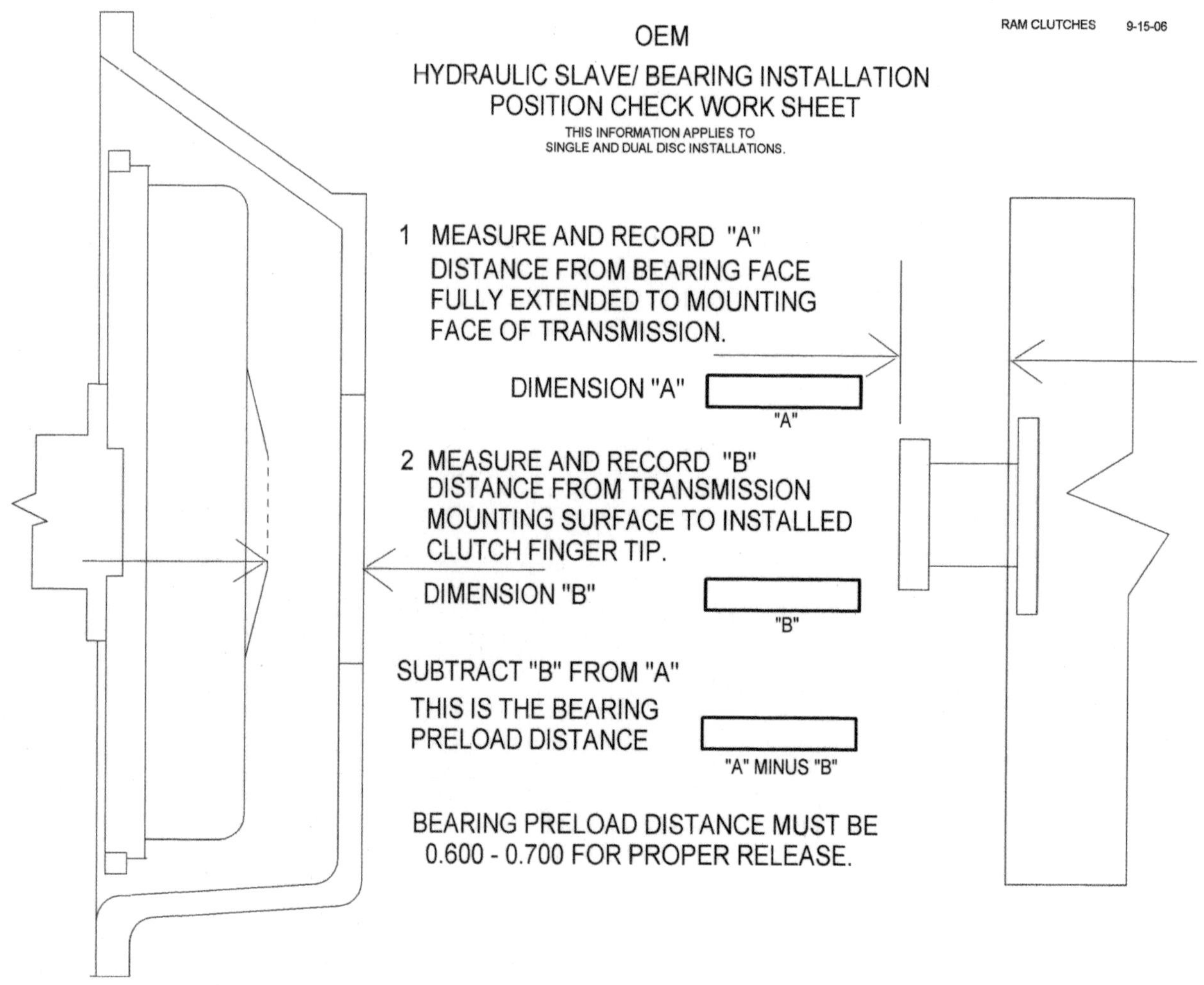

RAM CLUTCHES 9-15-06

OEM

HYDRAULIC SLAVE/ BEARING INSTALLATION
POSITION CHECK WORK SHEET

THIS INFORMATION APPLIES TO
SINGLE AND DUAL DISC INSTALLATIONS.

1 MEASURE AND RECORD "A"
DISTANCE FROM BEARING FACE
FULLY EXTENDED TO MOUNTING
FACE OF TRANSMISSION.

DIMENSION "A" [ ]
"A"

2 MEASURE AND RECORD "B"
DISTANCE FROM TRANSMISSION
MOUNTING SURFACE TO INSTALLED
CLUTCH FINGER TIP.

DIMENSION "B" [ ]
"B"

SUBTRACT "B" FROM "A"
THIS IS THE BEARING
PRELOAD DISTANCE [ ]
"A" MINUS "B"

BEARING PRELOAD DISTANCE MUST BE
0.600 - 0.700 FOR PROPER RELEASE.

Ram Clutches provides these clutch and hydraulic measurement worksheets (above and bottom of next page) with its clutches to help make sure the clutch measurements facilitate a full clutch release. If there is a problem with the measurements or slave cylinder position, it can be addressed before fully installing the transmission. Otherwise, with varying installations, you may have some issues due to the nonstock applications. *Ram Clutches*

Material variation alone is not the entire story; you can also add friction by increasing clamping pressure. Take a sheet of paper and drag it across a table. It's quite easy to do. Now put a brick on that same piece of paper on the same table. Not so easy anymore. You just increased friction without changing materials. Most of the aftermarket clutches are built around the F-body clutch dimensions and can interchange as far as flywheel and transmission input spline count (26 spline). Choosing the correct clutch for your project depends on the power level your LS engine produces. If your engine has been left stock, you can go with a mild clutch such as a Ram Powergrip. If you have some mild modifications and plan to hot rod your car around a bit, then the next one up would be a Ram Powergrip HD with a billet-steel flywheel. Now if you want something that can hold 1,000 horsepower and drive not much different than stock, you would want one of the Ram Dual Disc clutches, which feature stocklike drivability and an increase in holding capacity due to the extra clutch disc. This design offers a 100 percent increase in surface area over a single-disc clutch but not as drastic a clutch feel as far as pedal pressure. Still, it offers great holding power when you need it.

To mate a hydraulic master cylinder to a mechanical linkage setup, an adapter and hardware are necessary to mount a master cylinder to the firewall. Full kits are available from Keisler, American Powertrain, and Hurst Driveline. Most of the applications work best with a ¾-inch bore master cylinder, so if you have one of the vehicles that has no "bolt-on" master cylinder provisions, this gives you something to work from to build a custom setup. *Mast Motorsports*

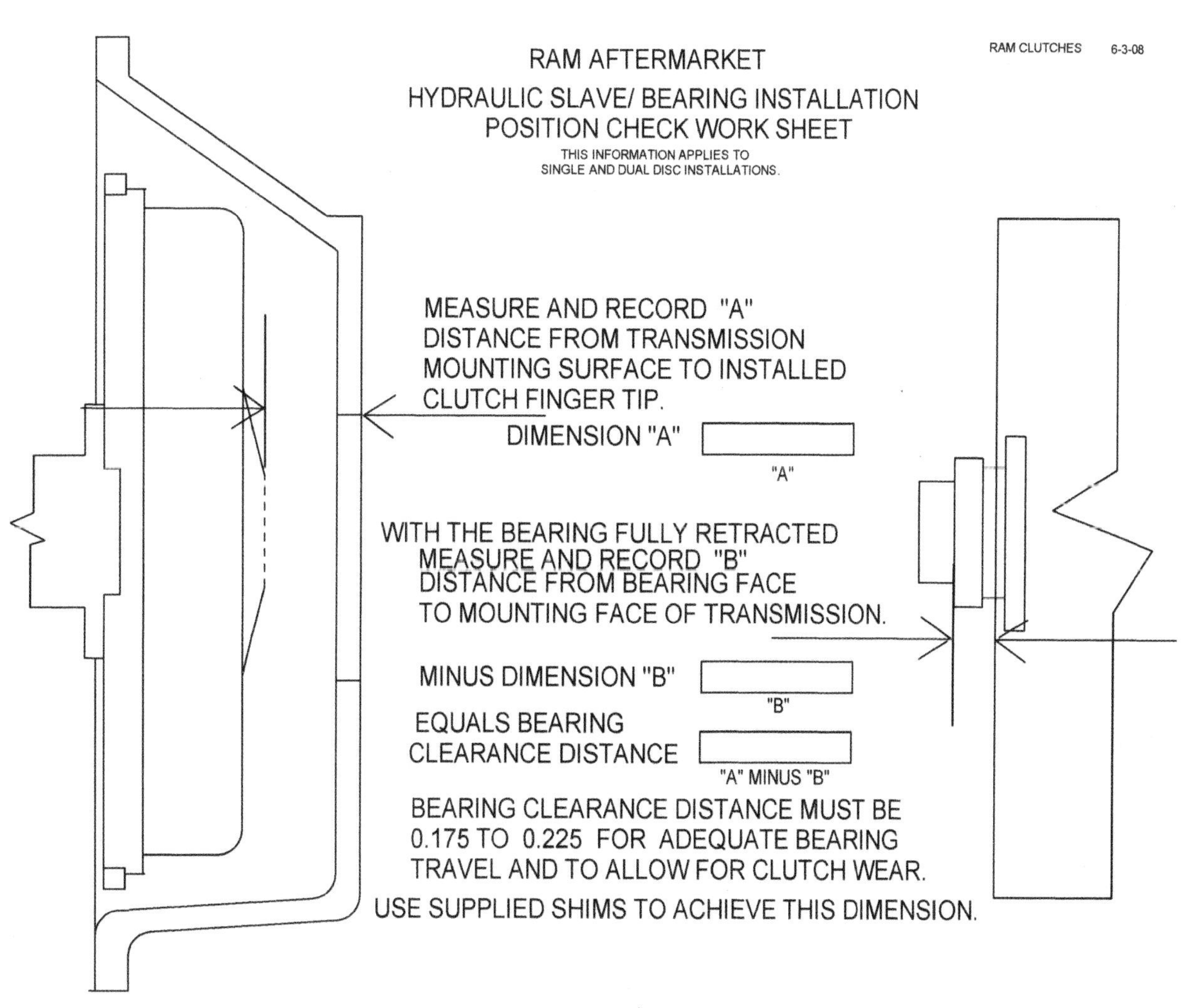

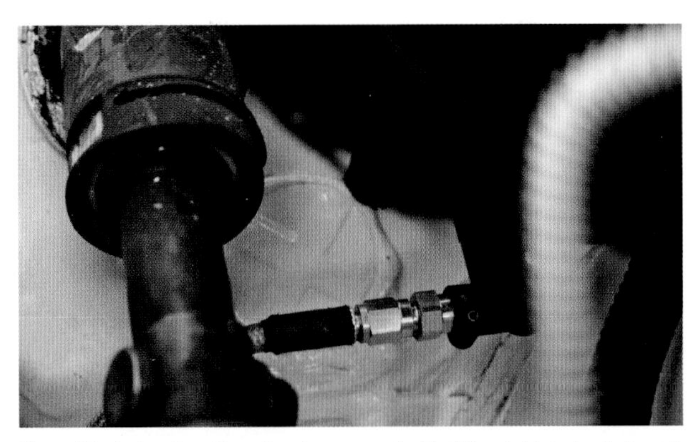

These AN adapters from Russell make easy work of building clutch hydraulic lines. In my 1991 Camaro swap, I used Russell P/N 640281 in the stock master cylinder, and then a 24-inch -3AN stainless hose from Inline Tube to connect to the clutch actuator. I also used a remote clutch bleeder kit to make easier work of bleeding the clutch hydraulics. These are available in kits from sources such as Tick Performance.

Hooker not only makes LS-swap headers, but they also make engine mounts and transmission crossmembers in the Blackheart lineup. This is the one needed for the 1982–1992 3rd-Gen F-body T56 installs, which provides a relocated torque arm mount as well (P/N 71222005HKR). Additionally, add-on brackets are available for other transmission choices. Note that the T56 shifter is located more rearward than the factory five-speed transmission and will need some clearance "modifications" to fit.

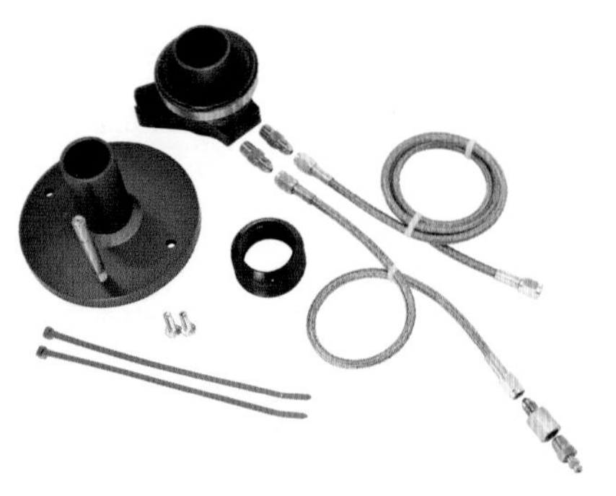

Ram Clutches offers a hydraulic clutch release bearing for both OEM T56 transmissions and nonstock applications and can even be used to retrofit a hydraulic clutch to any formerly mechanical linkage transmissions as well. *Ram Clutches*

The Ram hydraulic bearing has a bit more travel than the OEM clutch actuator. This can be quite beneficial when coupled with a nonstock application due to the extra travel and adjustably. *Ram Clutches*

CLUTCH HYDRAULICS

The main difference between the late-model six-speed transmission and your old four-speed transmission is the functional difference in clutch actuation. The older transmissions have mechanical linkage, whereas the LS type transmissions are hydraulically operated.

The easiest thing to do is reuse the hydraulic system from the donor vehicle by mating it to your older car or truck. The hydraulic clutch actuator is easy to retain; it just bolts on the front of the transmission as when in a stock vehicle.

The other side of the system is the clutch master cylinder. You can use an aftermarket master cylinder for the clutch, or you can adapt the GM F-body unit into an older car. Some companies such as BMR Fabrication and Keisler offer a plate that is sandwiched between the brake booster and firewall to create a mounting location for the clutch master, while in other applications you will have to fabricate a mount. The reservoir is the easiest portion as it just needs to be bolted somewhere above the clutch master cylinder.

Another transmission option for swaps via a billet adapter is the Aisin AR5 transmission. This is the manual transmission found in the Saturn Sky (pictured) and Colorado applications. The adapter is from FABbot Fabrication and bolts to the front of the AR5 to adapt to the LS 4L60E automatic bell housing. It comes with an adapter to use a T56 F-body clutch actuator, which you will need to install 180 degrees from stock, as shown, to have clutch line clearance. You also need longer -3AN clutch and remote bleeder lines with a minimum 36-inch length.

The stock master cylinder works best for stock replacement clutches and "like-stock" design clutches. When you use a different type of clutch, whether it be a billet flywheel and a puck clutch with a heavy-clutch pressure plate or a dual-disc clutch setup, sometimes the stock master cylinder is inadequate. Often they are on their last leg to begin with and finally expire when coupled with the sturdier clutch; like most stock reused parts when subjected to strains of heavier-duty items. There are some master cylinder options to go along with your clutch install. Adjustable master cylinders from shops such as Tick Performance are just that; simply adjustable. With most adjustable master cylinders, only the length of the pushrod can be changed when adjusted, in turn adjusting where your clutch pedal is in relation to the floor. This is a personal preference to suit your driving. You can adjust it to be even with the brake pedal or higher. This also adjusts the clutch's engagement point to a degree but not significantly. You have X amount of travel; changing the length of the pushrod does not lengthen available travel of the master cylinder plunger with a stock-type, unmodified master cylinder.

To significantly adjust the engagement point you need a longer travel master cylinder or a larger diameter, internally modified master cylinder such as the Tick Performance option. Anytime you have a nonstock application, you may need an adjustable master cylinder to get the clutch-engagement point correct.

Those with stock master cylinder disengagement issues should quickly get this resolved because clutch external components that do not work correctly can quickly accelerate clutch wear. If bleeding the clutch hydraulics thoroughly does not help, upgrading to an adjustable and higher volume master cylinder will usually help greatly. Be careful, though; if you set your pedal height incorrectly, you run the risk of overextending the slave cylinder, which in turn jeopardizes its hydraulics and rendering the car immobilized.

Adjust the master cylinder to the minimum height required for the clutch to operate. This usually means

Due to the many nonstock powertrain parts that end up on a retrofit buildup, more times than not a driveshaft length change will be required. If the old driveshaft is longer than necessary, you can save some coin and just have it shortened. To determine the new length measure from the output yoke to the rear U-joint centerline and take those measurements to your local driveshaft shop.

adjusting the pedal height so that midtravel is the engagement point for the clutch; not too low, not too high. The clutch adjustment point is under the dash near the clutch rod pivot point near where it attaches to the pedal. Clutch pedal adjustments are usually a set-aside-and-forget item after the clutch seat's in, but with significant clutch wear over time, a re-adjustment may be necessary to put the engagement point back where needed.

When using the Ram Hydraulic release bearing kit with both OEM and aftermarket transmissions, the most critical thing on hydraulic bearing installs is proper setup gap. This cannot be overstressed. Measure, measure, and when you think you have it, measure again! Overtravel can cause the bearing to extend to the snap ring stop, continue to pressurize, and blow out a seal. This happens *all* the time to people who don't measure. Also, measuring means precision measurement. Dial calipers recommended. No wooden yardsticks allowed.

The Ram retrofit hydraulic bearing kits include the release bearing and bleed line, but it is up to the customer to supply a master cylinder and feed line. You cannot just go and buy one without a new GM master cylinder, so it is best to find an aftermarket line kit. You can build your own clutch-line kit by using -3AN adapters from Russell Performance. They have adapters to mate the stock master cylinder to AN sizing with part number 640281. After adding the two adapter fittings, you can simply build your own -3AN line or have a -3AN line built to length at a local hydraulic hose supply store. Typically a ¾-inch bore master provides more than enough travel for the release bearing system to work. Some of the retrofit and transmission companies have devised masters and related mounting hardware to facilitate these installs. Some companies that offer specific aftermarket mounting brackets, master cylinders, and reservoirs are Keisler, American Powertrain, and Hurst Driveline. If you have the OEM hardware and can build your own mounting kit, you can always try this first rather than buying stuff you may not need. From the get-go, LS swappers have been installing F-body clutch master cylinders using pre-made adapter plates or fabbing adapter plates. The F-body master cylinder is readily available from GM, Ram, and other aftermarket sources. You will obviously need a clutch pedal assembly for your setup. If you do not have a clutch pedal option, you will have to procure a donor assembly. Once the pedal is in place, you can mount the clutch master. Find the leverage point for the clutch master by locating the old mechanical connection. Take a straightedge and put it against the left side of the clutch pedal flat section toward the engine firewall and mark a vertical line. This will be the centerline of the master cylinder. If using an aftermarket flat-mount master cylinder, this is where the pushrod of the master will be located. If using the F-body angled master cylinder, and fabricating from scratch, you have to take this angle into account by duplicating the master cylinder incline angle with your straightedge. You must match the angle, or the master cylinder rod could bind when being depressed.

Luckily, BMR Fabrication, Detroit Speed, and a few other companies have T56 swap kits that have an adapter bracket included for the most popular swaps such as Gen I Camaros. To use this bracket, you sandwich it between the brake master and the firewall. You simply bolt the master cylinder in the properly located mounting location. If your car is an automatic, you will have to hole-saw the firewall for the master cylinder pushrod. Since most older GM muscle cars used mechanical linkage, with this setup you will have to weld on a mounting tab to your clutch pedal for mounting of the master cylinder pushrod. Once the two ends of the clutch hydraulic system are in place and the clutch pedal mounted, the only thing left is to connect the two. You can use the stock quick-connect line if it reaches. Or if some more length or flexibility is required, you can get some -3AN adapter fittings from places such as McLeod

Clutches or Russell, which offer a more convenient method of line fabrication. If you use this setup, you cannot use the rubber-lined AN hose. You are required to use Teflon-lined high-pressure AN lines and fittings. Teflon-lined braided hoses are for higher pressures and are brake-fluid resistant. The rubber-lined hoses work great for oil and even fuel, but not for brake fluids. Once everything is sealed up, you can carefully fill the clutch master reservoir and bleed air out of the system. I've found it easiest to use a reverse pressure bleeder, forcing brake fluid and all air bubbles backward through the system. It can also be bled like a conventional brake system, with a buddy pumping the pedal and holding it down while you crack the bleed fitting.

Something that will help with this seemingly impossible hydraulic clutch task is a remote bleeder kit. Tick Performance has a great kit that matches the clutch actuator setup you have. It puts the bleeder screw in a more convenient location under the hood instead of under your car.

DRIVESHAFT CHANGES

As with many changes, sometimes the installation of new parts or different drivetrains change the measurements between two points. No better example is the length of the driveshaft. Depending on the design, you can vary the powertrain ½ inch either direction of stock and usually be ok. There is a minimum engagement point that will work just fine, but always check the measurements with the yoke and spline engagement, and if not enough then plan on a longer driveshaft.

Another reason a new driveshaft may be in order is due to transmission output shaft spline differences. All GM light-duty two-, three-, and four-speed RWD automatic transmissions use the same 27-spline output shaft as well as the GM four-, five-, and six-speed manual transmissions and aftermarket setups catering to GM installations. The 4L80, 6L80, and 6L90 do not share the same spline count as these other GM transmissions (6L90 two-wheel drive has a 36 spline, for example). In some cases, as with the 5th-Gen Camaro, the transmissions do not have a slip yoke, which means a 3-bolt-flange-to-U-joint yoke adapter and slip-shaft driveshaft must be built to address this issue. Some HD transmission designs also have a bolt-on transmission yoke requiring a two-piece driveline to be installed. Check with your driveshaft fabricator and see how they want you to take the driveshaft measurements. Some places want a measurement from the end of the splined output shaft to the centerline of the rear U-joint caps, while others want a measurement from the rear of the tailshaft and the output shaft to the rear U-joint cap centerline. If you do not have a local driveshaft shop, I recommend Denny's Driveshafts (dennysdriveshafts.com) for a mail-order driveshaft. They can build you a custom aluminum or steel driveshaft to your specifications for a fair price. If you measure right using the provided build sheet, it will fit perfectly the first time.

To order a new driveshaft, you also need to know your transmission output, shaft spline style, and the rear U-joint required dimensions (cap diameter and exterior width of U-joint or yoke), which vary between different GM yokes and Ford yokes, depending on axle style.

Denny's Driveshafts can custom build any driveshaft to your new specifications and ship it to your address if you have a few days to spare after getting your driveshaft measurements. Denny's Driveshafts can be built for anywhere from 400-horsepower applications on up in either aluminum or as bulletproof as possible using steel. *DennysDriveshafts.com*

Chapter 8
Performance Upgrades and Power-Adders

Power-adders increase engine power by cramming extra air and fuel into the cylinders. There are three ways to do this: nitrous, supercharger, or turbochargers. Each increases the power output well over what the engine can make on its own. LS engines respond well to power-adder systems. If you have grown tired of the power your engine makes, a power-adder setup can easily remedy that condition—but there are some quirks to address when building an LS-swapped power-adder vehicle.

The easiest to purchase and install is the nitrous setup, followed by a supercharger, and finally the turbocharger kit. The reason the turbo kit is the most complex and usually most expensive is because you are replacing the exhaust and induction and adding the turbo itself—the whole enchilada. The supercharged and nitrous setups are mainly an induction modification, complemented by your choice in exhaust. You can go fast with any setup as long as the rest of the vehicle's preparation reflects the increase in power and complements such power increase.

It comes as a no-brainer that with extra air and oxygen, more fuel is a requirement. If you insert a power-adder to your LS engine, plan on an improved fuel system upgrade as described earlier in the book and beyond the minimum required to run a stock LS engine. In addition to fuel system

Power-adders have always been a popular modification to add to the LS engines since their inception in the late 1990s. Superchargers, turbochargers, and nitrous are three items that build more power than the engine itself would make in the same configuration. Of the three, nitrous is the most economical as far as upfront costs. The superchargers are usually in the middle ground, and turbochargers are the most exotic and often most expensive of the group.

upgrades, a stand-alone aftermarket EFI is highly suggested. This should use speed-density tuning combined with air/fuel ratio to keep the engine safe. If you use the OEM ECU, adding an air/fuel monitoring wideband sensor setup should be way up on the list of complementing parts along with other monitoring systems, such as boost or fuel pressure gauges. Internal engine modifications include items such as camshafts, ported cylinder heads, upgraded valvetrain, higher compression, and often forged larger displacement engines, which have a variety of recipes to build. In this section, we delve into the ins and outs of what is recommended to add more power the old-fashioned way: by not giving the engine a choice and forcing it to perform.

WHAT IS BOOST?

Boost is the result of adding artificial atmosphere into the engine by pressurizing the intake system with greater than static atmospheric pressure. Naturally aspirated engines run with only the air we breathe around us; boosted engines compress that air so that more air and fuel can be introduced into engine combustion. More air means more power as long as the engine can handle it and does not detonate. This should not be thought of as more psi equaling more air, as this is true only up to a point. Always remember that the intake manifold psi is a measurement of a restriction; it is the amount of air entering the cylinder itself that matters. With boost, significant gains can be had. The restriction could be the intake valves, cylinder heads, or actual engine displacement; all factor into the measurement of psi. With free-flowing intake, exhaust, and cylinder heads, the psi settings can affect the power output when increasing boost within a conservative range. Typically, if math on paper correlated to engine math, each 7.35 pounds of boost would be worth a 50 percent increase over the N/A horsepower numbers. As is often the case, this seldom happens, but it provides a general idea of what to expect.

There are some calculations that can get close estimates to how much power is gained by X amount of boost, but there are too many variables unless you build a cookie-cutter setup, matching a proven combination, in both engine specifications and boost equipment.

Typically, boost is related to air pressure, most often measured as psi, but sometimes as bars. The atmospheric pressure we live in is 14.7 psi at sea level, or about 1-bar. A psi of 14.7 equates to the weight of a column of air from sea level to the top of our atmosphere in a 1-square-inch cross-sectional measurement. Each additional 14.7 psi is another bar added; 2-bar would be about 29 psi or two times the amount of ambient pressure. That 14.7 psi is what we breathe at sea level, and that is what naturally aspirated engines draw air from. The highest oxygen level is at sea level due to the weight of the air, with less and less available as altitude is increased. This is why even normal cars will have a huge performance loss when traversing at high-elevation. They have to work more to get the air in the engine, and the oxygen content is decreased as compared to the lower elevations.

This is the reason boosted engines operate so well; the engine does not have to draw in its own air supply. Rather, it is pushed into the cylinder. Compressed air is also more dense, so there is more powerful combustion taking place. You have many factors at work; simply adding extra air quantities alone does not increase power. To go along with that, air in a proportionately equivalent requirement is what the now increased fueling demands. Anytime airflow is doubled, the fuel quantity must be increased a complementing amount. We get power from the fuel, not from the air alone. The power gained from forced induction is not by the psi settings. It is due to how much of that air gets into the cylinders, truly based on cfm alone. Psi is often used because at the power levels we deal with, this can often tell how efficient the setup is. Although boost as we read it in the intake manifold is purely a measurement of restriction, whether it be in the cylinder head, camshaft events, or just the cubic inches of the engine, all are contributing factors to the "restriction" amount.

There are numerous approaches to add a blower, turbos, or nitrous to your LS swap. Honestly, entire books have been dedicated to these subjects, so you will need to dig deeper than I offer here. Much of my information will speak to feasibility and compatibility, and which brackets and such are needed for the different examples.

BOOST OPTIONS

The two common ways to compress air in an engine are the supercharger or the turbocharger. Many people have installed Magnacharger/Eaton Rotor style and Whipple Twin-Screw blowers into their swap projects. While some have gone full-fab mode and installed twin turbos during their LS swap installation, other resourceful builders have scavenged the factory 1.9-liter TVS supercharger from the 6.2-liter LSA and applied it to their swap project, making for a great retrofit package. With intake adapters from places like ICT Billet, the LSA blower can fit other types of cylinder heads than just the LSA/LS3 style.

Beyond the Eaton or Twin-Screw design supercharger kits, builders can consider a centrifugal belt-driven design, where the supercharger functions directly off engine power via a crankshaft hub. The more engine rpm, the more boost. Supercharger kits rely on a tight belt grip without slip to deliver proper and desired airflow through the system. If the belt is slipping, the blower is not turning the desired speed, whereas turbochargers are exhaust heat and flow driven, not driven directly by mechanical means. They rely fully on the combustion and exhaust process to operate. In this way, they are not directly driven by the engine itself. They are a byproduct of a normal running engine.

While to an inexperienced hot rodder "boost is boost," both types of systems do not have the same attributes or have boost at the same time or rpm. Turbos have boost earlier than blowers but can experience turbo lag before boost builds. Centrifugal blowers need rpm to build boost by function: more boost at higher rpm. For example, you have two cars, one with a turbo, and one with a blower. Each is set for 10 psi. The turbo car would typically achieve full boost at about 3,000 to 3200 rpm and keep it there the whole time. The centrifugal blower would start building boost early on but will incrementally build more and more until full boost is hit at the top end of the rpm band, basically at the rpm cutoff level. Each system has it advantages and disadvantages—is there a disadvantage to more power?—which will be discussed more in their own topics.

CENTRIFUGAL SUPERCHARGERS

Superchargers come in a few different designs. Some of the most popular types of LS-series blower kits are centrifugal belt-driven compressors, such as from ProCharger. These have a similar compressor housing to a turbocharger and similar compressor wheels but are belt-driven versus exhaust gas-driven, as are turbochargers. Much like the gears on a bicycle, you have your powered gear and your driven gear. The powered gear on blower kits is the crankshaft pulley hub, while the driven gear would be the equivalent to the blower pulley itself. The fixed diameter of each determines how much rpm the blower is spinning at. Further, the blower itself has its own internal gear-up components to spin the impeller much faster. These are normally in the 1 to 4 ratio, and each single pulley rotation, will spin the impeller four times.

The way centrifugal blowers work is by increasing the air speed by slinging it outward from the central inlet opening and then forcing it out one direction. This increase in air speed and cfm starts stacking up at the nearest restriction

Centrifugal blowers are a belt-driven form of a turbocharger, although they only make peak boost and peak power at peak rpm. Below the engine redline the boost gain is progressive throughout the rpm curve, so while making more power, it is often a lot more controlled when kept mild. If the pulleys are changed to spin the blower faster, then look out, as these have a lot of potential. A Procharger D1SC C6 kit is shown.

Magnacharger roots blowers are a popular modification to add to any LS engine. You will want one that meets a similar application as your accessory drive and matches your cylinder heads. There are numerous variations of the Magnacharger blower as far as applications, so there is likely something that already fits your engine. You can choose between a black discrete blower or a chromed blower, depending on the look you are going for.

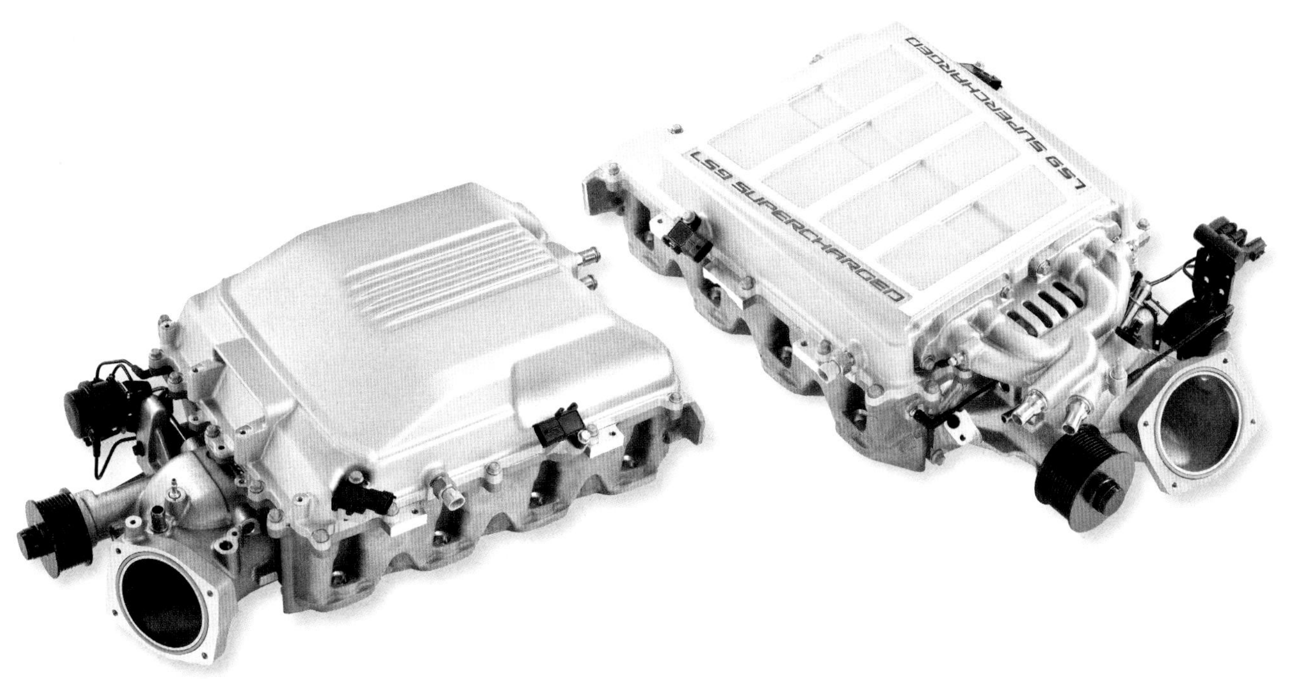

The blower that comes with the Cadillac CTS-V helps to pump out 556 horsepower on a conservative boost level using a roots-style 1.9-liter blower. GM Performance Parts released the bigger brother 2.3-liter ZR1 blower as a kit for the 6.2-liter engines using square port LS3-style heads. One other difference is the LSA uses its own blower belt (a third belt) while the ZR1 due to dimension constraints shares an 11-rib belt with all accessories. *Courtesy of General Motors*

and starts to compress, due to the air not having a sufficient escape path other than flowing through the engine itself. Since centrifugal blowers rely on impeller speed to make boost at a calculated rpm, making either the crankshaft pulley diameter bigger or the blower pulley smaller increases boost output and shifts the rpm from which boost starts building down a little, as long as the belt can provide proper grasp.

Peak boost is normally at the rpm limiter with centrifugal blowers. The higher the blower is spun = the most boost. Even though these can bolt on much like an alternator, they are often less popular for swaps. They definitely have a cool factor, and hearing a centrifugal supercharger whine as the car sits there and idles with a choppy cam is a very distinct sound—as is the car taking off during a spirited run.

A roots blower works by compressing air underneath the two rotors by way of the interlocking rotor design. This varies from a twin-screw setup that would compress air in the blower "screws" then release it into the lower intake. The GM blowers have a quiet operation and are virtually maintenance free for the lifetime of the engine. *Courtesy of General Motors*

The Magnacharger kits also share the same belt drive as the remaining accessories. You can get an intercooled setup from Magnacharger with applications fitting just about any engine combination and accessory drive there is. It's likely something that will work for your swap project. In this case, an aftermarket top-mount A/C compressor was adapted by Mast Motorsports. *Mast Motorsports*

TVS-, ROOTS-, AND SCREW-TYPE SUPERCHARGERS

Roots- and screw-type blowers fit in place of the intake manifold and also, much like the centrifugal, are a belt-driven blower. The difference between a screw-type blower and a roots-type are in the design of the gears and rotors internally. Screw-type blowers compress air pockets inside the blower itself, while roots-type compress the air under the blower and are more of an air pump. Screw-type blowers also move airflow internally, turn internally the complete opposite direction than Roots-type, which diverts air around the perimeter of the blower housing, the screw-type because of its reverse direction directs air in and out straight through. Whipple and Kenne-Bell offer Eaton TVS-style, twin-screw superchargers for LS engines, while Magnacharger offers roots-type blowers. Additionally, the LSA (1.9-liter) and LS9 (2.3-liter) are both OEM TVS blowers. Both styles perform great, but the production 1.9 TVS is found more often in the used marketplace.

The blowers that use the Eaton TVS rotors (LSA, LS9, Magnacharger, etc.) are not compressing air while you are doing your "normal" point-A-to-point-B driving. The blower is just functioning as an intake manifold at this point and it's mostly just along for the ride. This means that the air is not compressed while cruising, merely recycled internally until the loud pedal is pressed, then, as the bypass shuts off, the air pressure builds instantly and the car goes *vroom*.

Because the air is not compressed until boost is built, the air temperature does start off pretty close to ambient outside air, and then of course as it is compressed the temperature rises. The Twin-Screw blowers work similarly, but they compress the air as the rotors turn, so that pockets of compressed air are pumped under the screws. These are great for power and have more potential, but if you are already in a hot climate, there is a small chance of detonation if the air starts getting too hot. This hot compressed air requires a solution to cool back off.

This is the rationale for giving these types of superchargers internal intercooler bricks, which remove the heat from compressing the air. These are not 100 percent efficient, but they do a good job of keeping things inside the blower more chill. Designing an adequate heat exchanger setup is crucial for the engine's health: the cooler the air, the better. Hot air from the blower is bad for parts, especially the pistons. They will thank you for the cooler air by enjoying a much longer lifespan.

Magnacharger is possibly the most flexible of the engine-mounted blowers, due to having many OEM applications. If using F-body or GTO accessories, you need the GTO kit. If using C6 accessories, you need the C6 kit. If using LS3 cylinder heads, you will have to use a setup for the relocated intake ports. If so, then the LS3/L99 Camaro is the best choice, although it will require the 2010–2015 Camaro accessories spacing and crankshaft damper. Magnacharger kits are available in shiny chrome or in a black OEM-appearing finish. These are also available in a hot rod–style kit, made specifically for swap projects.

TURBOCHARGERS

Turbochargers are nothing new to the automotive world, although they have an exotic ring, especially when coupled with an earlier GM vehicle. How great would it sound to have a Twin Turbocharged LS2 Chevy II? Turbochargers operate and are driven from exhaust gas speed and also from leftover unburned fuel; they are often likened to having an extra cylinder. The hot-side (piping from exhaust port to turbo) of the turbo system directs all exhaust through the turbine side of the turbocharger, which drives the connecting compressor side of the turbo. The exhaust turbine side can be likened to the pulley of a blower, although it is not truly dependent on engine rpm to dictate turbo rpm. Many factors come into play to determine the boosting point of a turbo system, such as exhaust turbine diameter, turbo housing sizes, and then downpipe and muffler restrictions. Since the turbo is not directly connected to the engine other than by exhaust flow itself, boost is not a constant number per rpm like that of a supercharger. For instance, you can be cruising at 3,500 rpm and have no boost, but you can be at a lower 3,200 rpm holding the car from moving with the brake and have 12 pounds of boost. Turbo systems are more load-based systems versus rpm-based, although a minimum load or rpm is usually required to get things going.

Turbocharger installations in LS swaps often require much fabrication and dedication to the project. You can't just go out and buy a turbo kit for any car; it often needs to be built. Often, the turbocharger fabrication can be much harder to build than the actual LS swap itself, and it takes a lot of time to build. This is because the turbo setups are integrated with the entire engine package, more so than with other bolt-ons such as blowers and nitrous, which essentially are only on the intake tract of the engine. Turbocharger systems consist of

Hooker Blackheart has a ton of great products and solutions for LS swap applications, and their cast-iron turbo manifolds are no exception, such as this P/N 8510HKR, which functions as a manifold, turbocharger mount, and merge pipe all in one. A variety of crossover pipes are available for different transmission selections, which helps builders who don't have fabrication skills to have a simple yet effective turbocharger build on their swap project. *Holley*

The super-clean engine bay of this Chevy II is home to a forged piston/rod iron 6.0-liter engine. Features shown include TrickFlow 220cc as-cast heads, a custom BTR camshaft and spring kit, an effective twin-turbo setup using TSR-modified truck manifolds for feeding combustion exhaust into the Borg-Warner turbos, the hot-and-cold side piping were fabricated by Cambo Built. To handle cooling the compressed intake air, a 417 Motorsports 2,000-horsepower intercooler is sandwiched between the two halves of the Holley Low-Ram intake manifold.

the exhaust headers, exhaust system, induction, and often the intercooler to cool the compressed air back down. Additional components such as wastegates to control turbocharger speed and blowoff valves are necessary to bleed intake pressure when you let off the throttle.

For LS swaps, there are several ways to start a turbo kit. Some companies such as Hooker (P/N 8510HKR) have cast-iron turbo manifolds and universal crossover pipe (P/N 8517HKR) that allow a solid starting point for turbo kit fabrication. There are other crossover pipes available that fit specific transmission choices. This takes the legwork out of fabricating custom headers and provides the complete hot-side system (other than the downpipe). If you have ample underhood room, these types of kits should fit with little effort.

If you want to build a twin turbo kit, a different method must be employed. Look at the Flowtech (P/N 11535FLT) for a pair of up-and-forward 304 stainless-steel turbo headers. Like the cast-iron example, this provides a great place to start fabrication. While these won't fit every car, for the sake of saving several days of fab work on headers they're worth trying first.

Speed Engineering also has a similar up-and-forward, stainless-mandrel, bent twin-turbo-designed header (P/N 25-1023). The header kit is a 1¾-inch primary tube, a 3-inch collector, and comes with V-band clamps. These give you a good entry point for your turbo fabrication to build from.

NITROUS OXIDE INJECTION

Nitrous oxide offers significant horsepower gains via simple chemistry. Composed of 2 parts nitrogen and 1 part oxygen (36 percent by weight), nitrous oxide is stored as a compressed liquid. While not flammable by itself—it doesn't explode in the way portrayed in movies—nitrous oxide is classified as an oxidizer. A few things associated with nitrous oxide translate directly and indirectly into power production. Upon injection into the motor, we notice a significant drop in air-intake temperature. The drop in temperature is a direct result of a phase change from a compressed liquid to a gaseous vapor state. This indirectly translates to a colder more dense air charge and any time that more oxygen can occupy the same space, we can add more fuel, resulting in higher power generation.

Most LS engines feature only a press-fit to keep the pulley in place. When you include a power-adder such as a blower, the pulley can work its way loose due to the extra strain on the accessory belt. To counteract your pulley flying off the car, all blower kits require pinning the crankshaft pulley. Pinning the crankshaft pulley is a good idea for any power-adder variations to stop the pulley from spinning off the car and getting lost forever, sometimes resulting in hidden engine damage.

While these gains may be slightly insignificant, the chemistry that follows as nitrous and fuel are inhaled into the cylinders will produce generous amounts of power. Once inside the cylinder, the piston moves toward the top of the chamber on the compression stroke, and pressure and heat begin to build within. Once 572 degrees F is reached, the nitrous oxide molcecule breaks down and releases the oxygen particle. This release directly allows additional fuel to be burned, and, in turn, generate power. Another indirect result is the release of nitrogen. Nitrogen in the combustion process helps to absorb heat, which indirectly translates to aiding in the prevention of detonation, but there is a trick here: You need enough fuel and octane to sufficiently complement the extra oxygen produced as well, or you can cause detonation even with the much cooler air.

Nitrous kits come from a variety of manufacturers; the more specific LS-oriented kits come from Nitrous Outlet. There are economical kits, and there are complete kits that offer premium prices and features. Dry kits inject nitrous only through a metered jet and nozzle assembly—and rely on the vehicles' ECM to supply extra fueling via the engines' fuel injectors. Wet kits supply both metered nitrous and required supplemental fueling through a nozzle. Double- and triple-check jet sizing when changing the nitrous shot amount. I cannot stress the requirement of having plenty of fuel and matching ECM tuning to not hurt parts; so many people add nitrous without complementing fueling upgrades and hurt engine parts. Make sure your fuel system can cover your total power consumption, with a surplus.

Nitrous oxide injection can be used to gain substantial power when you get bored of your naturally aspirated engine. Stock engines can safely take an extra 150 horsepower with fuel system upgrades, whereas engines built for nitrous can take much more. Talk to your engine builder or performance shop for specific capabilities. Shown is a Nitrous Outlet direct port kit on top of a FAST 102mm intake manifold for a forged LS7 427ci engine.

Nitrous oxide kits inject the intake tract with a supply of N_2O and extra fueling that makes more power by releasing more oxygen in the cylinder to create a harder combustion hit to the pistons. Nitrous also has the benefit of cooling the intake air stream. To use N_2O, you will need a great fuel system to keep up.

Nitrous oxide injection can be an easy addition to any LS setup. Both carbureted intake manifolds and composite intakes can be fitted with plate kits that mount between the throttle body and intake manifold. Direct port kits take a bit more piping and plumbing work to initially set up in each intake runner, but they offer more N_2O adjustment and tuning per each cylinder.

Camshaft swaps in LS engines are quite easy to perform as compared to other engines. For one, you do not need to remove the intake or cylinder heads to swap camshafts, as the lifters can either stay in place or they can be manually held into place with lifter dowel rods while the camshaft is extracted and reinstalled. Use plenty of camshaft lube for the initial startup.

INTERNAL ENGINE MODIFICATIONS

Camshafts or cylinder heads are each popular items to replace before, after, or during the installation of engine exterior bolt-on components. These can take your car as far as possible for naturally aspirated or power-adder applications. A camshaft package alone would consist of the camshaft, valve springs, and pushrods. Depending on mileage, I would recommend a new timing chain and replacement oil pump to prevent future problems, especially if the condition of the engine is unknown. These items can be replaced at the same time as the camshaft for not much extra work, and you are likely in the engine already for an oil pan swap, so you're halfway to the oil pump anyway. The camshaft is what would give your LS engine its rumpity idle chop, so if that is your thing, the camshaft is what you will want to do. Although many builders do camshafts for the power increase, the tone of the exhaust note is also a wonderful thing.

Ported heads alone are not a good sole provider of performance gains when dealing with the stock camshafts. There will be improvements, of course, but if comparing a cam-only install to a heads-only install, the camshaft will almost always come out ahead in the performance department. Once a camshaft is in place, then cylinder head performance gains are highlighted much more than if by themselves, although a package that works together is always best. This is why it is popular to do a heads and cam package at the same time. If the choice to be made is one or the other, always do the camshaft if your goals are lofty, but keep in mind that the "cam-only" setup is usually a good 25 to 40 rear-wheel horsepower behind the ported heads with the camshaft option.

Ideal camshaft choices tend to follow the cylinder-head design to be optimized; 4.8-liter/5.3-liter/6.0-liter cathedral-head truck intake engines like different cams than the 6.2-liter LS3-headed engine, while the 5.7-liter LS6 car-version intake likes different camshafts than the cathedral truck engines, even with the same heads. Then you have the 427ci LS7, which is entirely different than its next closest relative, the 6.2-liter LS3, and for optimum performance these traits must be taken into account.

If you have one of the common truck engine donors, such as the 5.3-liter truck engine, you have the option to run one of the low-lift camshaft grinds from places like BTR, TSP, or Cam Motion. These cams provide a bump in power and sound, while being friendly on the valvetrain. While there are cams that reuse the stock springs, I would recommend going with at least LS6 springs, which handle the 0.550–0.560-inch valve lift of these camshaft families. Expect gains of 45–75 horsepower on average with a camshaft swap of this type.

There are some general rules related to the 5.3-liter truck engine. The smaller camshaft durations (208–218 intake duration) will drive more nicely and have strong torque curves—overall they are stock torque-converter friendly—while the longer-duration cams (above around 220 intake duration) will be more top-rpm happy and have less low-end grunt torque. These higher-rpm camshafts usually require an aftermarket torque converter and gears to optimize the setup.

LS3 engines (more specifically, the heads) do better with higher-lift camshafts, which does require an aftermarket valve spring swap. Several companies make a 229/244-duration-size camshaft that sounds good, makes great power, and does not sacrifice a ton of low-end torque, though it's a middle-of-the-road camshaft that works for most N/A builds. LS3 engines like any cam, of course, but 218/230 would be the smallest upgrade recommended, while up to the 235/248-size camshafts can barely fit with stock pistons—albeit providing max power curve.

If external engine modifications don't keep you satisfied, or you want other complementing items inside the engine, there are even more aftermarket things available that you can do to build up your engine internally. One thing you can do is install a performance camshaft package. GM and several aftermarket camshaft companies have many choices from mild camshafts that you barely notice to camshafts that rattle every trim piece in your entire car. *Courtesy of General Motors*

There are basically three main options for ported OEM heads intended for the stock displacement LS1 or 4.8/5.3-liter engines. Your choices in castings are porting the stock 5.7-liter LS1 heads, porting the stock 4.8/5.3-liter truck version heads, or starting with the upgraded LS6/LS2 heads. When building up Gen IV engines, these usually have great heads from which to work. LS2 heads are based on the LS6 heads. LS3 are similar but not the same as the LS7 head design, and with LS7, while great performing heads in stock form, there is plenty left on the table for porting and modifying.

Each head has its advantages, such as if porting your own 241 casting LS1 core heads. If you already own these, you save on replacement core charges up front. This would be the most economical route, but also the stingiest for performance. While you do gain in airflow, with any significantly sized camshaft, you are limited on the cylinder head milling to boost compression, with each point of compression worth about mid-20s in horsepower and a good chunk of torque. This may not be the most desirable cylinder head for your goals, most LS1 builders swap to the ported 5.3-liter heads or 243 casting LS6 heads.

The truck 4.8- and 5.3-liter heads you would think are not good performing heads, and in OEM form, they are less stellar than the stock LS1 heads. Where they shine is that when ported they flow really well and provide increased compression without extravagant milling being necessary as they start out with a 61cc chamber. The truck heads are plentiful and inexpensive as cores; hence, they have become popular and are ideal for builders with stock-sized engines and anywhere from mild camshafts to large behemoth camshafts. Valve upgrades are a necessity with these heads, as the stock 1.89-inch intake valve is a restriction. You can upgrade these to the stock 2.0-inch intake valve easily with a valve job. The 2.02/1.57-inch intake/exhaust valve upgrade fits onto the machined-out stock intake/exhaust seats as well.

LS6/LS2 heads are one of the great stock heads to come out in OEM form. These can be improved by quality porting work, either by hand or by CNC machines.

Along with about any camshaft swap, a valve spring replacement is necessary to cover the added lift and engine rpm the camshaft may operate with. There are drop-in beehive springs that reuse the stock valve seals, spring seats, and spring retainers, or there are dual-spring packages that replace everything from the valve guide to the valve tip. Check with your camshaft designer to see which setup is required with each specific camshaft.

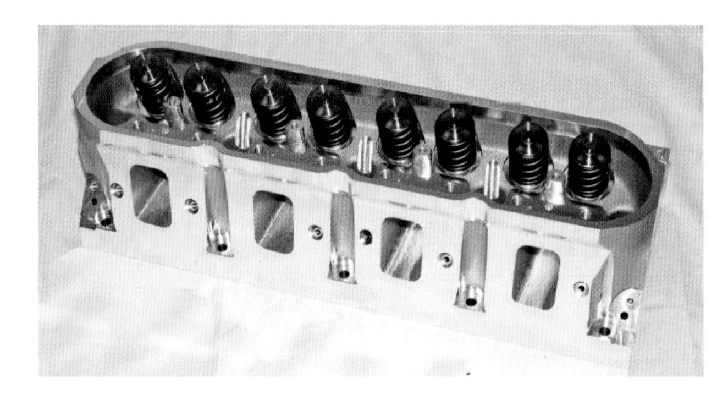

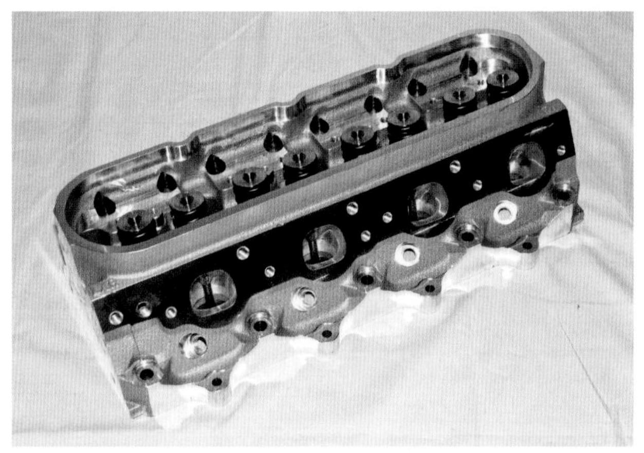

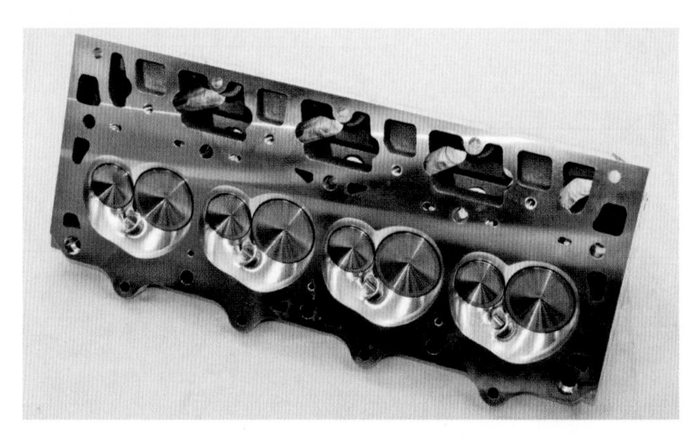

Ported cylinder heads along with a camshaft choice will really wake up the engine as much as possible without a power-adder or extra displacement being added. Typical cylinder head power gains are anywhere from 25 to 35 horsepower when a matching camshaft is already in place. Both heads and camshaft together can add anywhere from 80 to 100 horsepower when equipped with headers and a good induction and intake tract. Six-bolt Performance Induction heads are shown here. These are CNC ported by Precision Race Components, and LS7-style heads are ideally suited for large displacement engines (427-cubic inch or larger), but there are also many stock-replacement ported heads that are direct replacements for 5.7-, 6.0-, and 6.2-liter heads.

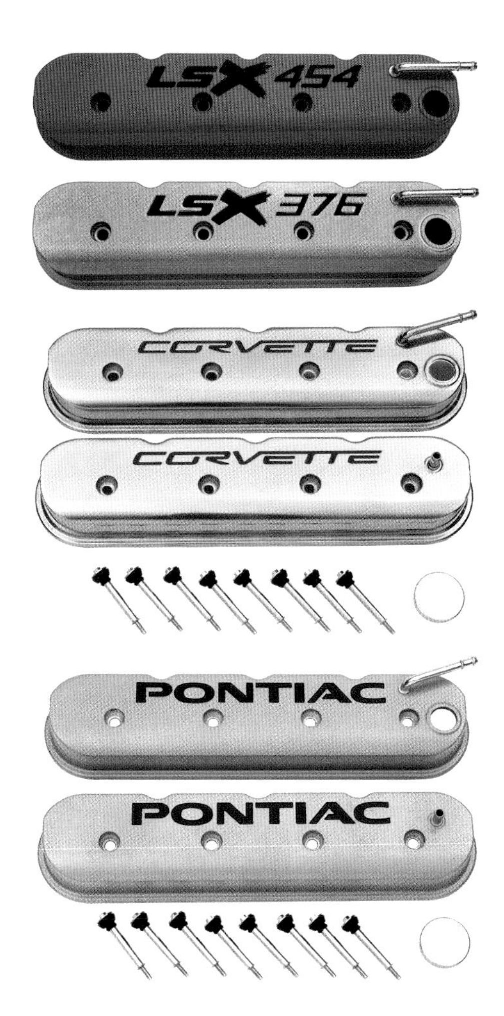

Something you may want on your new engine is a better set of valve covers. There are no aftermarket valve covers for the 1997–1998 perimeter valve cover heads but quite a growing selection for the 1999 and newer center-bolt designs, including cast aluminum and billet options. There are various aftermarket offerings, but be leery of the extremely inexpensive covers. Some have inadequate sealing provisions, especially the fabricated ones. The GM valve covers reuse factory sealing designs and gaskets. The only drawback to using aftermarket valve covers is the coil pack mounting. You will need to relocate the coil packs. You can use items such as the Katech coil pack mounting solutions for this. *Courtesy of General Motors*

CNC machines allow repeatable results head after head, duplicating a master port. The LS6/LS2 heads were factory heads on the 2001–2004 Z06 Corvette (with lightweight valves) and later on the 2005–2007 Corvette C6 LS2 and Trailblazer LS2 engine. They were also available on certain Gen III later-model 5.3-liter truck engines and on all 2007–2012 Gen IV 5.3-liter engines. These heads have great potential rivaling aftermarket casting heads.

Engines based on the 6.2-liter LS3 have some great cylinder heads out of the box, and if you are after every last morsel of horsepower, you can have these heads ported as well. The gains are not as good as the earlier cathedral-head port work, but they do respond to cylinder-head porting. Once you have a good LS3 camshaft and port the heads, expect around 25 horsepower extra in the mid-upper-rpm range. Unless you have a forged engine with intake/exhaust valve reliefs, you cannot mill the LS3 heads very much, as there is not a lot of real estate left due to their big intake valves. Avoid milling LS3-based heads when using aftermarket camshafts with OEM pistons.

As you can see, there are loads of options to make more power with your LS swap project. I always like to start in the middle of the road with a fresh build—nothing that would be annoying to drive. Then, once the car is sorted, I'll plan the next step, if needed. Once the starting point is set, you can decide which direction you want to go. From mild to wild, anything's possible.

The 6.2L LSA engines have made their way into many swap projects, even non-GM vehicles such as this classic Dodge Coronet. The 6.2-liter LSA in its rare stock form makes 556 horsepower and 551 torque, which is 130 horsepower more than a stock 426ci Dodge Hemi.

The 6.2L LSA is right at home in many classic GM vehicles such as this early C10 patina truck. This one has the CTS-V lid and Cadillac engine cover to dress up the engine a little more. I'm sure it surprises many people when the hood is opened or when the throttle is opened up next to you.

Chapter 9
Instrumentation and Auxiliary Add-Ons

This chapter will cover more specific gauge control and optional accessories such as GM factory and aftermarket cruise control. While tech info is sprinkled throughout the manual, this chapter will be dedicated to information on various ways to make these add-ons operate, whether you want the OEM gauges to work seamlessly, or you're going completely aftermarket with equipment like Dakota Digital gauges, or your project is to replace the OEM gauge cluster and use an aftermarket touchscreen display. Additionally, there may be other things you want to operate, such as monitoring fuel pressure with your Holley EFI or lighting up your LED PRNDL display correctly on your OBS truck LS swap.

OEM GAUGES

In many instances, a user may wish to keep their factory gauges intact, and in some vehicles there may be no option to change them. In GM vehicles, you can address this by keeping segments of the old engine harness and wiring sensors/senders directly, specifically the oil pressure and coolant temperature. Simply put, you wire these in just like the old engine, but usually in the new designated spot using metric-to-NPT adapters. There are instances where LS sending units can operate similar scale gauges, such as in GM SBC to GM LS swaps. This is the case with 1980s and 1990s GM vehicles.

First-Gen Camaros can retain their classic styling with modern amenities by using tasteful modern modifications, such as new electronic gauge kits and cruise and climate control.

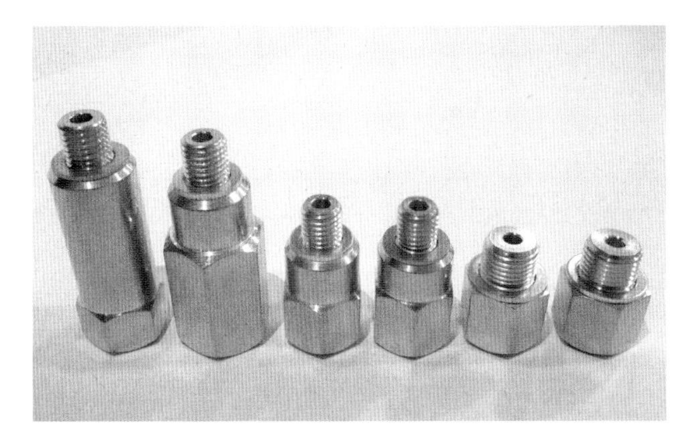

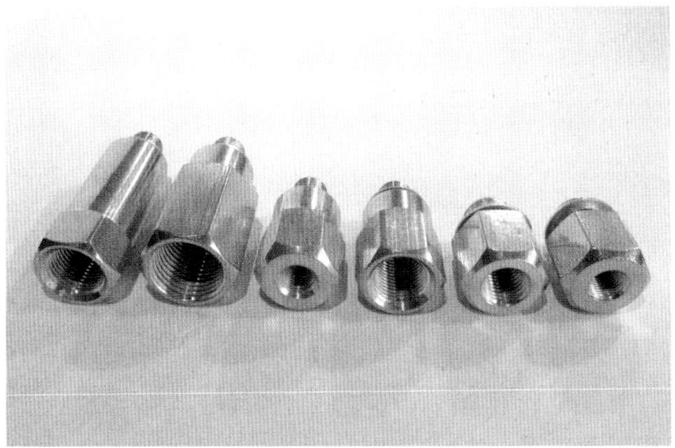

If you don't know what style fittings you need to adapt, ICT Billet offers this convenient master fitting adapter kit P/N 551156 for your gauge sending units to adapt NPT fittings to the metric LS engine ports. While not required to run the LS engine itself, they're often required for items like OEM or aftermarket oil pressure and coolant temperature gauges, aftermarket cooling fan controls, or additional sensors for stand-alone ECUs.

My 1991 Camaro used a 4th-Gen Camaro LS1 tall plastic housing oil sending unit M16x1.5mm (P/N 12551640), and the three-pin coolant temperature sensor (CTS) along with the correct wiring pigtails. Two of the coolant sensor wires go to the ECM to report engine coolant temperature for fuel and spark calibration, and the added third wire is connected to the coolant temp gauge. You can also use the CTS separately without the ECM portion in use, similar to a single-wire sensor, by using the third leg of the sensor only and on the back of the passenger-side cylinder head.

Operating the tachometer and speedometer is a different story. This depends on which engine control system you are using. If you are using factory ECMs and have speedometer and tachometer outputs, you can usually calibrate these to work the equipment factory GM electric speedo and tach if these options are included in your harness. Dakota Digital has a module solution that reads OBD II data and can output/calibrate the speedometer and tachometer using data from the OBD II port; you wire this module to your factory (or aftermarket) electronic gauges (P/N STA-1000). Moreover, this module can also illuminate the check engine light if any important diagnostic trouble codes pop up and you don't have it wired in separately.

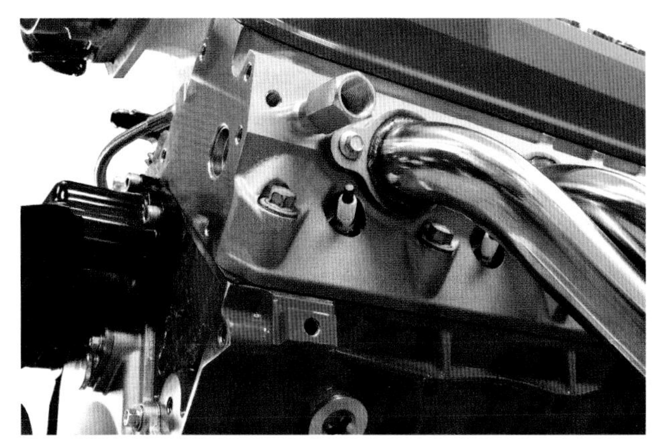

Example placement of the ICT Billet adapters: There are multiple places where these adapter fittings can be applied, such as for cylinder head coolant temperature or engine oil pressure locations. You can also buy the necessary fittings individually if you know which fitting sizes you need for your exact application. *ICT Billet*

If you have a GM vehicle with working OEM gauges, you can often use 1998 F-body LS1 sending units to operate the GM OEM gauges, as these have identical resistance values. This is useful on a slew of 1970s through late-1990s GM vehicles with factory gauges. ICT Billet can supply the Gen III oil sending unit (P/N SEN035) and the three-wire coolant sensor I sourced from GM (P/N 12551708). If you need the correct wiring pigtails, ICT Billet can supply these also. *General Motors*

If you are using a Holley EFI system such as the Terminator X, it is only controlling the engine. While it does have a tachometer output signal, you often need to recalibrate this signal to read the correct rpm. The base Terminator X does not do vehicle speed, so you'll need to wire in a third-party module to recalibrate the speedometer signal. In the 1991 Camaro swap, I used the Dakota Digital Speedo/Tach interface (P/N SGI-100BT), which wires into the T56 vehicle speed sensor and Holley Tachometer output and allows calibration of both to the OEM 1991 Camaro gauges. This module also has Bluetooth-capable calibration using a Dakota Digital app as the interface.

If you need to use sensor data that the OEM GM ECU uses in the OBD II data stream, you know that interpreting this data for use with aftermarket gauges is a challenge. An OBD II interface like this one, from Dakota Digital (P/N STA-1000), can help. Plug it into the OBD II port, attach your gauge input wires to the correct terminal, and the interface will provide the correct signals for the speedometer, tachometer, and your aftermarket cruise control. This is purely for OEM OBD II data, as aftermarket ECUs require different methodology. *Dakota Digital*

My 1991 Camaro with LS6 and T56 uses the Holley Terminator X ECU and factory Camaro gauges. I needed a way to operate and calibrate the speedometer and tachometer, which for this vehicle came from Dakota Digital (P/N SGI-100BT). The Universal Speedometer and Tachometer Interface wires to the unused speed sensor in the T56, providing the proper signal to the speedometer. In addition, the tachometer signal from the Holley can go through this module for calibration using either a smartphone app or the buttons on the module.

MECHANICAL CABLE SPEEDOMETERS

More modern technology and more user-friendly devices have been introduced since the first edition of this book appeared. This topic is for builders who want to reuse the exact OEM speedometer their car came with (or apply aftermarket mechanical speedometers). If you have a nonelectronic transmission such as a Powerglide, TH350, or TH400, you just connect your OEM speedometer cable to the transmission speedometer adapter and—*voilà*—it'll work as it did before. If you did a gear swap or anything like tire diameter change, though, you'll still have to calibrate it the same way as before the LS swap (with different speedometer drive/driven gears). For non-GM LS swap projects using mechanical speedometers (probably pretty rare), you may have to get a speedometer cable modified by adapting it to the non-GM setup; in this case, there are several companies that can build custom cables. Bobsspeedometer.com and Speedoservice.com are some that have received good feedback.

If you have a mechanical speedometer but a modern electronic transmission design, again Dakota Digital comes to the rescue with a mechanical cable drive adapter module. This can be used with OEM LS engine controllers via the OBD II connector, which reads the engine data stream for speedometer information, or as a stand-alone if you have a speed sensor in your transmission. The OBD II connector takes the electronic digital signal and changes it to analog via the speedometer cable; it's extremely useful for vehicles from the 1980s era or before that employed speedometer cables. This electronic to mechanical speedometer module comes with a 36-inch speedo cable and can also be calibrated wirelessly via Bluetooth using the Dakota Digital app to have the speedometer read correct speeds. The fully sealed module can be mounted anywhere that's convenient; passenger cabin or underhood options are available.

Dakota Digital Speedometer Cable Drive:

- ECD-200BT-1: GM/Ford thread-on, ⅝ inch
- ECD-200BT-2: GM clip-on
- ECD-200BT-5: Ford clip-on

You may want to keep and use your OEM mechanical speedometer for your classic car, or you may just want to have a sleeper incognito effect where you can still operate the mechanical gauge with the dash working like it was factory stock. The challenge here is not having a direct connection for your electric transmission vehicle speed sensor (VSS) and a mechanical speedometer. Using an electronic cable drive that can convert the signal and drive the cable, such as this ECD-200BT, will solve the problem. It takes the digital VSS signal and connects it to your mechanical speedometer. *Dakota Digital*

CAN BUS–CONTROLLED GAUGES

Electronic gauges have come a long way in the last decade. Swaps for non-GM vehicles like the Jeep JK and Jeep JL, or for import vehicles such as Toyotas, Nissans, and other foreign swaps, offer no option for running analog/electric gauges directly. These vehicles transmit the data to the gauges via a CAN bus network, though each manufacturer uses a different language for the data. To communicate over a given data stream, many swap companies will perform CAN bus sniffing, a complicated mapping out of the entire vehicle that may be beyond the expertise of a typical home user. This helps them develop and program a turn-key wiring harness that simply plugs in and works as if the LS engine came factory in that application.

Let's not go down the rabbit hole with details. Basically, all the gauge information is communicated via two wires through the entire vehicle. That means you need a way to send the relevant data over this network if you want to keep the OEM functions intact.

Many modern vehicles require the LS engine controller to function as a piggyback system, where both the OEM "insert whatever car here" engine controller is still present and powered up, but there is no engine for it to control; it's essentially a pass-through for sensors. In the early Jeep JK swap example, where we reused the Jeep coolant temp sensor and oil pressure switch, these still went through the Jeep ECM and then to the gauges.

The Jeep speedometer derives data from wheel speed, but the tachometer information comes from the Jeep ECM. Now, with the V-6 being tossed in the Dumpster, the tach signal must be replicated. Novak Adapters makes a tach emulator that adapts and interprets the LS signal (both OEM GM and Holley outputs) and outputs this to the Jeep crankshaft sensor input, which in turn makes the tachometer operate as if the V-6 was still there. This is a much better option than fabricating a crankshaft trigger wheel and V-6 crank position sensor to run the tach, although on some swaps that can still be done.

In newer Jeeps, a CAN bus gateway module must be used from MRS Electronics. Since both ECMs speak different languages, the gateway module functions as a translator between the GM ECM and the Jeep ECM. Note that these do not come programmed, except when purchased from Jeep-specific LS swap shops, where they are available in kit form and come with a complete PnP harness and programming solution. At this level these are not budget-friendly swap kits, but they do offer OEM-like functions and everything still works as though from the factory, such as cruise control, ABS, A/C, and so on.

The high-level LS Jeep swap builders like Bruiser Conversions will program the Jeep ECU with a V-8 Hemi calibration and make the LS engine run with the Jeep ECM—next-level engineering indeed. This takes a lot of R&D and, at that point, you've graduated beyond working with someone reading a book to do a swap.

While we're discussing Jeep LS swaps, I should say that Bruiser Conversions provides an at-home option for performing LS swap work. This is available as a Junkyard Dog Series swap kit that comes with the equipment to run the GM E38 ECM with the swap, so that the gauges and all normal Jeep things operate as a factory-equipped setup.

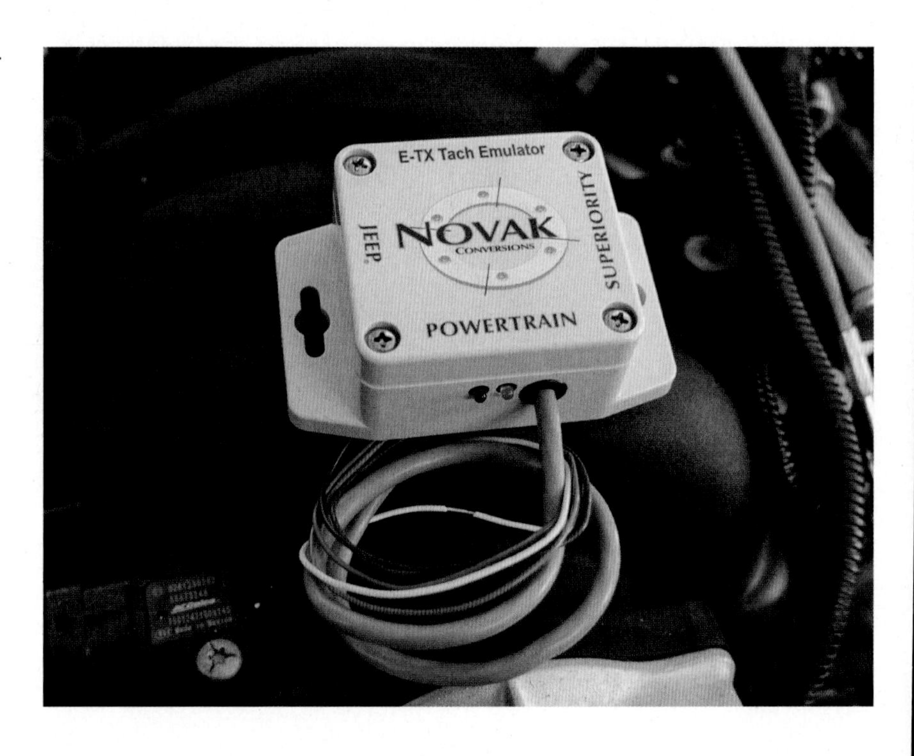

Shown here, my 2010 Jeep JK is getting an LS swap. Vehicles like this require some solutions in order to operate the gauges. In this swap, I need to retain the factory ECM in addition to the LS ECM. The factory ECM is along for the ride and will mainly operate the gauges, so I'll reuse the sending units from the Jeep into the Jeep ECM. For the tachometer to function, the LS signal has to be interpreted and integrated into the crankshaft position wire of the Jeep. There are a few ways to do this, but here I am using this module P/N E-TX62 from Novak Conversions.

Some Terminator X kits come with the 3.5-inch interface screen, but you can also buy it separately if needed. You can datalog, tune/tweak, and view sensor data in real time when this is plugged into the Holley CAN connection. This is more of a handheld-style scanner that you can leave in the glove box until needed, rather than a legitimate gauge cluster, although it can also be used for this purpose with the included mounting bracket. *Holley*

HOLLEY DIGITAL DASH

As mentioned in the EFI section, Holley has something for everyone, with widely expanded swap component offerings. They've built several Holley Digital Dash touchscreens, from the 3.5-inch mini size that comes optional with the Terminator X kits to the big 12.3-inch Pro Dash screens that replace the entire gauge cluster, along with several sizes in between that offer various options.

The base 3.5-inch LED Screen (P/N 553-108) is a compact solution to integrate with the Holley EFI system. It has one simple CAN bus connection for communication and power. You can view sensor data for diagnostics or use the small screen as a mini-auxiliary gauge cluster configured with everything the Holley can see. The 3.5-inch unit can do internal data logging to the SD card, while tuning changes can be accomplished using the screen and stylus—without having to lug a laptop everywhere you go. While a laptop affords the widest range of control for major changes, if you're out driving, then this option is a great way to leave the laptop behind and still make some changes on the road.

You can use the handheld screen to make changes on the fly, view data, and tweak things like fan-on speeds. When you don't need it, you can disconnect it and put it in the glove box until needed. While this offers a lot of mobility, some builders will use custom 3D-printed mounts with a screen for full access to data and settings.

If digital touchscreen dashes are what you require for your LS swap using Holley EFI, Holley has you covered with several options. The 7-inch Digital Dash has multiple screens that can be programmed as needed, along with switches, alarms, shift light, tuning, and data logging/playback. The 6.86-inch and huge 12.3-inch are the Pro Dash versions, which have additional features for external inputs, GPS speedometer, and shift lights. It also provides an additional connector to use the included 34-pin harness for thirteen separate inputs and four output controls that are separate from the ECU control. All Holley dash kits connect to the ECU data stream easily via a CAN bus connection. *Holley*

(ABOVE AND BELOW) The Holley Dash choices connect easily to the main Holley harness via a CAN bus connection. This is a 3.5-inch connection that does not require external power to operate, though the larger screens require additional power and ground sources and also have a Y-splitter CAN bus connection for future expansion needs.

(RIGHT) To supply sensor data to the Holley EFI or the Holley Dash, pressure sensors can be added and configured for readouts such as fuel, oil, boost, transmission, and nitrous pressures, among other data. The pressure sensor available ranges are from 100psi to 3000psi, so choose the pressure sensor range you need, then match the calibration in the ECU or Holley Dash as required. There are other sensors that read temperatures or speed, or can be configured to any other customization needs. Being able to program and calibrate sensor data yourself is one of the perks of having a stand-alone ECU and display screen. *Holley*

The larger, more visible screen sizes—such as the 6.86-, 7.0-, and 12.3-inch options—are intended for full-time dash and gauge cluster permanent replacement. Each has slight option variations, but they're all configurable for use as a complete replacement gauge solution. While these are mostly known as a being compatible with the Holley EFI, they can also be configured as a stand-alone gauge cluster with or without Holley EFI ECUs. The layout is completely adjustable and programmable, in addition to monitoring your engine conditions, speed, tach, oil pressure, and so on. You can have LED indicators or switches put on the screen that light up when your cooling fan is commanded on, or you can turn on the fan with the touchscreen. The options and layouts are comprehensive, if color digital dashes are your thing. These integrate well with the Holley EFI when paired together, and are expandable to include things like fuel level, turn signal indicators, outside air temperature, GPS speed, data logging, and tuning, all without requiring a laptop.

While the base Holley EFI does not need data like fuel pressure to operate the engine, you can add these pressure sensors and monitor fuel pressure from the screen. The regular pressure transducers from Holley are a 0–100 psi sensor and, while the fuel pressure is plug and play, you can easily add more sensors for the Holley Dash to configure and monitor. Holley has pressure transducers up to 3,000 psi, so get the one closest to the pressure range you need and then add the wiring and channel to monitor as appropriate. Data points like transmission pressure and nitrous bottle pressure are available with this technology.

(TOP LEFT, TOP RIGHT, ABOVE) Reversion Raceworks builds their Holley screen–compatible dash bezels by first 3D-scanning the original cluster, then digitally reengineering it for the application and Holley screen size before having it 3D printed. The 3D print, reinforced with carbon fiber, is a rigid unit; depending on which version you have, you can customize it further by color matching it with interior paint. Here you can see some examples in their ever-expanding catalog, such as the 1978–1988 G-body for the 6.86-inch Holley Pro Dash and the 1994–2004 Mustang design for the Holley 7-inch Digital Dash. *Reversion Raceworks*

(ABOVE) To conveniently mount the Holley Dash into popular applications, Holley has several bezel choices. This one is for the 7-inch Holley Dash into the 1st-Gen Camaro (P/N 553-300). It offers a clean and simple mounting solution you can use instead of fabricating a mount using scrap metal and zip ties. *Holley*

These Digital Dashes have mounting holes to mount to a panel. Holley has designed many dash bezels for specific applications, such as GM trucks, all common muscle cars, and many other early domestic vehicles. There are also flat-panel cut-to-fit mounts if your vehicle is not supported. If you want to be fancy, you can implement a custom 3D-printed gauge cluster mount; check out Reversion Raceworks for 3D-printed options.

DAKOTA DIGITAL GAUGES

Dakota Digital has been at the forefront of modernizing classic cars with new gauge technology for years. They are the absolute best at what they do. While gauges in later model cars are already modern (mostly), the instrumentation in older cars just does not match the modern powertrain for what would be considered a high-tech EFI LS swap.

If you're like me, new cars cannot capture the classic car heart, soul, or timeless styling of 1960s and '70s vehicles. Take the 5th-Gen Camaro: portions of it resemble its '69 Camaro counterpart, such as the rear window, small trunk lid, grill and lights, side scallops, and general shape of the car. But look at them side by side and it's a night-and-day difference. A design can be new and improved and still completely lose the soul of the original creation.

Enter Dakota Digital, which says, "Hold my beer and watch this," taking modern high-tech electronic gauge technology and making it look like it came right out of the original car. I'm talking about the RTX (Retrotech Analog/Digital) gauge kits, which are available for all mainstream muscle and performance cars and trucks. These gauges have original styling that looks right out of the era of the original vehicle. If you showed someone these "modern" gauges, the initial appearance would give no hint that they've been changed. Until you turn the key, that is, at which point the LED screen and auxiliary gauges come into view. All are programmable for backlight colors, or they come with additional gauges with the addition of optional add-on bus interface modules (BIM). You can see automatic transmission gear position with the GSS-3000 connected, or you can see ambient outside air temperature and compass position with BIM-17-2, tire pressure monitoring, GPS speed, EGT and wideband sensors, and more.

The Dakota Digital RTX gauge sets offer a nod to classic styling, with an OEM retro design of original gauges and cluster. They also provide modern features and technology, such as electronic stepper motors, RGB colored backlighting, and LED indicator lights. RTX gauges have LED screens that feature animated, original-style odometers; also available are information screens for other data, expandable with different accessory modules. With the ignition off, a bystander wouldn't even know you've changed your gauges: turning the key brings many of the features to life. *Dakota Digital*

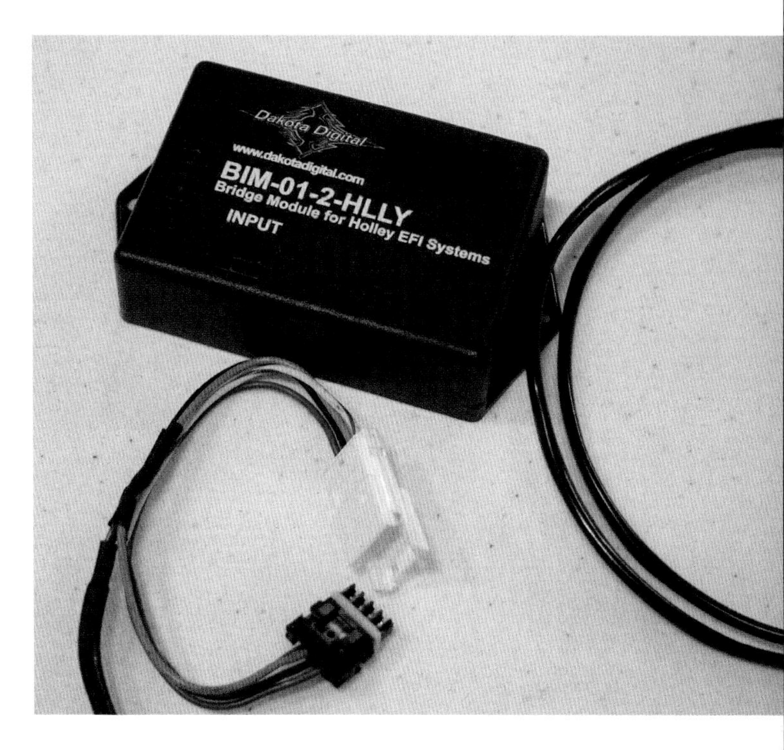

While wiring in each gauge to the individual component is standard procedure, Dakota Digital gauge systems can optionally be used with the OEM data stream like a modern car. With their BIM modules, Dakota Digital allows for easy data stream gauge integration. These modules interpret either OEM or aftermarket CAN bus data and feed that to the corresponding gauges. Basic data such as speed, tachometer, oil pressure, and coolant temperature are all used by the ECM to begin with, so why not tap into it and use this as a source for your gauge displays? OBD II (P/N BIM-02-2) and Holley (P/N BIM-01-2) modules shown.

Now for the cool factor: the Dakota Digital gauges can also read your data stream with the BIM-01-02 module connected to the GM OBD II port—and use this data as the gauge data. They took it a step further and have modules that interface with the Holley EFI systems, FAST EFI, J1939 CAN communication, AEM, Edelbrock EFI, Megasquirt, etc. Anything communicated over the CAN bus network can be implemented as gauge data, which means that, if your ECM already has oil pressure and coolant temperature, there's no need to use a separate coolant temperature and oil pressure sensor.

The RTX, like most Dakota Digital gauge kits, operates with a central control module. This is the key to their integrated setup. Dakota Digital gauges come with pre-wired harnesses for each sensor: oil, coolant, and speedometer for versatility in various applications (such as SBC/BBC/LS-Series), though some of this is not utilized with the LS swap instrumentation. The Dakota Digital central control box settings can be accessed via Bluetooth and a Dakota Digital app through your phone for convenient calibration and customization.

PRNDL Display

I installed a 6L80 six-speed auto into a 1998 OBS truck and quickly found that the PRNDL LED display did not operate, though everything else in the swap worked perfectly. I hadn't expected this, much less imagined it could happen. The cause? With the 6L80 transmission using CAN bus for everything, it has its own T43 transmission controller that handles all these duties. No one told the 1998 truck about this, and technically there was no park/neutral safety switch for startup or any similar options.

I ended up taking a Dakota Digital gear indicator kit (P/N GSS 3000) and modifying it for use in the OBS truck 6L80 (without a guarantee that this would work). This helped with a few problems. First, the reverse light operation: I could now power up the reverse lights. Secondly, it functioned as a neutral safety switch, so now the truck would start in park and neutral only. Finally, I was able to use the GSS-3000 to simulate the sequence of ground wiring that went to the old PRNDL switch on the side of the old 4L60 transmission. I did this using four relays and twelve diodes (three per relay) and a lot of schematic studies and wire mapping. I'm sharing the pictures for reference if it helps someone do the same. It may be that using only diodes to ground the four different sequence of wires and programming the DD GSS-3000 to ground path control would work, but I also was using the relays themselves because I knew it would work easier in my situation as simple on/off remote switches. Of course, an electrical engineer could probably design a solid-state module that replicates this approach to capturing CAN bus data, but this example shows how being resourceful and knowing a thing or two about wiring can accomplish your lofty goals.

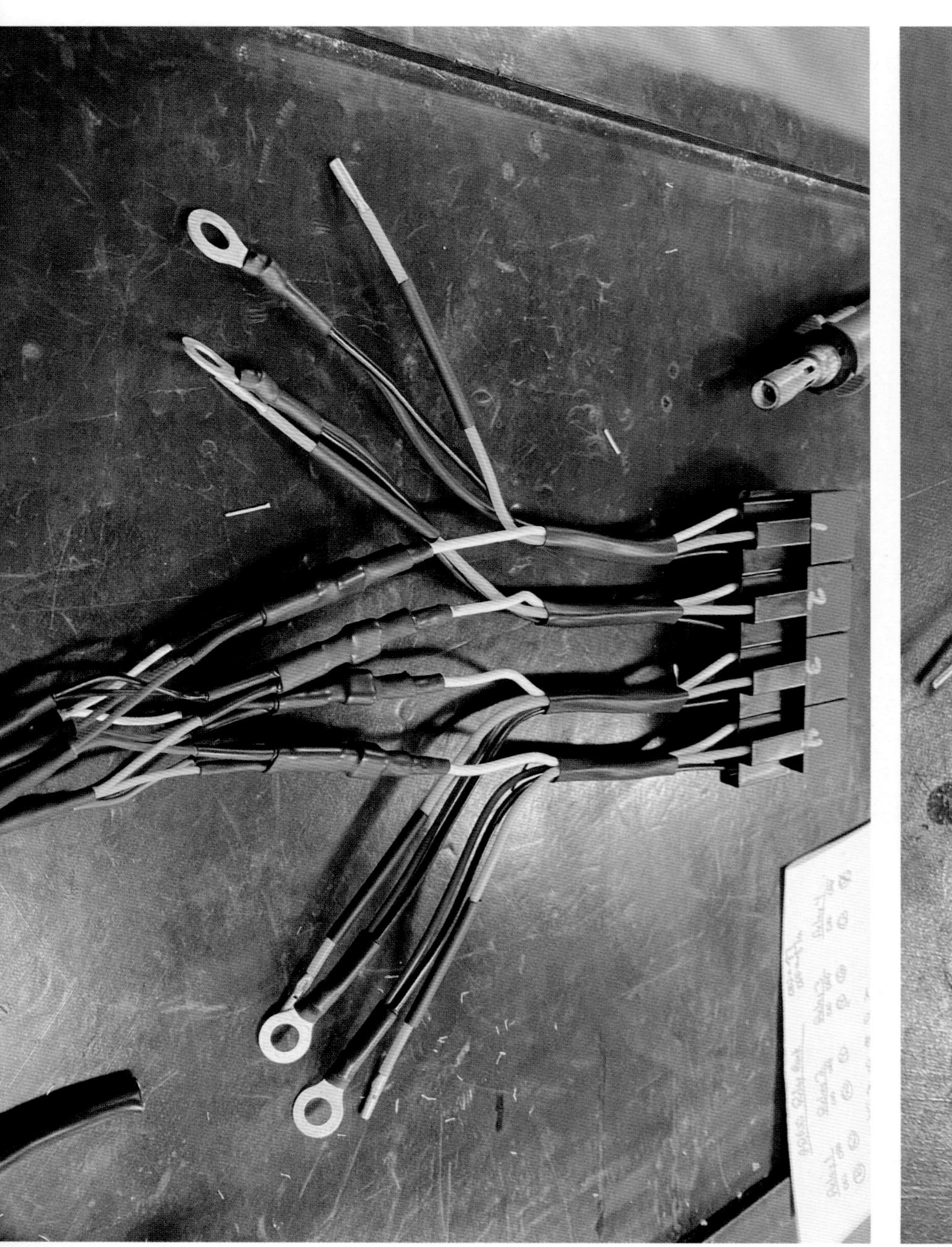

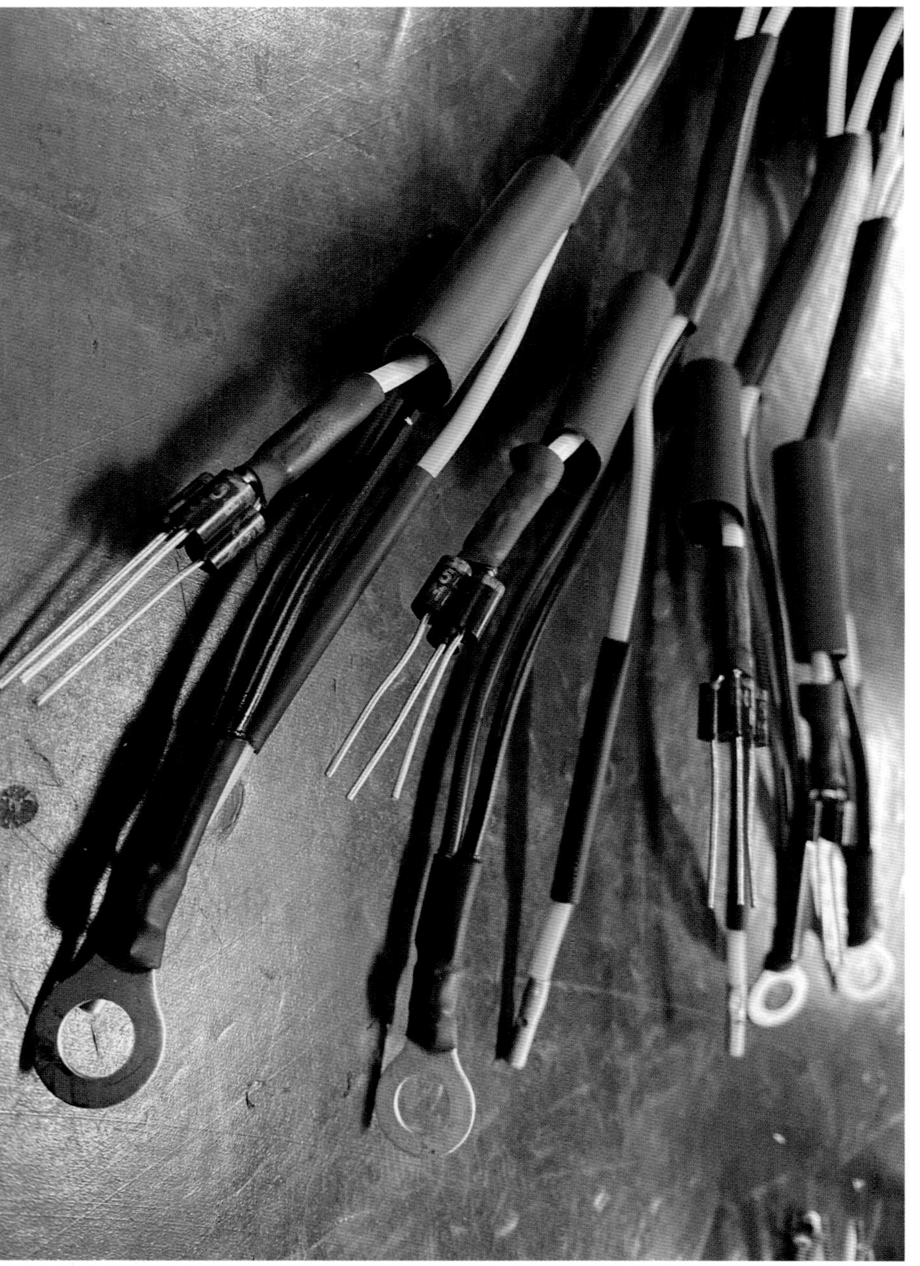

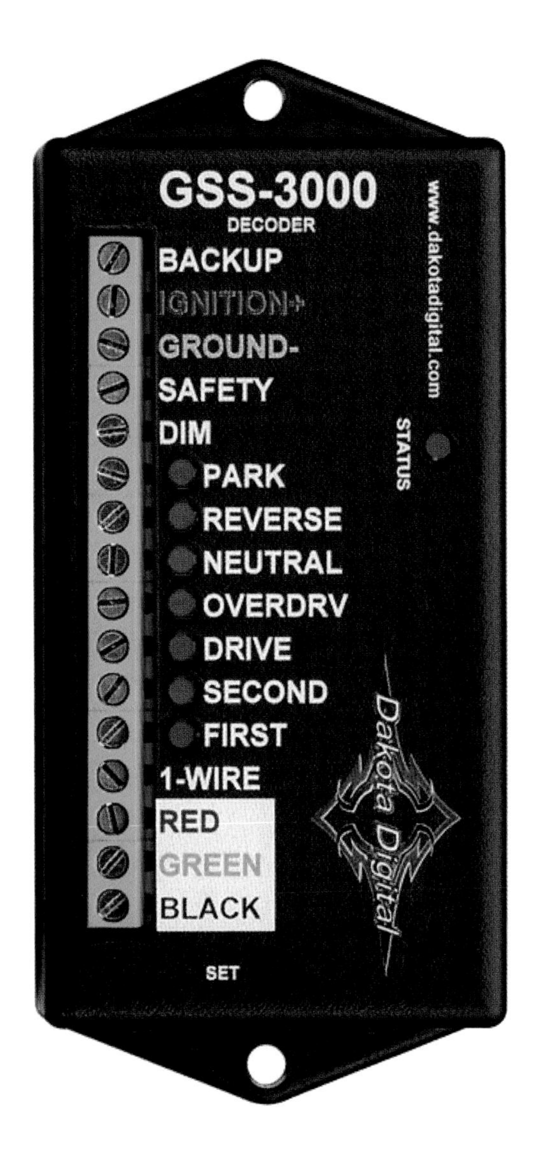

(ABOVE) The GSS-3000 from Dakota Digital is a gear position module that has a transmission-mounted potentiometer on the transmission shift lever. This is helpful as a dash-mounted aftermarket gear indicator, but it also serves as a park/neutral position safety switch for startup. This kit can be used creatively in CAN bus–equipped transmissions for the same purposes. *Dakota Digital*

(OPPOSITE) Chevy trucks manufactured from 1995 to 2000 with LED shift indicator lights will not work with a 6L80/90 transmission, as the 4L60/80E neutral safety switch can no longer be used and all electronics run through the CAN bus network. If you're resourceful and can work with the electrical system, you can use the Dakota Digital GSS-3000 to output the required gear position in these trucks. But because combinations of multiple wires activate various indicator LEDs, you must use diodes and relays for the setup to operate. In the future, maybe someone will make a module that does this, but currently this was the best way to create a working PRNDL display.

LED	*PRNDL Wires to Ground*
P	White and Black
R	Yellow and Black
N	Yellow and White
D	Yellow and Gray
3	All Wires to Ground (I did not use this position with 6L80)
2	Black and Gray
1	White and Gray

Knowing the sequence and wire pairing required told me I needed four relays, one for each wire color, so I named the relays as follows where the number corresponds to a wire color. Each wire has three possible locations, but because it took combinations of six different positions to light up six LEDs using the four wires, that added up to twelve diodes. Three wire locations per relay × 4 relays = 12 diodes needed to not backfeed and just light up all seven LEDS.

Relay	*Wire Color*
1	White
2	Black
3	Yellow
4	Gray

Now I would take the GSS-3000 and wire it to control two relays per output location via diode 3 into 1 trident setup. The relays would then provide a ground path, lighting up the specific PRNDL-corresponding LED.

Park	Activate Relay 1 and 2
Reverse	Activate Relay 2 and 3
Neutral	Activate Relay 1 and 3
Drive	Activate Relay 3 and 4
3	Activate Relay 1, 2, 3, 4 (not used on 6L80)
2	Activate Relay 2 and 4
1	Activate Relay 1 and 4

I don't want to downplay how involved this was. It literally took a day of work to figure this out and to test, assemble, and make this PRNDL work. If you are paying shop labor for something like this, at conservatively $100 an hour, you would now owe the shop $800 to do the same thing, plus the $120 for the GSS-3000 module. Now you can see the value in purchasing this book. This information just is not available online for free.

If your project didn't come with cruise control, you can add this function using either OEM or aftermarket kits. Dakota Digital has these kits, which also include convenient replacement handles that match the classic car looks. Because almost all cruise control systems use standard wiring schematics, you can usually find what you need for cruise control switches, or even add in a dash-mounted control panel.

CRUISE CONTROL

Now that you've built the perfect modern-powertrain weekend cruiser, what are you going to do with it? Drive it to church on Sunday? Attend the local car meets? Maybe you want to show up for one of the cross-country rallies like the Hot Rod Power Tour? To do so comfortably, first put some good seats and A/C in your car and, second, implement cruise control on your new build. Cruise control is easy to add, and can be added later on after the build is complete, if you want to break up your workload.

Gen III Cruise Control

There are a couple ways to do this with Gen III controllers:

- For a DBC source, take a factory cruise control box from a 4th-Gen Camaro/Firebird, or an early model DBC truck/van (either without traction control) and wire it in just like your car was a 4th-Gen F-body. You can mount the cruise control module anywhere underhood as long as your cable reaches the mechanical throttle body. If you source one of these from a donor vehicle, make sure you get as much of the wiring as possible; the ten-pin connector and wiring can be spliced into your vehicle setup. You'll still need a four-wire cruise control switch from an analog setup that has On/Off, Set/Coast, and Resume/Accel. This may be on your multifunction turn signal switch.
- If you have a DBW Gen III, you have the makings of cruise control without additional hardware. Provided you have the TAC module and cruise control switch already, you only need to implement three wires needed to activate controls on the TAC module for cruise control: On/Off, Set/Coast, and Resume/Accel. This means that four wires, including the power to the switch, must be utilized for the control switch itself. This is simply wired into connector C1 of the TAC module.

For either, it is possible to use the Dakota Digital HND-2 On Dash replacement control panel, which works with the OEM Gen III cruise control—if you don't have a factory switch available. These use a 4,000-pulse-per-mile signal from the ECM to keep a constant set speed and, for safety, a brake and clutch switch must be implemented.

Gen IV Cruise Control

The cruise control is a little trickier on the Gen IV. The E38 engine controllers rely on CAN bus cruise control functions through a body control module. This requires wiring in the compatible "new car" multifunction switch and matching BCM to use cruise control. No, thank you.

If you have a DBW system such as from the Gen IV LS engines, you will have a DBW pedal. You can easily add cruise control to these using the CRC-1000 or CRC-2000 Dakota Digital module. They tap into the pedal connection itself to replicate the proper cruise-speed throttle angle. There are a few other things that need to be wired in, such as speed input, brake pedal switch, and of course the cruise switch itself. These kits offer a great solution that otherwise might not be possible for DBW-equipped vehicles.

With the E67 engine controller, this ECM does have the ability to handle cruise control via direct analog connection signals. So you can use your 1960s–1990s-style analog multifunction switch or any of the Dakota Digital cruise handles/switch panels because they use standardized analog wiring. The main trick here will be integrating the firmware in the ECM: Calibrate the ECM with a 2009 Trailblazer SS segment file to make sure everything jibes, but because the control signals are the same as they have been for forty years, this should be easy to wire in following the pin locations.

Dakota Digital DBW Cruise Control

I cannot say enough good things about the people at Dakota Digital and their commitment to hot rodders. No surprise that they have a cruise control solution for DBW Gen IV engines. There are two versions: the P/N CRC-1000, for integrating and reading speed on the OBD II CAN bus data stream, and the P/N CRC-2000, for a separate speedometer pulse-per-mile input that you add yourself, such as when using the P/N SGI-100-BT or comparable speedometer modules.

Both systems connect to the DBW electronic pedal, require a brake input, clutch pedal switch input, and the three control wires from the switch. If you are like me and appreciate things that look factory, you will be pleased to know that the add-on switch looks right at home in the 1967 Camaro build, matching the era of the car. A more modern switch would look out of place.

The CRC-1000 and CRC-2000 are known to function on the following engine controllers and pedals:

- 2005–2013 Corvette
- 2008–up GM full-size truck/van/SUV with throttle pedal contain inline six-pin connector
- 2010–2015 Camaro
- 2008–2010 Pontiac G8 GT (verify pedal connector type first)
- GM Connect & Cruise packages

The Dakota Digital is the aftermarket solution for OEM-level cruise control functionality with Gen IV engine controllers such as the E38 and E67. It's also reasonably priced for what it is, if you compare the alternative choices. Check the images and captions for more information.

Chapter 10
Startup and Troubleshooting Guide

Throughout this manual we have discussed all of the major points of the LS-series engine swap: engines, transmissions, accessories, ECM/engine harnesses, cooling systems, and fuel systems. In this last chapter we will cover a basic checklist to go over before startup and things to do after startup. Also, if there are startup issues, we cover what to look for in when diagnosing the problem. Before starting up a new engine for the first time, or even a used engine, there are certain things that you need to be sure of, and it also helps to have a buddy standing by with a fire extinguisher anytime this amount of work is done, just in case. Much of the diagnosis part is the beginning steps and just touching on common subject matters to help get to the bottom of any issues. If things get too complicated, the reliance of a good service manual and having someone at hand who is experienced in such diagnosis goes a long way in helping find and fix your issues.

INITIAL CHECK OVER

Before even hooking up the battery, go over *all* of your work one last time. Check hose clamps and the oil drain plug, make sure you have secure fuel connections, and ensure all electrical connectors and grounds are good and tight. Make sure the engine is full of oil, the oil filter is in place, transmission and cooling systems are as full as possible and that no moving parts rub on nonmoving parts. Also, be absolutely sure that no wiring is against the exhaust. It may be ok now, but when the engine fires up, the wiring won't be put in a good position for longevity. This is also a good time to make sure fresh fuel is available for the engine.

Finally, your swap is completed! If you did everything correctly, you can prepare for initial startup. *KWiK Performance*

Obvious mandatory steps before startup include filling the fuel tank, radiator, adding or changing engine oil, and maybe power steering or transmission fluid top-offs, if these are new items. Filling the radiator on LS-series engines can be tedious if an air bubble gets trapped in the coolant passages. Using a radiator funnel seems to help, but loosening the engine coolant sensor will help equalize the radiator and engine block coolant levels if needed.

One thing I like to do when filling up the oiling system on a new engine or one that has sat for a long time is pre-lubricate the valvetrain. Remove both valve covers and take 1 quart of engine oil and coat the rocker arms and valve springs on each cylinder head. Use about ½ quart on each side. This helps keep everything lubricated while the oiling system builds up pressure and engine oil traverses to the top end of the engine.

I have watched it take up to five minutes from initial startup for the rocker arms and springs to get oil on an idling engine—on an engine that only sat overnight. If you hear a squeaky noise under the valve covers at half the pace of the engine rpm, that is a lack of oil to the rocker arms. Verify oil pressure, then increase engine speed 500 rpm to quickly take care of this. I typically run brand-new engines at 1,500 to 1,800 rpm to avoid this situation, and it helps lubricate the cylinder walls at this elevated rpm. Connect battery cables (positive first) and observe that there is no excessive arcing. A little spark is fine. Quickly look over the car to make sure there is no burnt wiring or funny smells. After ensuring that the electrical system is safe, you can crank the engine over after disabling the ECM or coil packs so that the engine does not start. Crank it over in park or neutral while listening for any odd noises. If everything is deemed all right, you can enable the ignition or ECM and prepare to fire her up for the first time.

Once the ECM is back in place, turn the ignition on, but do not start the car. Make sure you hear the fuel pump running for the few seconds during prime mode. Check the fuel lines from the fuel tank to the engine completely for any leaks; this is where you would find them and address any issues before startup. It would be helpful to monitor fuel pressure if you suspect an issue, but I know you are ready to fire this thing up. It has been long enough, I'm sure.

As a side note, I prefer to start up new engines on the factory fuel injectors, as these tend to be easier to keep the ECM calibration to match with, and it's less likely that you'll wash the oil off the cylinder walls (which is bad for ring seal). That being said, if you have some dirty old injectors that have been plugged up with old fuel crud, you obviously cannot use those. Keep this tip in mind, and do not run your new engine with the wrong fuel injector calibration—have some patience until the injector data matches up.

With the car still in park or neutral and tires chocked or on jack stands, attempt to start the engine. It may take a bit for air to be purged through the fuel lines, but it should start up rather quickly. It may need some assistance from your throttle pedal. If it doesn't start up, we need to check a few things out. Make sure the VATS is disabled, fuel pressure is good, and that the ignition system is working.

Assuming the engine does start up as expected, make sure there are no oil, coolant, transmission, or fuel leaks first of all. Then triple-check nothing is against the exhaust and melting (the burning plastic smell is very noticeable). This is where it pays to have a friend or helpful neighbor present to alert you. Four eyes are better than two, especially when you are trying to look at 15 different things at once.

Second, make sure you have oil pressure within about 8 to 10 seconds. If not, shut it off and figure out why. When you first start up a fresh motor or motor that has been sitting, there will also be quite a bit of engine noise from the valvetrain until oil pressure fills the hydraulic lifters with oil. The majority of this will go away when oil pressure is present, but sometimes you will have one stubborn lifter that just won't get quiet until the engine warms up. Don't panic, just hold the engine rpm above 1,000,

If you have an adjustable fuel pressure regulator such as the one needed with an Aeromotive fuel system, you will be required to set the fuel pressure before startup. Out of the box this Aeromotive EFI regulator was set to a hair above 20 psi, which is well below the minimum requirement of 58 psi. Using a fuel pressure gauge and a few wrenches, the fuel setting should be adjusted to 58 psi before startup. It may need another tweak back up to specifications to be exactly the fuel pressure needed while the engine is running. Using the C5 fuel filter and regulator combo will keep the fuel pressure at the correct specification, although checking fuel pressure is always a great idea with all new components.

and let it warm up a bit while watching coolant temp and levels. Often these just need some heat to free up.

It's worth mentioning again to not let a new engine idle, as the valvetrain needs some rpm to get a steady supply of engine oil to the top-end. This does not happen quickly at engine idle. Bump it up and let the oil flow. No matter how good your idle chop is, never let a fresh engine idle for very long. The tolerances are tighter on a brand-new engine and the cylinder walls and pistons do not get pressure-fed oil supply; running them without adequate lubrication can be detrimental to piston wear. Oil slung off the connecting rod bearings hits the cylinder walls and pistons, so it typically takes the 1,500–1,800 rpm range to get some oil splashed around. You can idle the engine again after a few heat cycles during the break-in period.

DIAGNOSING A NO-START CONDITION

So you have spent the better part of the last three to six months performing the LS-swap in your project car. You have all your buddies around, the neighbors come out to see what all the fuss

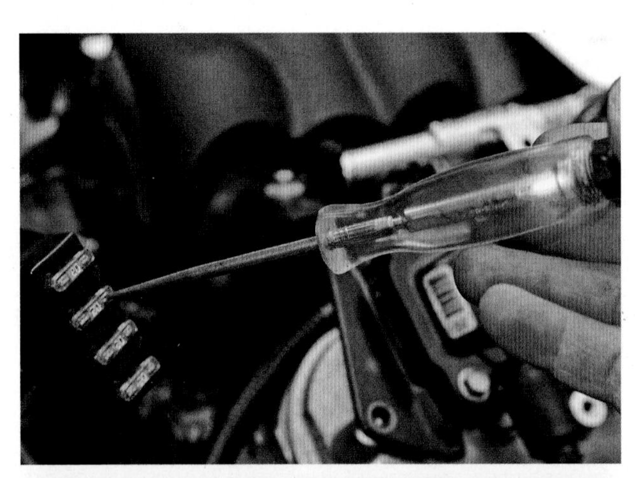

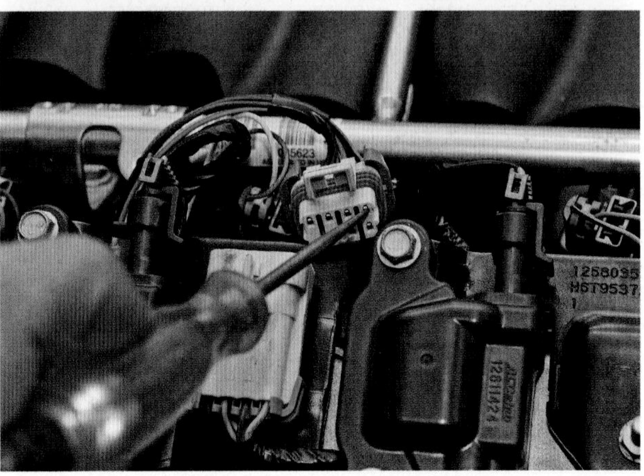

If you try to start your car and it just cranks over and over without a hint of starting, you may need to put on your mechanic's hat and start checking your work. If you know you have fuel pressure, you will need to check other areas related to the engine. Test lights come in handy to check simple things like blown fuses or power to coil packs. If the fuel pump is not working, check for power and ground on the fuel pump terminals and back trace the wiring if an issue is present.

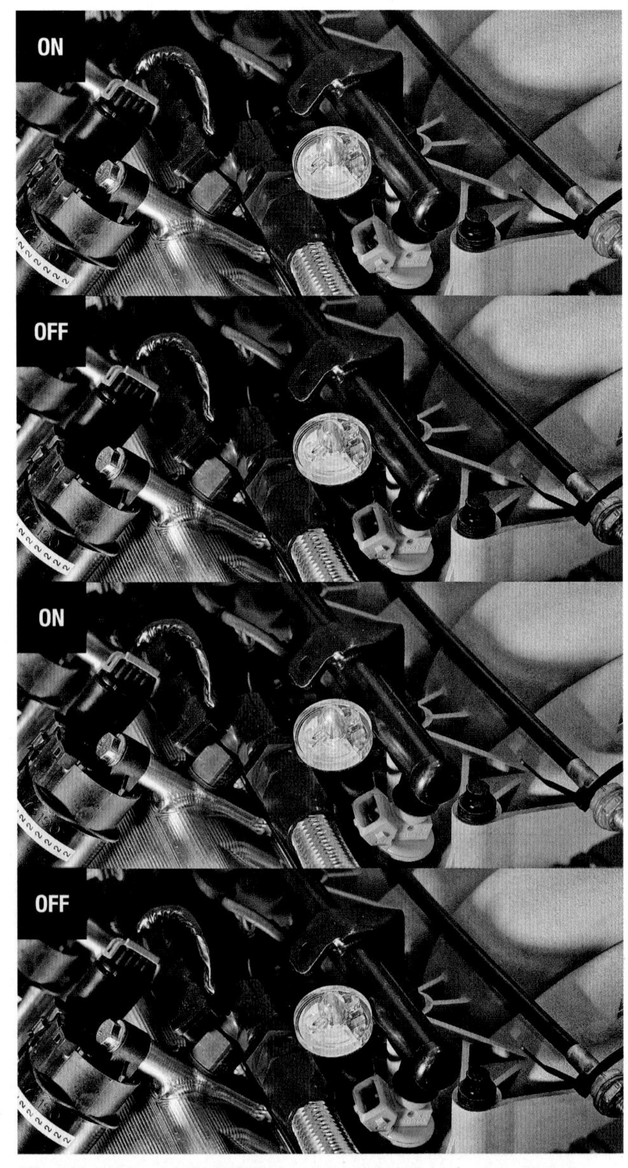

Often with an LS swap, the theft deterrent determines if your engine will start. Usually the theft deterrent rears its head by disabling the fuel-injector pulse. You can diagnose that issue by using an injector noid light and cranking over the engine. The light will pulse on and off as shown, indicating an injector signal. Keep in mind that the injectors can be getting a signal and still not open if old fuel or debris plugged or stuck the injectors. In that case, either new injectors must be procured or you may want to have the injectors cleaned.

is, the kids are in their best attire for the occasion, and the wife has the video camera recording the occasion. Like a child on Christmas morning opening his new presents, you gleefully turn the ignition switch to start the engine—and something quickly happens (or doesn't happen, rather) to turn the best day ever into the worst day ever. The longer it cranks (or does not crank) the sicker you feel.

This is often the case with many swaps, but this is no reason to go arsonist on your car or start throwing screwdrivers across the garage into the sheet rock. It could just be something simple. Once the commotion dies down and you have some time—and a clear head to think—now is the time to diagnose the no-start condition. Sometimes it is better to just let the car sit while you think about what you are going to check. A clear head goes a long way when working on engines and cars.

So you know it won't start, but the diagnosis procedure for testing varies a bit depending on exactly what is going on. Is the starter turning the motor over? Is the car trying to start but not quite making it? Is it backfiring? The answer to these questions decides which direction you need to start checking items.

Many wiring problems are self-inflicted. Usually these are ground related, caused for example by relocating a battery to your trunk and relying on the chassis for your ground path. You need a clean metal-on-metal contact path from the engine to the battery. If every panel is freshly painted, it is likely a bad ground path. In this case, you need to run a ground wire to your engine directly from the battery: bigger is better on the battery voltage and ground cables when remote mounted. For reference, I use 1/0 sized battery cable. Additionally, the ECM ground points need to be clean engine grounds. Learning what is not allowing the engine to start will gives the diagnostic starting point, but never underestimate the importance of engine and chassis grounding.

First, check for diagnostic trouble codes in the ECM. This should automatically be the first thing you check, should you have issues. There is some stuff that the ECM can help you with, but it cannot tell you if the starter, fuel pump, or injectors are working. It can tell you if something is not plugged in, though, which is sometimes helpful. If the engine does not crank over, use a test light or voltmeter check for power at the starter ignition wire (small wire on starter) while someone is cranking the engine. If there is power, but the starter still does not crank over, it could likely be a bad starter. If there is no power to that little wire, you will need to retrace the steps backward from the starter to the ignition switch or starter relay if used. It could be something like a misadjusted neutral safety switch not allowing the engine to crank over, or maybe an aftermarket car alarm interrupting the starter signal. Whatever it could be, the test light or volt meter will help isolate it.

If you have power coming out of the ignition switch but none at the starter ignition wire, it obviously is a broken connection between the two. Count the VATS system out on a retrofitted engine in an older car, but if swapping into a vehicle

Checking the ignition side of things is easy also. Rather than use a screwdriver connected to a spark plug wire to check for spark as many people do, find or borrow a spark tester. This simply plugs inline to the spark plug and wire and lights up while cranking if the ignition system works. You likely only have to check one to determine if the ignition system is functional or not.

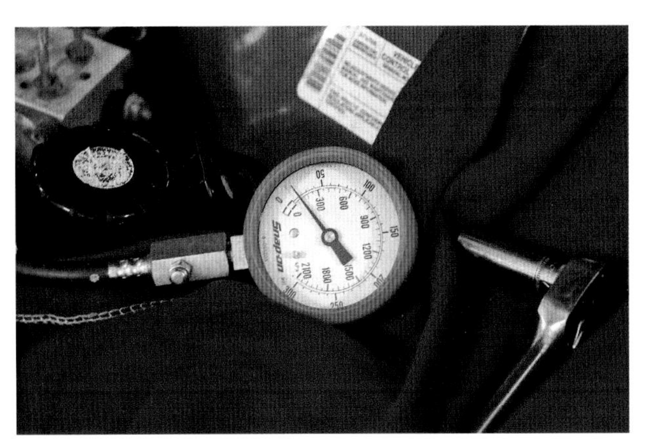

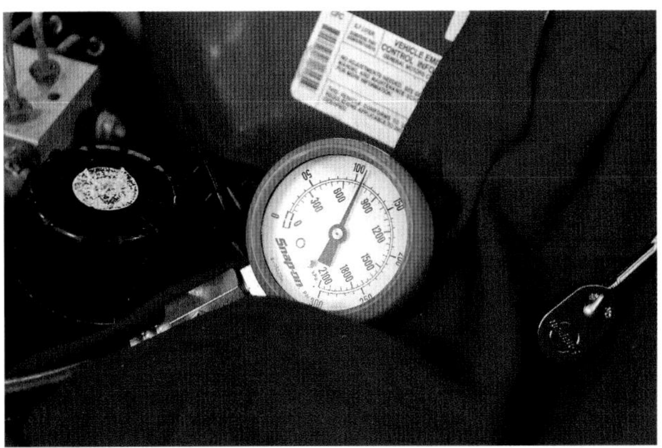

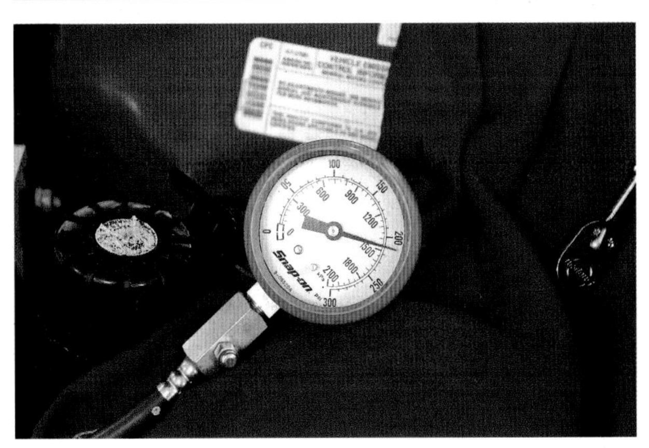
If you have a low-powered, rough-running or misfiring engine, the next thing to check is the engine health to eliminate possible internal issues. The easiest thing to do is a compression test on all cylinders. The compression test results depend on actual compression ratio, engine displacement, and camshaft duration. Typical good numbers are 160 psi and up. The first image shows a bad cylinder (25 psi). In this case, the piston and rings were bad. The second image is the same engine, just with slightly better compression (110 psi) but still not ideal. The last image shows compression test results on a normal cylinder (210 psi).

with VATS, you may have to bypass the system to start up the LS engine since so much has been changed in the wiring system and engine bay. I ran into this in the 1991 Camaro swap: I had to bypass the Passlock module starter signal for the starter to crank.

Now, if your engine does crank over and over and over and does not attempt to startup, it could be related to the fuel or ignition system. Begin with ignition first so as to not have any fuel vapors nearby.

The ignition system is the easier of the two to verify proper operation by using a spark tester connected to a spark plug wire. A screwdriver would work, but do not handle it at all while cranking the engine. Stock coils have 40K volts of power each.

Instead, safely zip-tie it near a metal object like the exhaust manifold while cranking and stand back. If it jumps spark while cranking, then the ignition system is working and no more tests are needed at this time. Reconnect everything as it was beforehand and move on to the fueling.

No-start conditions due to fuel system issues can stem from hose misrouting, miswiring, plugged fuel injectors, or even from being out of fuel. Anytime you are working on a fuel system, you must exert careful techniques so as not to become a fire hazard. I recommend de-pressurizing the fuel system when working on it, even if you know there is no fuel in it, treat it as such.

In this situation you would listen for the fuel pump to come on and prime when the ignition switch is first turned on. If you hear the pump, you will want to check fuel pressure next. You can check it at the fuel rail in the engine bay. Connect the fuel

A fancier way to check engine health and to isolate the problem area is with a leak-down test. In a leak-down test, you add 100 psi of air pressure to the problematic cylinder with the piston at top dead center and both the intake and exhaust valves closed. The second gauge shows how much air the cylinder holds. On a warm stock engine you would typically see less than 5 percent leak down; on a cold engine you could see 15 percent. This one (reference above low-compression cylinder) has 85 percent leak down, which is really unfortunate. The engine requires repair or replacement.

GM CRATE ENGINE STARTUP AND BREAK-IN PROCEDURES

Safety first. If the vehicle is on the ground, be sure the emergency brake is set, the wheels are chocked, and the car cannot fall into gear. Verify everything is installed properly and nothing was missed.

1. **Oil and fluid fill:**This engine assembly may need to be filled with oil or have oil added. After installing the engine, ensure the crankcase has been filled with the appropriate motor oil to the recommended oil fill level on the dipstick. The LS crate engine requires a special oil meeting GM Standard GM4718M (this will be specified on the oil label). Mobil 1 is one such recommended oil. Other oils meeting this standard may be identified as synthetic. However, not all synthetic oils will meet this GM standard. Look for and use only oil that meets GM Standard GM4718M. Also, check and fill as required any other necessary fluids such as coolant, power steering fluid, and so on.
2. **Oil system prime:**
   a. The engine should be primed with oil before starting. Install an oil pressure gauge (the existing oil pressure sensor location at the upper rear of the engine may be used) and disconnect the engine control system (removing power from the engine control module is generally recommended, but check your engine control system information for additional details). *Note: Disconnecting only ignition or fuel injector connectors is not recommended. Make sure the control system will not provide ignition or fuel to the engine.*
   b. Once the engine-control system has been disconnected, crank the engine using the starter for 10 seconds and check for oil pressure. If no pressure is indicated, wait 30 seconds and crank again for 10 seconds. *Repeat this process until oil pressure is indicated on the gauge.*
3. **Initial engine start:** Reconnect the engine control system. Start the engine and listen for any unusual noises. If no unusual noises are noted, run the engine at approximately 1,000 rpm until normal operating temperature is reached.
4. **Engine warm-up recommendation:** When possible, you should always allow the engine to warm up prior to driving. It is a good practice to allow the oil sump and water temperature to reach 180 degrees F before towing heavy loads or performing hard acceleration runs.
5. **First 30-mile break-in period:** The engine should be driven at varying loads and conditions for the first 30 miles or one hour without wide open throttle (WOT) or sustained high rpm accelerations.
6. **Medium accelerations for break-in:** Run five or six medium throttle (50 percent) accelerations to about 4,000 rpm and back to idle (0 percent throttle) in gear.
7. **Hard accelerations for break-in:** Run two or three hard throttle (WOT 100 percent) accelerations to about 4,000 rpm and back to idle (0 percent throttle) in gear.
8. **Change the oil and filter:** Replace the oil per the specification in step 1, and replace the filter with a new PF48 AC Delco oil filter. Inspect the oil and the oil filter for any foreign particles to ensure that the engine is functioning properly.
9. **500-mile break-in period:** Drive the next 500 miles (12 to 15 engine hours) under normal conditions. Do not run the engine at its maximum rated engine speed. Also, do not expose the engine to extended periods of high load.
10. **Change the oil and filter after 500-mile break-in:** Again, inspect the oil and oil filter for any foreign particles to ensure that the engine is functioning properly.

(Source: GMPP Crate Engine Instructions)

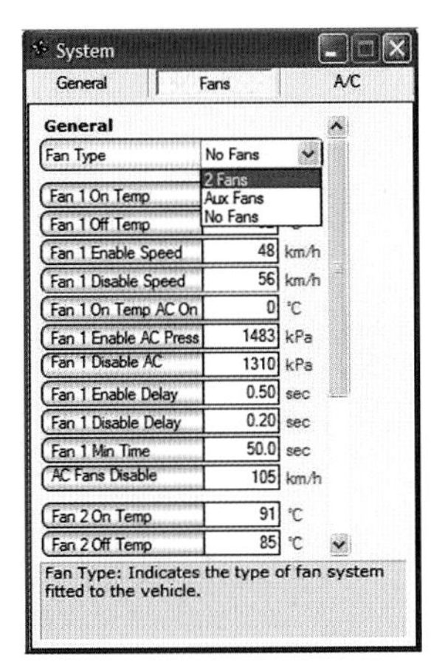

If your cooling fans are not turning on, there could be a problem with the wiring, or it could just be as simple as turning them on in the ECM software. Truck ECMs that come with a mechanical cooling fan would be the most prone to needing the fan settings programmed. In addition to the fan option, the cooling fan temperature settings may be recalibrated, especially so if you have a lower opening temperature, such as a 160 degrees F.

pressure tester to the test port on the fuel-injector rail. If there is any fuel pressure at all, be careful when connecting the fuel pressure tester so as to not dump fuel everywhere. Use some shop rags to absorb the fuel spray or puddles. To test fuel pressure after connecting the gauge, energize the fuel pump to pressurize the fuel system. While it is running, check the needle on the fuel pressure gauge and note the readings, if any. The fuel pressure should be set to 58–60 psi if you have an adjustable pressure regulator; if you have a C5 Corvette fuel filter/regulator, verify that the pressure is the same 58–60 psi.

If there is no fuel pressure, but you hear the pump, it could be crossed lines at the fuel filter or regulator. This happens a lot. What you would think is the inlet to the filter because of being located in the middle of the filter is actually the return line. The fuel inlet is offset and the larger of the two fittings on that side of the filter. The front outlet is by itself so unlikely that that is misrouted. Some aftermarket fuel pressure regulators have different spring pressures and come with a low-pressure spring and a high-pressure spring. If your fuel pressure is low, make sure you did not use the wrong FPR spring.

If the pump is running, but you suspect that it is not pumping fuel, you can first of all check polarity of the wiring. Make sure positive is to positive and ground is to ground. Also, check to make sure the pump is oriented correctly and not backward. If you have an inline fuel pump, that is.

Finally, if you see nothing wrong and the pump is verified to turn on, make a test hose to connect to the outlet of the fuel pump long enough to reach outside far from the car and flammable materials. Zip-tie the test fuel line into something sturdy like a fuel jug and cycle the key to on, stand clear, and observe. If no fuel, or little, is delivered, then you have a problem in the suction side of the fuel pump or delivery system.

If you want to check fuel volume at this time to evaluate fuel system health, perform the same technique but for a set amount of time. I use a clear 5-gallon fuel jug, then power up the fuel pump using a momentary starter button for fifteen seconds. Use a timer and see how much fuel pumps into the fuel jug. Log your measurement and multiply this by 4 to get the minute measurement. If your fuel pump is rated by gallons per minute, compare this to your pump specs. If your pump is liters per hour, take this measurement and convert to liters (3.78 liters per gallon), but you also have to multiply by 60 to get the liters-per-hour (LPH) measurement.

- GPM × 3.78 × 60 = LPH
- LPH/3.78/60 = GPM
- 450 LPH pump would deliver 2.0 gallons per minute
- 340 LPH pump would deliver 1.5 gallons per minute
- 255 LPH pump would deliver 1.1 gallons per minute

If you are using the stock carbureted fuel delivery line and no sump fitting to gravity feed the fuel pump, then this could likely be your problem. The fuel pump does not like to pull fuel. It likes to push fuel that is fed to it, and a sumped fuel tank with bottom-feed fittings is almost always required for an inline fuel pump.

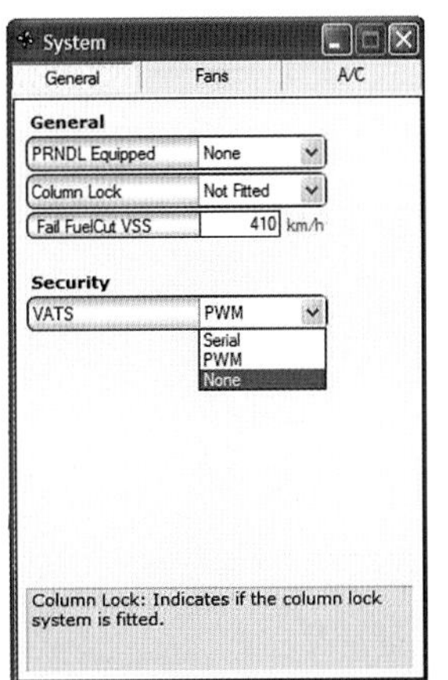

Another nuance when using OEM ECMs is the theft deterrent system. You are likely to not reuse the OEM theft deterrent, as it is more of a car feature than an engine feature; but you do need to "turn off" the theft deterrent in the ECM so that you can start the car. In the OEM vehicle, the starter and fuel injectors are disabled. In your retrofit, only the injectors would likely be disabled as the starter would normally be part of the car's existing wiring system.

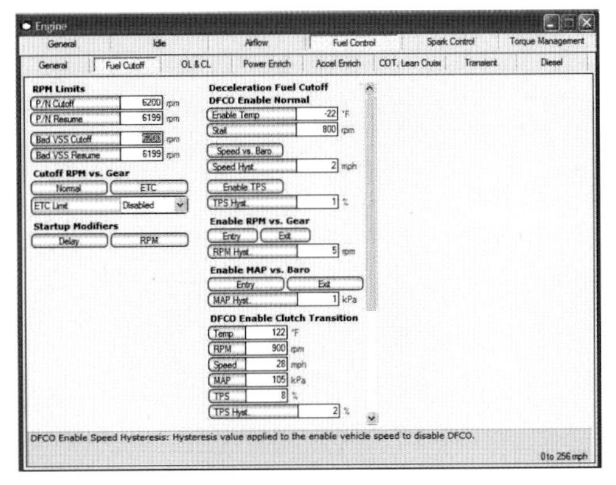

If you are using a mechanical speedometer with no speed-sensor inputs into the ECM, you will be required to raise the rpm limiter for a missing vehicle speed sensor (VSS) as the ECM will use the lower "Bad VSS" rpm rev-limiter. Simply raise both numbers to your desired value, usually 200 rpm past your ideal shift point.

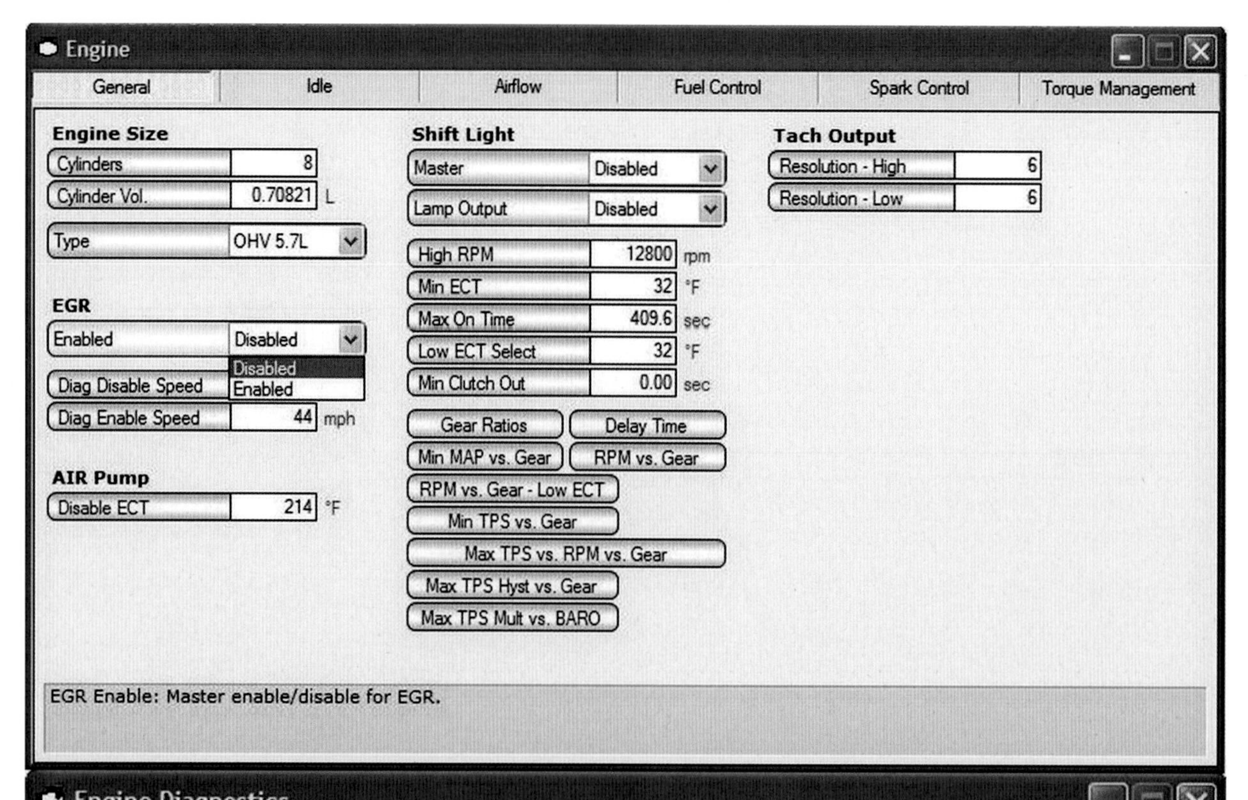

Engine Diagnostics

General | Misfire | Airflow | DTCs

	Description	SES Enable	Error Mode
P0131	HO2S Circuit Low Voltage Bank 1 Sensor 1	☑	1 - MIL on Second Error
P0132	HO2S Circuit High Voltage Bank 1 Sensor 1	☑	1 - MIL on Second Error
P0133	HO2S Slow Response Bank 1 Sensor 1	☑	1 - MIL on Second Error
P0134	HO2S Circuit Insufficient Activity Bank 1 Sensor 1	☑	1 - MIL on Second Error
P0135	HO2S Heater Performance Bank 1 Sensor 1	☑	1 - MIL on Second Error
P0137	HO2S Circuit Low Voltage Bank 1 Sensor 2	☑	3 - No Error Reported
P0138	HO2S Circuit High Voltage Bank 1 Sensor 2	☑	3 - No Error Reported
P0140	HO2S Circuit Insufficient Activity Bank 1 Sensor 2	☑	3 - No Error Reported
P0141	HO2S Heater Performance Bank 1 Sensor 2	☑	3 - No Error Reported
P0147	HO2S Heater Performance Bank 1 Sensor 3	☐	3 - No Error Reported
P0151	HO2S Circuit Low Voltage Bank 2 Sensor 1	☑	1 - MIL on Second Error
P0152	HO2S Circuit High Voltage Bank 2 Sensor 1	☑	1 - MIL on Second Error
P0153	HO2S Slow Response Bank 2 Sensor 1	☑	1 - MIL on Second Error
P0154	HO2S Circuit Insufficient Activity Bank 2 Sensor 1	☑	1 - MIL on Second Error
P0155	HO2S Heater Performance Bank 2 Sensor 1	☑	1 - MIL on Second Error
P0157	HO2S Circuit Low Voltage Bank 2 Sensor 2	☑	3 - No Error Reported
P0158	HO2S Circuit High Voltage Bank 2 Sensor 2	☑	3 - No Error Reported

Tampering with emissions control devices can be illegal. Please check your local laws as well as EPA rules and regulations for legal modification. Disabling emissions controls should be used on off road use only vehicles. It is illegal to modify diagnostic test results in order to pass emissions testing. FOR OFF ROAD USE ONLY!

Finally, if you are building a vehicle that doesn't have to adhere to the late-model cars' EPA requirements, you can disable the emissions sensors that are not used. The sensors not required for the engine to run are the rear O_2 sensors, EGR system, EVAP solenoid, and the AIR injection system. You simply check off the sensors you are not using. This is probably much better than the original engine EPA requirements.

If everything checks out thus far, you will have to dig deeper. The LS ECMs turn off the fuel injectors when theft deterrent is activated. This means that while you can crank the engine, have fuel pressure, and have spark, you still actually have no fuel getting to the engine.

To test this, you will want to use a noid light to check for a fuel injector electrical pulse. These are really inexpensive testers and useful for such diagnosis. I have a complete set of them with a spark tester that cost just $30. You simply take an injector connector loose and connect the proper noid light into the injector harness connector and crank the engine. If there is an injector pulse, then your fuel injectors are being pulsed by the ECM just fine.

Just because you have fuel pressure and an injector pulse signal doesn't mean that they are working. They could be clogged from fuel varnish, rust, or are simply sticking. Usually there are a few that work, so the motor will try to start but acts like it is out of fuel. If you have these symptoms and your used engine has sat for longer than six months, this is likely the situation. You need to take your injectors to be backflush-cleaned and flow-tested, or you can also replace them with better ones.

The best course of action is to get new injectors, and preferably not new OEM injectors, as these will drain your bank account quickly. Rather, look for aftermarket injectors or recent take-out injectors of the same size and dimensions from someone who is upgrading.

If you change injector ratings, dimensions, or connector designs, you will need a new fueling calibration in the ECM, the addition or removal of spacers, and the possibility of changing injector harness ends or needing adapter connectors. If replacement is necessary, try to get what you already have if buying used. If buying new injectors from Injector Dynamics, Holley, SVO, or FAST, then you can get the correctly dimensioned injector, but still likely need an injector recalibration in the ECM for both the injector flow rate and the fuel pressure setting.

DIAGNOSING ROUGH RUNNING

A rough running or stalling engine is usually a sign of incorrect ECM calibration, incorrect fuel pressure, intake vacuum leaks, or a bad MAP, MAF, or O_2 sensor reading. Rough running can also be a sign of bad fuel or a missing coolant temperature sensor signal.

The first thing to do is have someone knowledgeable with an engine scanner check over the sensor data stream to look for anomalies. Sometimes it could even be related to bad spark plug wires or plugs. Either way, check the data stream first and then go from there.

Sometimes problems are obvious; sometimes you have to think a little. Often it is related to just the engine swap itself, so if all else fails, find a competent LS engine tuner and have your ECM recalibrated better.

Remember that if you performed internal engine modifications such as replacing a camshaft or cylinder heads, or added a blower or turbo, the engine may not purr like a content kitten on your lap. Anytime internal engine modifications are made, you have to compensate in the ECM tuning a bit for the difference in engine vacuum and camshaft overlap. Whether this means adding more idle rpm or adding or taking away idle fuel or ignition timing, you can usually adjust just about any quirkiness out of the ECM with the use of an observant and capable LS engine tuner.

DIAGNOSING NO OR LOW OIL PRESSURE

Low or no engine oil pressure is cause for alarm. Do not run the engine for long if you don't see oil pressure rise within the first 10 seconds of engine run time. Usually you can tell when oil pressure is building, as the engine will have noisy lifters; then all at once the lifters will lose 90 percent of their noise once oil pressure comes up. The engine will even run a little differently due to having more camshaft duration and lift from the oil pressure stabilization. If there is no oil pressure, the engine will

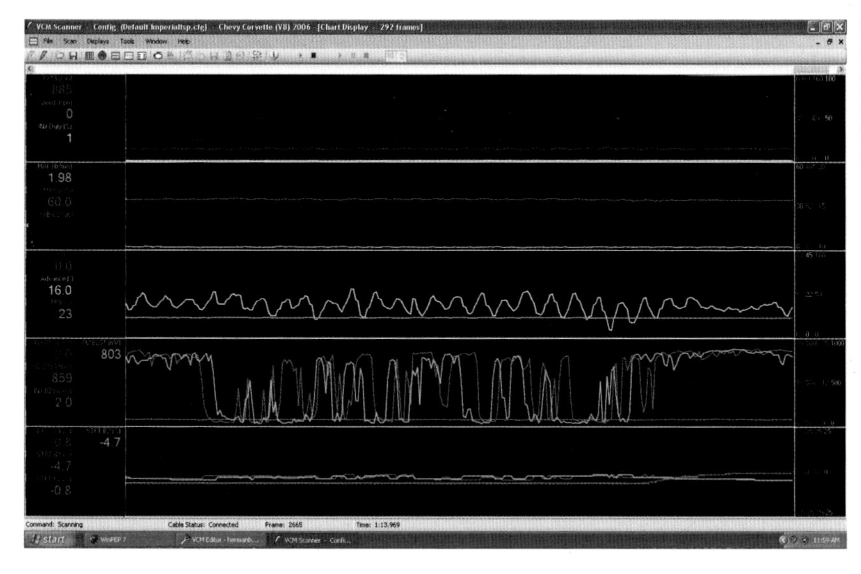

Scanning the data stream and logging the history of the engine sensors can provide some excellent data if sensor problems arise or the engine runs bad. Using the scan data tells you exactly what each sensor is doing and if there is a problem. Where the engine runs funny, you can usually see it when the data is graphed.

still sound like there is no oil pressure due to the lifters causing valvetrain clacking.

No oil pressure can be traced to a few problem areas. First of all, the stock oil pumps are known to have sticking pressure regulator valves. Usually this is apparent by the engine having normal oil pressure at first and then on a restart the oil pressure is nonexistent. Also, since the engine oil pump is not located in the oil pan, it uses a long pickup tube with an O-ring seal to seal inside the oil pump. If this seal was disrupted and not replaced, or cut, or misaligned on reassembly, there will be insufficient oiling. Last, if this was a brand-new engine buildup, the engine assembler could have left out the front or rear galley plug or dog-bone restrictor, which would leak engine oil volume internally within the engine. You will never get oil pressure until these items are properly fitted.

Gen IV engines with Active Fuel Management have oil towers in the block that must be plugged off when doing an AFM delete. If these plugs are not plugged (whether with a non-AFM alley plate or physical plugs), they will leak oil pressure. Additionally, the stock 5th-Gen Camaro and truck oil pans (and other Gen IV oil pans) have an internal pressure regulator near the oil filter to keep the AFM lifters happy. If you are reusing these oil pans with the bypass and AFM delete, remove this plug and use a M14 × 1.5 Honda oil drain plug to seal off this small nuisance.

These are the common things to look at. GM specifications call for low oil pressure at the minimum end of the scale. In real-world conditions, the GM specifications are never seen with a healthy engine. Typically, aluminum LS engines have 25 to 35 psi of oil pressure at warm idle, and around 50 to 60 psi on a cold startup. An iron-based motor may have a little less, but it will typically vary less when cold versus hot due to the lessened expansion rate of iron compared to aluminum.

When all problematic areas are taken care of you are left with a modern high-tech engine package in a car that has timeless looks and appearance. Sure you can install any older SBC or BBC engine into your Gen I Camaro, and it still will be a Gen I Camaro. But if you install an LS-series engine into any vehicle, the entire powertrain can borrow all of the late-model EFI aspects such as reliability, performance, and even obtain increased fuel economy. Enjoy your newfound LS-engine features.

DIAGNOSING OVERHEATING

Engine overheating is never any fun and can be more frustrating than the other issues because you know the car is close to being drivable, but just needs one little problem fixed. The most common issue with overheating is that the engine develops a huge air pocket that it cannot burp out.

For an easy fix, refer to the Accessory and Cooling Systems chapter that shows a thermostat housing modification that will help bleed an air-locked cooling system. If you have a heater hose loop, this is also helping coolant flow internally. Another thing that helps is to remove the coolant temperature sensor and the coolant air-bleed pipes. Further, if none of these work out, you can pull a vacuum on the engine side of the coolant bleed pipe hose fitting while filling the radiator or coolant reservoir with fluid.

LS-series cooling systems are hard to bleed, but using common sense will go a long way. If the cooling system takes only a gallon of coolant but seems "full" and you know capacity is 2.5 gallons, then figure out a way to get the rest in there. If it takes 2.5 gallons, and it is still overheating, check and see if the fans are on and the fan settings are right, then make sure the thermostat is opening by testing the radiator hose temperatures. If one is cold and one is hot, then the thermostat could be malfunctioning or could just need some extra time to open for the first time. New thermostats always seem to take a little extra coolant heat to open initially, from then on they seem to operate completely fine.

IN CLOSING

Once all the installation-related problems go away, you are left with a modern high-tech and high-performance engine that in stock form blows most mildly modified SBC engines out of the water. Looking at the first LS-series engine made, what other small-block engine from a stock Camaro can make 400 horsepower at the flywheel with just the addition of headers? The newer engines are more impressive, with the LS3 making 480 horsepower flywheel and 600 horsepower with the addition of a ported heads and cam package. No other engine has that capability, and it is no wonder that the LS-series engines are so popular to scavenge from newer cars. They have everything desirable to a car builder. Huge power numbers, widespread aftermarket support, efficiency, flexibility, ease of installation, and that is without the engine even working up a sweat. The larger LS engines rival big-block power, in a package dimensionally smaller than a SBC.

Best of all, by performing an LS-series engine swap or retrofit, you have this highly desirable engine now in the car of your choice.

Index